Introduction to the Theory of Thermal Neutron Scattering

Since the advent of the nuclear reactor, thermal neutron scattering has proved a valuable tool for studying many properties of solids and liquids, and research workers are active in the field at reactor centres and universities throughout the world.

This classic text provides the basic quantum theory of thermal neutron scattering and applies the concepts to scattering by crystals, liquids and magnetic systems. Other topics discussed are the relation of the scattering to correlation functions in the scattering system, the dynamical theory of scattering and polarisation analysis. No previous knowledge of the theory of thermal neutron scattering is assumed, but basic knowledge of quantum mechanics and solid-state physics is required.

The book is intended for experimenters rather than theoreticians, and the discussion is kept as informal as possible. A number of examples, with worked solutions, are included as an aid to the understanding of the text.

G. L. SQUIRES (1924–2010) was a Lecturer in Physics at the University of Cambridge and a Fellow of Trinity College Cambridge from 1956. He published two other books with Cambridge University Press: *Practical Physics* and *Problems in Quantum Mechanics with Solutions*, wrote an article on quantum mechanics for the *Encyclopaedia Britannica* and contributed extensively to leading scientific journals. From his retirement in 1991 until his death in 2010, Dr Squires was the curator of the Cavendish Laboratory Museum and wrote a number of historical articles on scientists and scientific discoveries in Cambridge.

Introduction to the Theory of Thermal Neutron Scattering

G. L. SQUIRES (1924–2010)
Lecturer in Physics at the University of Cambridge and Fellow of Trinity College Cambridge

CAMBRIDGE
UNIVERSITY PRESS

CAMBRIDGE UNIVERSITY PRESS
Cambridge, New York, Melbourne, Madrid, Cape Town,
Singapore, São Paulo, Delhi, Mexico City

Cambridge University Press
The Edinburgh Building, Cambridge CB2 8RU, UK

Published in the United States of America by Cambridge University Press, New York

www.cambridge.org
Information on this title: www.cambridge.org/9781107644069

First published by Cambridge University Press 1978
Published in paperback by Dover Publications Inc. 1996
First published in paperback by Cambridge University Press 2012
Reprinted 2012

A catalogue record for this publication is available from the British Library

ISBN 978-1-107-64406-9 Paperback

CONTENTS

PREFACE

This book arose from some lectures given in Cambridge in 1973 at a Summer School organised by the Neutron Scattering Group of the Institute of Physics and the Faraday Society. It is intended for experimenters in the field of thermal neutron scattering who wish to see the theoretical ideas developed in a not too formal manner. But I hope it may be of interest to students and research workers in related fields.

I assume no previous knowledge of the theory of thermal neutron scattering, but a familiarity with the basic concepts of quantum mechanics and solid state physics is necessary for a proper understanding of the text. The required results in these subjects are summarised in appendices. The latter also contain proofs of some of the mathematical results.

Some problem examples have been given at the ends of chapters. They are intended to illustrate the text, and the reader is advised to glance at them even if he is not inclined to try to solve them. Their purpose is partly, as with some of the appendices, to remove some mathematical material from the main body of the text, and partly to stimulate the reader to a more active understanding of the subject.

Thermal neutron scattering is being applied in more and more areas of science. But this book does not attempt to cover the theory of all the applications. Instead I have confined myself to the basic ideas of the theory. My aim is to bring the reader to the point where he can tackle the theoretical expositions of the specialised branches given in more advanced textbooks and in theoretical papers.

It is a pleasure to thank Mr Cole, Dr Dore, Dr Howie, Professor Joannopoulos, Dr de Vallera, and Dr Zeilinger for reading parts of the manuscript and making useful comments.

Cambridge, July 1977 G. L. SQUIRES

1

Introduction

1.1 Basic properties of the neutron

With the advent of nuclear reactors, thermal neutrons have become a valuable tool for investigating many important features of matter – particularly condensed matter. The usefulness of thermal neutrons arises from the basic properties of the neutron. These are listed in Table 1.1.

The value of the mass of the neutron results in the de Broglie wavelength of thermal neutrons being of the order of interatomic distances in solids and liquids. Thus, interference effects occur which yield information on the structure of the scattering system.

Secondly, the fact that the neutron is uncharged means, not only that it can penetrate deeply into the target, but also that it comes close to the nuclei – there is no Coulomb barrier to be overcome. Neutrons are thus scattered by nuclear forces, and for certain nuclides the scattering is large. An important example is light hydrogen which is virtually transparent to X-rays but which scatters neutrons strongly.

Thirdly, the energy of thermal neutrons is of the same order as that of many excitations in condensed matter. So when the neutron is inelastically scattered by the creation or annihilation of an excitation, the change in the energy of the neutron is a large fraction of its initial energy. Measurement of the neutron energies thus provides accurate information on the energies of the excitations, and hence on the interatomic forces.

Fourthly, the neutron has a magnetic moment, which means that neutrons interact with the unpaired electrons in magnetic atoms. Elastic scattering from this interaction gives information on the arrangement of electron spins and the density distribution of unpaired electrons. Inelastic magnetic scattering gives the energies of

Table 1.1 Basic properties of the neutron and values of physical constants

Basic properties of neutron	
mass	$m = 1.675 \times 10^{-27}$ kg
charge	0
spin	$\frac{1}{2}$
magnetic dipole moment	$\mu_n = -1.913\ \mu_N$
Values of physical constants	
elementary charge	$e = 1.602 \times 10^{-19}$ C
mass of electron	$m_e = 9.109 \times 10^{-31}$ kg
mass of proton	$m_p = 1.673 \times 10^{-27}$ kg
Planck constant	$h = 6.626 \times 10^{-34}$ J s
Boltzmann constant	$k_B = 1.381 \times 10^{-23}$ J K^{-1}
Avogadro constant	$N_A = 6.022 \times 10^{23}$ mol^{-1}
Bohr magneton	$\mu_B = 9.274 \times 10^{-24}$ J T^{-1}
nuclear magneton	$\mu_N = 5.051 \times 10^{-27}$ J T^{-1}

magnetic excitations, and in general permits a study of time-dependent spin correlations in the scattering system.

It is convenient to develop the theories of nuclear and magnetic scattering separately. Thus Chapters 1 to 6 of the book are concerned mainly with nuclear scattering, though the definitions in Chapter 1 and the theoretical development in Chapter 2 give basic results that apply to both types of scattering. Chapters 7 and 8 are devoted to magnetic scattering. Chapter 9 deals with polarisation effects and includes both nuclear and magnetic scattering.

1.2 Numerical values for velocity, energy, wavelength

At present the source of thermal neutrons in most scattering experiments is a nuclear reactor. In the thermal region, the velocity spectrum of the neutrons emerging from the reactor is close to Maxwellian, with the temperature T that of the moderator.

The Maxwellian distribution for flux is

$$\phi(v) \propto v^3 \exp(-\tfrac{1}{2}mv^2/k_B T), \tag{1.1}$$

where $\phi(v)\,dv$ is the number of neutrons through unit area per second with velocities between v and $v+dv$, m is the mass of the neutron, and k_B is the Boltzmann constant. The maximum of the function $\phi(v)$ occurs at

$$v=\left(\frac{3k_BT}{m}\right)^{1/2}, \tag{1.2}$$

which corresponds to a kinetic energy

$$E=\tfrac{1}{2}mv^2=\tfrac{3}{2}k_BT. \tag{1.3}$$

It is conventional to say that a neutron with energy E corresponds to a temperature T, given by

$$E=k_BT. \tag{1.4}$$

The de Broglie wavelength of a neutron with velocity v is

$$\lambda=\frac{h}{mv}, \tag{1.5}$$

where h is the Planck constant. The wavevector $\boldsymbol{k}$ is defined to have magnitude

$$k=\frac{2\pi}{\lambda}, \tag{1.6}$$

its direction being that of $\boldsymbol{v}$. The momentum of the neutron is

$$\boldsymbol{p}=\hbar\boldsymbol{k}. \tag{1.7}$$

We thus have

$$E=k_BT=\tfrac{1}{2}mv^2=\frac{h^2}{2m\lambda^2}=\frac{\hbar^2k^2}{2m}. \tag{1.8}$$

Inserting the values of the constants m, e, h, k_B in Table 1.1 in (1.8) gives the following relations between the wavelength, wavevector, velocity, energy, and temperature for thermal neutrons:

$$\begin{aligned}\lambda&=6.283\frac{1}{k}=3.956\frac{1}{v}=9.045\frac{1}{\sqrt{E}}=30.81\frac{1}{\sqrt{T}},\\ E&=0.08617T=5.227v^2=81.81\frac{1}{\lambda^2}=2.072k^2.\end{aligned} \tag{1.9}$$

In these equations, λ is in Å, k in $10^{10}\,\mathrm{m}^{-1}$, v in $\mathrm{km\,s^{-1}}$, E in meV, and T in kelvin.

The value $v=2.20\ \mathrm{km\,s^{-1}}$ is conventionally taken as a standard velocity for thermal neutrons. For example, the absorption cross-section is usually proportional to $1/v$, and its value is then quoted for this value of v. For the standard velocity

$$v = 2.20\ \text{km s}^{-1}, \qquad \frac{1}{v} = 455\ \mu\text{s m}^{-1},$$
$$E = 25.3\ \text{meV}, \qquad T = 293\ \text{K}, \tag{1.10}$$
$$\lambda = 1.798\ \text{Å}, \qquad k = 3.49 \times 10^{10}\ \text{m}^{-1}.$$

For most reactors designed to produce high thermal neutron flux, the temperature of the moderator is about 300 to 350 K, and the resulting velocity spectrum of the neutrons is suitable for many experiments. However, in some experiments the velocities required for the incident neutrons lie on the low-energy tail of the thermal spectrum, while in other experiments neutrons on the high-energy side are required. It is therefore desirable to be able to change the temperature of the velocity distribution. This is done by placing in the reactor a small amount of moderating material at a different

Fig. 1.1 Maxwellian flux distribution $\phi(v) \propto v^3 \exp(-mv^2/2k_B T)$ for $T = 25$, 300, 2000 K. The curves are normalised to have the same area.

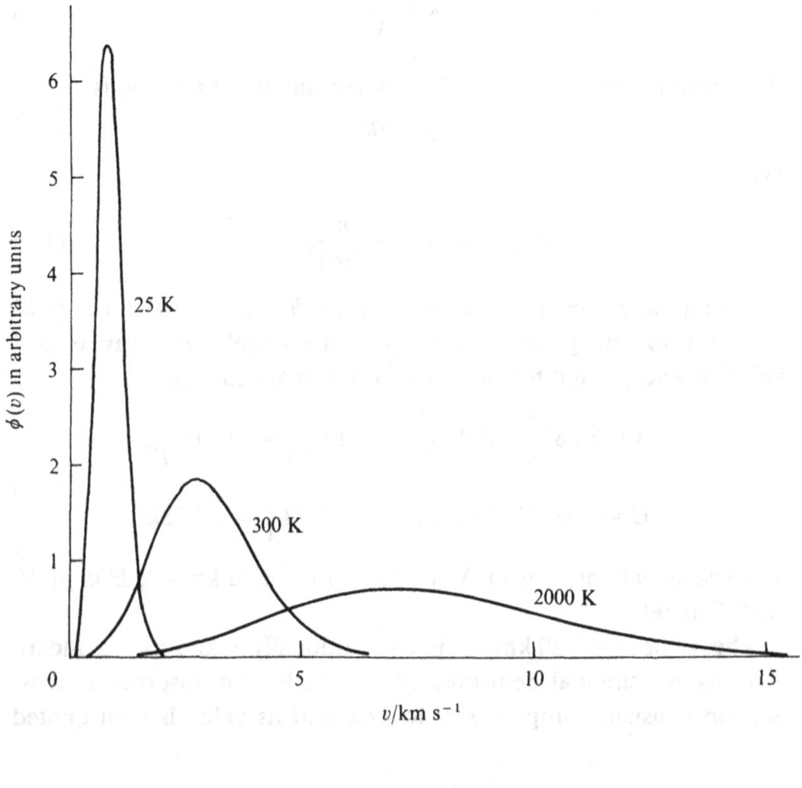

temperature. In the high-flux reactor at the Laue–Langevin Institute in Grenoble, a vessel containing 25 litres of liquid deuterium acts as a cold moderating source, providing a velocity distribution with $T \sim$ 25 K, while a block of hot graphite of about half this volume provides a distribution with $T \sim 2000$ K. Curves of the velocity distribution for $T = 25$, 300, and 2000 K are shown in Fig. 1.1. The approximate ranges for energy, temperature, and wavelength of the neutrons for the three types of source are given in Table 1.2.

Table 1.2 Approximate values for the range of energy, temperature, and wavelength for three types of source in a reactor

Source	Energy E/meV	Temperature T/K	Wavelength $\lambda/10^{-10}$ m
cold	0.1– 10	1– 120	30–3
thermal	5–100	60–1000	4–1
hot	100–500	1000–6000	1–0.4

1.3 Definitions of scattering cross-sections

Consider a beam of thermal neutrons, all with the same energy E, incident on a target (Fig. 1.2). The target is a general collection of

Fig. 1.2 Geometry for scattering experiment.

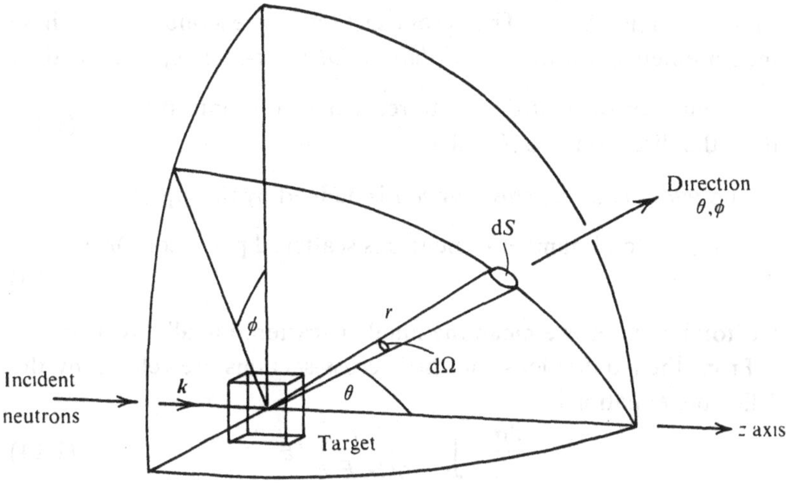

atoms – it may be a crystal, an amorphous solid, a liquid, or a gas. We shall call it the *scattering system.* Various types of measurement can be made on the neutrons after they have interacted with the scattering system. The results in each case can be expressed in terms of a quantity known as a *cross-section.*

Suppose we set up a neutron counter and measure the number of neutrons scattered in a given direction as a function of their energy E'. The distance of the counter from the target is assumed to be large compared to the dimensions of the counter and the target, so the small angle $\mathrm{d}\Omega$ subtended by the counter at the target is well defined. To specify the geometry of the scattering process we use polar coordinates, taking the direction of the incident neutrons as the polar axis. Let the direction of the scattered neutrons be θ, ϕ. The *partial differential cross-section* is defined by the equation

$$\frac{\mathrm{d}^2\sigma}{\mathrm{d}\Omega\,\mathrm{d}E'} = \begin{array}{l}\text{(number of neutrons scattered per second into a}\\ \text{small solid angle } \mathrm{d}\Omega \text{ in the direction } \theta, \phi \text{ with final}\\ \text{energy between } E' \text{ and } E'+\mathrm{d}E')/\Phi\,\mathrm{d}\Omega\,\mathrm{d}E',\end{array} \tag{1.11}$$

where Φ is the flux of the incident neutrons, i.e. the number through unit area per second, the area being perpendicular to the direction of the neutron beam. Note that the dimensions of the numerator on the right-hand side of the equation are [time^{-1} energy]. The dimensions of flux are [area^{-1} time^{-1}]. Thus the dimensions of the cross-section are [area], as we would expect from the name.

Suppose we do not analyse the energy of the scattered neutrons, but simply count all the neutrons scattered into the solid angle $\mathrm{d}\Omega$ in the direction θ, ϕ. The cross-section corresponding to these measurements, known as the *differential cross-section,* is defined by

$$\frac{\mathrm{d}\sigma}{\mathrm{d}\Omega} = \begin{array}{l}\text{(number of neutrons scattered per second into } \mathrm{d}\Omega \text{ in}\\ \text{the direction } \theta, \phi)/\,\Phi\,\mathrm{d}\Omega.\end{array} \tag{1.12}$$

The *total scattering cross-section* is defined by the equation

$$\sigma_{\text{tot}} = \text{(total number of neutrons scattered per second)}/\,\Phi. \tag{1.13}$$

By 'total number' we mean the number scattered in all directions.

From their definitions the three cross-sections are related by the following equations

$$\frac{\mathrm{d}\sigma}{\mathrm{d}\Omega} = \int_0^\infty \left(\frac{\mathrm{d}^2\sigma}{\mathrm{d}\Omega\,\mathrm{d}E'}\right)\mathrm{d}E', \tag{1.14}$$

$$\sigma_{\text{tot}} = \int_{\text{all directions}} \left(\frac{d\sigma}{d\Omega}\right) d\Omega. \tag{1.15}$$

If the scattering is axially symmetric, i.e. if $d\sigma/d\Omega$ depends only on θ and not on ϕ, the last equation becomes

$$\sigma_{\text{tot}} = \int_0^{\pi} \frac{d\sigma}{d\Omega} 2\pi \sin\theta \, d\theta. \tag{1.16}$$

The cross-sections are the quantities actually measured in a scattering experiment. The basic problem, with which this whole book is concerned, is to derive theoretical expressions for these quantities. Experimental cross-sections are quoted per atom or per molecule, i.e. the cross-sections defined above are divided by the number of atoms or molecules in the scattering system.

The present cross-sections do not take account of the initial and final spin states of the neutron. The definitions are extended to do this in Chapter 9 when we consider polarisation experiments.

1.4 Scattering of neutrons by a single fixed nucleus

The definitions of cross-sections apply to any kind of scattering. We now consider a simple case – nuclear scattering by a single nucleus fixed in position. The nuclear forces which cause the scattering have a range of about 10^{-14} to 10^{-15} m. The wavelength of thermal neutrons is of the order of 10^{-10} m, and is thus much larger than the range of the forces. In these circumstances the scattering, analysed in terms of partial waves, comes entirely from the S waves ($l = 0$). The angular distribution for S-wave scattering is spherically symmetric. We do not need the theory of partial waves to see that the angular distribution has this form. It is a basic result from diffraction theory that if waves of any kind are scattered by an object small compared to the wavelength of the waves, then the scattered wave is spherically symmetric.

We take the origin to be at the position of the nucleus, and the z axis to be along the direction of $\boldsymbol{k}$, the wavevector of the incident neutrons (Fig. 1.2). Then the incident neutrons can be represented by the wavefunction

$$\psi_{\text{inc}} = \exp(ikz). \tag{1.17}$$

As the scattering is spherically symmetric, the wavefunction of the scattered neutrons at the point $\boldsymbol{r}$ can be written in the form

$$\psi_{sc} = -\frac{b}{r}\exp(ikr), \tag{1.18}$$

where b is a constant, independent of the angles θ, ϕ. The minus sign in the equation is arbitrary and corresponds to a positive value of b for a repulsive potential (see Section 2.3).

Note that the magnitude of the wavevector is the same for the scattered and incident neutrons. The energy of thermal neutrons is too small to change the internal energy of the nucleus. And since we are taking the position of the nucleus to be fixed, the neutron cannot give the nucleus kinetic energy. Thus the scattering is elastic. The energy of the neutron, and hence the magnitude of $\boldsymbol{k}$, is unchanged.

The quantity b in ψ_{sc} is known as the *scattering length*. We may distinguish two types of nucleus. In the first type the scattering length is complex and varies rapidly with the energy of the neutron. The scattering for such nuclei is a resonance phenomenon and is associated with the formation of a compound nucleus (original nucleus plus neutron) with energy close to an excited state. Examples of nuclei which show this behaviour are ^{103}Rh, ^{113}Cd, ^{157}Gd, and ^{176}Lu. Since the imaginary part of the scattering length corresponds to absorption, such nuclei strongly absorb neutrons. The majority of nuclei are of the second type, in which the compound nucleus is not formed near an excited state. The imaginary part of the scattering length is small, and the scattering length is independent of the energy of the neutron. We shall confine the discussion to such nuclei and take the scattering length to be a real quantity.

The value of the scattering length depends on the particular nucleus (i.e. nuclide), and the spin state of the nucleus–neutron system. The neutron has spin $\frac{1}{2}$. Suppose the nucleus has spin I (not zero). Then the spin of the nucleus–neutron system is either $I+\frac{1}{2}$, or $I-\frac{1}{2}$. Each spin state has its own value of b. So every nucleus with non-zero spin has two values of the scattering length. If the spin of the nucleus is zero, the nucleus–neutron system can only have spin $\frac{1}{2}$, and there is only one value of the scattering length.

If we had a proper theory of nuclear forces we would be able to calculate or predict the values of b from other properties of the nucleus. But we do not have such a theory, so we have to treat the scattering lengths as parameters to be determined experimentally. If the nuclides are arranged according to their Z, N values the values of b vary erratically from one nuclide to its neighbour. The actual values have important practical consequences, which we shall see later. A selection of values is given in Table 1.3.

Table 1.3 Values of scattering lengths

Nuclide	Combined spin	b/fm	Nuclide	Combined spin	b/fm
^1H	1	10.85	^{23}Na	2	6.3
	0	−47.50		1	−0.9
^2H	$\frac{3}{2}$	9.53	^{59}Co	4	−2.78
	$\frac{1}{2}$	0.98		3	9.91

The values for H, Na, and Co are from Koester (1977), Abragam *et al.* (1975), and Koester *et al.* (1974) respectively. The spin values refer to the nucleus–neutron system.

We can readily calculate the cross-section $d\sigma/d\Omega$ for scattering from a single fixed nucleus, using the expressions for ψ_{inc} and ψ_{sc} in (1.17) and (1.18). If v is the velocity of the neutrons (the same before and after scattering), the number of neutrons passing through the area dS per second is

$$v\, dS|\psi_{sc}|^2 = v\, dS\frac{b^2}{r^2} = vb^2\, d\Omega \tag{1.19}$$

(see Fig. 1.2). The flux of incident neutrons is

$$\Phi = v|\psi_{inc}|^2 = v. \tag{1.20}$$

From the definition of the cross-section

$$\frac{d\sigma}{d\Omega} = \frac{vb^2\, d\Omega}{\Phi\, d\Omega} = b^2, \tag{1.21}$$

and in this simple case

$$\sigma_{tot} = 4\pi b^2. \tag{1.22}$$

2

Nuclear scattering – basic theory

2.1 Introduction

We now start on the theory proper and consider the nuclear scattering by a general system of particles. We first derive a general expression for the cross-section $d^2\sigma/d\Omega\, dE'$ for a specific transition of the scattering system from one of its quantum states to another. Although the calculation relates to nuclear scattering there will be no difficulty in applying the basic formula (2.15) to the magnetic case. We start by ignoring the spin of the neutron. This means that the state of the neutron is specified entirely by its momentum, i.e. by its wavevector.

Suppose we have a neutron with wavevector $\boldsymbol{k}$ incident on a scattering system in a state characterised by an index λ. Denote the wavefunction of the neutron by $\psi_{\boldsymbol{k}}$ and of the scattering system by χ_λ. Suppose the neutron interacts with the system via a potential V, and is scattered so that its final wavevector is $\boldsymbol{k}'$. The final state of the scattering system is λ'.

We set up a coordinate system with the origin at some arbitrary point in the scattering system. Denote the number of nuclei in the scattering system by N. Let $\boldsymbol{R}_j$ $(j=1,\ldots N)$ be the position vector of the jth nucleus, and $\boldsymbol{r}$ that of the neutron (Fig. 2.1).

2.2 Fermi's golden rule

Consider the differential scattering cross-section $(d\sigma/d\Omega)_{\lambda\to\lambda'}$, representing the sum of all processes in which the state of the scattering system changes from λ to λ', and the state of the neutron changes from $\boldsymbol{k}$ to $\boldsymbol{k}'$. The sum is taken over all values of $\boldsymbol{k}'$ that lie in the small solid angle $d\Omega$ in the direction θ, ϕ, the values of $\boldsymbol{k}$, λ, and λ' *remaining constant* (Fig. 2.2). *From the definition of $d\sigma/d\Omega$ given in*

(1.12) we have

$$\left(\frac{d\sigma}{d\Omega}\right)_{\lambda\to\lambda'} = \frac{1}{\Phi}\frac{1}{d\Omega}\sum_{\substack{\boldsymbol{k}' \\ \text{in } d\Omega}} W_{\boldsymbol{k},\lambda\to\boldsymbol{k}',\lambda'}, \tag{2.1}$$

where $W_{\boldsymbol{k},\lambda\to\boldsymbol{k}'\lambda'}$ is the number of transitions per second from the state $\boldsymbol{k}$, λ to the state $\boldsymbol{k}'$, λ', and Φ is the flux of incident neutrons.

To evaluate the expression on the right-hand side of (2.1) we use a fundamental result in quantum mechanics, known as *Fermi's golden*

Fig. 2.1 Coordinates of nucleus and neutron.

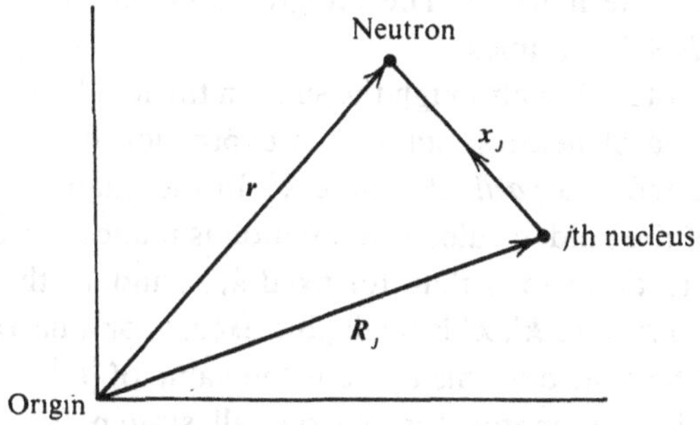

Fig. 2.2 Calculation of $\rho_{k'}$. The points represent $\boldsymbol{k}'$ values permitted by box normalisation. The spheres corresponding to neutron energies E' and $E'+dE'$ are shown.

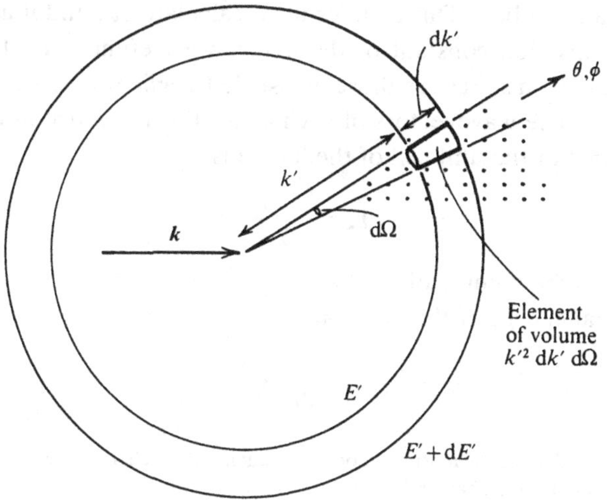

rule.† This is

$$\sum_{\substack{\boldsymbol{k}' \\ \text{in d}\Omega}} W_{\boldsymbol{k},\lambda\to\boldsymbol{k}',\lambda'} = \frac{2\pi}{\hbar}\rho_{\boldsymbol{k}'}|\langle\boldsymbol{k}'\lambda'|V|\boldsymbol{k}\lambda\rangle|^2. \tag{2.2}$$

In this expression $\rho_{\boldsymbol{k}'}$ is the number of momentum states in dΩ per unit energy range for neutrons in the state $\boldsymbol{k}'$. The matrix element is given explicitly by

$$\langle\boldsymbol{k}'\lambda'|V|\boldsymbol{k}\lambda\rangle = \int \psi^*_{\boldsymbol{k}'}\chi^*_{\lambda'}V\psi_{\boldsymbol{k}}\chi_\lambda \, \mathrm{d}\boldsymbol{R}\, \mathrm{d}\boldsymbol{r}, \tag{2.3}$$

where

$$\mathrm{d}\boldsymbol{R} = \mathrm{d}\boldsymbol{R}_1\, \mathrm{d}\boldsymbol{R}_2 \ldots \mathrm{d}\boldsymbol{R}_N. \tag{2.4}$$

$\mathrm{d}\boldsymbol{R}_j$ is an element of volume for the jth nucleus, and $\mathrm{d}\boldsymbol{r}$ is an element of volume for the neutron. The integral is taken over all space for each of the $N+1$ variables.

It may be noted that although the sum on the left-hand side of (2.2) is over a range of neutron states, the expression on the right-hand side is evaluated at a *particular* state $\boldsymbol{k}'$. In the quantum mechanical derivation of the golden rule, a summation is made over all values of $|\boldsymbol{k}'|$. The derivation shows that, for fixed $\boldsymbol{k}$, λ, and λ', the probability of a transition $\boldsymbol{k}$, λ to $\boldsymbol{k}'$, λ' is negligible except for a narrow range of $|\boldsymbol{k}'|$ values. The centre of this range is the value of $|\boldsymbol{k}'|$ corresponding to conservation of energy for the overall system of neutron plus scattering system, and it is this particular $\boldsymbol{k}'$ which must be inserted on the right-hand side of (2.2).

We adopt a standard device in quantum mechanics, known as *box normalisation*, which is to imagine the neutron and scattering system to be in a large box. This enables us to calculate $\rho_{\boldsymbol{k}'}$, and it also fixes the normalisation constant of the neutron wavefunctions. The only allowed neutron states are those whose de Broglie waves are periodic in the box. The wavevectors of such states form a lattice in $\boldsymbol{k}$ space. The volume of the unit cell of the lattice is

$$v_{\boldsymbol{k}} = \frac{(2\pi)^3}{Y}, \tag{2.5}$$

where Y is the volume of the box.

The final energy of the neutron is

$$E' = \frac{\hbar^2}{2m}k'^2 \tag{2.6}$$

† The rule is derived in most textbooks on quantum mechanics. See for example Merzbacher (1970), Chapter 18.

with
$$\mathrm{d}E' = \frac{\hbar^2}{m} k'\,\mathrm{d}k'. \tag{2.7}$$

By definition $\rho_{k'}\,\mathrm{d}E'$ is the number of states in $\mathrm{d}\Omega$ with energy between E' and $E'+\mathrm{d}E'$, which is the number of wavevector points in the element of volume $k'^2\,\mathrm{d}k'\,\mathrm{d}\Omega$ (Fig. 2.2). Thus

$$\rho_{k'}\,\mathrm{d}E' = \frac{1}{v_k} k'^2\,\mathrm{d}k'\,\mathrm{d}\Omega. \tag{2.8}$$

Eqs. (2.5) to (2.8) give

$$\rho_{k'} = \frac{Y}{(2\pi)^3} k' \frac{m}{\hbar^2}\,\mathrm{d}\Omega. \tag{2.9}$$

We now consider the wavefunction $\psi_{\boldsymbol{k}}$. It is a plane wave and has the form $\exp(i\boldsymbol{k}\,.\,\boldsymbol{r})$. There is one neutron in the box of volume Y. So the neutron density is $1/Y$. Thus

$$\psi_{\boldsymbol{k}} = \frac{1}{\sqrt{Y}} \exp(\mathrm{i}\boldsymbol{k}\,.\,\boldsymbol{r}). \tag{2.10}$$

The matrix element in (2.3) is

$$\begin{aligned}\langle \boldsymbol{k}'\lambda'|V|\boldsymbol{k}\lambda\rangle &= \int \psi^*_{\boldsymbol{k}'}\chi^*_{\lambda'} V \psi_{\boldsymbol{k}}\chi_\lambda\,\mathrm{d}\boldsymbol{R}\,\mathrm{d}\boldsymbol{r} \\ &= \frac{1}{Y}\int \exp(-\mathrm{i}\boldsymbol{k}'\,.\,\boldsymbol{r})\chi^*_{\lambda'} V \exp(\mathrm{i}\boldsymbol{k}\,.\,\boldsymbol{r})\chi_\lambda\,\mathrm{d}\boldsymbol{R}\,\mathrm{d}\boldsymbol{r}.\end{aligned} \tag{2.11}$$

We shall write the last line as $\langle \boldsymbol{k}'\lambda'|V|\boldsymbol{k}\lambda\rangle/Y$, so from now on the neutron wavefunction in the matrix element is $\exp(\mathrm{i}\boldsymbol{k}\,.\,\boldsymbol{r})$.

The flux of the incident neutrons is the product of their density and velocity, i.e.

$$\Phi = \frac{1}{Y}\frac{\hbar}{m}k. \tag{2.12}$$

2.3 Expression for $\mathrm{d}^2\sigma/\mathrm{d}\Omega\,\mathrm{d}E'$

We now derive a basic expression for the cross-section $\mathrm{d}^2\sigma/\mathrm{d}\Omega\,\mathrm{d}E'$. We start by substituting (2.2), (2.9), (2.11), and (2.12) into (2.1), which gives

$$\left(\frac{\mathrm{d}\sigma}{\mathrm{d}\Omega}\right)_{\lambda\to\lambda'} = \frac{k'}{k}\left(\frac{m}{2\pi\hbar^2}\right)^2 |\langle \boldsymbol{k}'\lambda'|V|\boldsymbol{k}\lambda\rangle|^2. \tag{2.13}$$

All the Ys have cancelled out in (2.13), as they must do since the volume of the normalisation box is arbitrary.

$(\mathrm{d}\sigma/\mathrm{d}\Omega)_{\lambda\to\lambda'}$ is the cross-section for neutrons scattered into $\mathrm{d}\Omega$ in the direction of $\boldsymbol{k}'$. However, as already mentioned, since $\boldsymbol{k}$, λ, and λ' are fixed, the scattered neutrons all have the same energy, determined by conservation of energy. If E and E' are the initial and final energies of the neutron, and E_λ and $E_{\lambda'}$ are the initial and final energies of the scattering system, then

$$E+E_\lambda = E'+E_{\lambda'}. \qquad (2.14)$$

In mathematical terms the energy distribution of the scattered neutrons is a δ-function (see Appendix A). So the expression for the partial differential cross-section is

$$\left(\frac{\mathrm{d}^2\sigma}{\mathrm{d}\Omega\,\mathrm{d}E'}\right)_{\lambda\to\lambda'} = \frac{k'}{k}\left(\frac{m}{2\pi\hbar^2}\right)^2 |\langle \boldsymbol{k}'\lambda'|V|\boldsymbol{k}\lambda\rangle|^2 \delta(E_\lambda - E_{\lambda'} + E - E'). \qquad (2.15)$$

This follows from (1.14) and the result

$$\int \delta(E_\lambda - E_{\lambda'} + E - E')\,\mathrm{d}E' = 1. \qquad (2.16)$$

Fourier transform of the potential function

The first step in evaluating the matrix element is to integrate with respect to $\boldsymbol{r}$, the neutron coordinate. The potential of the neutron due to the jth nucleus has the form $V_j(\boldsymbol{r}-\boldsymbol{R}_j)$. So the potential for the whole scattering system is

$$V = \sum_j V_j(\boldsymbol{r}-\boldsymbol{R}_j). \qquad (2.17)$$

Put

$$\boldsymbol{x}_j = \boldsymbol{r}-\boldsymbol{R}_j. \qquad (2.18)$$

Now

$$\langle \boldsymbol{k}'\lambda'|V|\boldsymbol{k}\lambda\rangle$$

$$= \sum_j \int \chi^*_{\lambda'} \exp(-\mathrm{i}\boldsymbol{k}'.\boldsymbol{r}) V_j(\boldsymbol{r}-\boldsymbol{R}_j)\chi_\lambda \exp(\mathrm{i}\boldsymbol{k}.\boldsymbol{r})\,\mathrm{d}\boldsymbol{R}\,\mathrm{d}\boldsymbol{r} \qquad (2.19)$$

$$= \sum_j \int \chi^*_{\lambda'} \exp\{-\mathrm{i}\boldsymbol{k}'.(\boldsymbol{x}_j+\boldsymbol{R}_j)\} V_j(\boldsymbol{x}_j)\chi_\lambda \exp\{\mathrm{i}\boldsymbol{k}.(\boldsymbol{x}_j+\boldsymbol{R}_j)\}\,\mathrm{d}\boldsymbol{R}\,\mathrm{d}\boldsymbol{x}_j \qquad (2.20)$$

$$= \sum_j V_j(\boldsymbol{\kappa})\langle\lambda'|\exp(\mathrm{i}\boldsymbol{\kappa}.\boldsymbol{R}_j)|\lambda\rangle, \qquad (2.21)$$

where

$$V_j(\boldsymbol{\kappa}) = \int V_j(\boldsymbol{x}_j)\exp(\mathrm{i}\boldsymbol{\kappa}.\boldsymbol{x}_j)\,\mathrm{d}\boldsymbol{x}_j, \qquad (2.22)$$

$$\langle\lambda'|\exp(\mathrm{i}\boldsymbol{\kappa}.\boldsymbol{R}_j)|\lambda\rangle = \int \chi^*_{\lambda'}\exp(\mathrm{i}\boldsymbol{\kappa}.\boldsymbol{R}_j)\chi_\lambda\,\mathrm{d}\boldsymbol{R}, \qquad (2.23)$$

and
$$\boldsymbol{\kappa} = \boldsymbol{k} - \boldsymbol{k}'. \tag{2.24}$$
$\boldsymbol{\kappa}$ is known as the *scattering vector*. The logic in going from (2.19) to (2.20) is that, for each j term in the sum, integrating with respect to $\boldsymbol{r}$ and to $\boldsymbol{x}_j$ give the same result, because both integrations are done at fixed $\boldsymbol{R}_j$ over all space. We see from (2.22) that $V_j(\boldsymbol{\kappa})$ is the Fourier transform of the potential function for the jth nucleus.

Fermi pseudopotential

The next step is to insert a specific function for $V_j(\boldsymbol{x}_j)$. To find a suitable mathematical function we make an apparent digression and calculate $\mathrm{d}\sigma/\mathrm{d}\Omega$ for a single fixed nucleus using the present formalism. Consider (2.19). There is only one term, $j=1$, in the sum over j. Since the nucleus is fixed at the origin, $\boldsymbol{R}_1=0$, and $\lambda'=\lambda$. Thus
$$\begin{aligned}\langle \boldsymbol{k}'\lambda'|V|\boldsymbol{k}\lambda\rangle &= \int \chi_\lambda^*\chi_\lambda\,\mathrm{d}\boldsymbol{R}_1 \int V(\boldsymbol{r})\exp(\mathrm{i}\boldsymbol{\kappa}\,.\,\boldsymbol{r})\,\mathrm{d}\boldsymbol{r}\\ &= \int V(\boldsymbol{r})\exp(\mathrm{i}\boldsymbol{\kappa}\,.\,\boldsymbol{r})\,\mathrm{d}\boldsymbol{r},\end{aligned} \tag{2.25}$$
since χ_λ is normalised. Inserting this result in (2.13), together with $k'=k$, gives
$$\frac{\mathrm{d}\sigma}{\mathrm{d}\Omega} = \left(\frac{m}{2\pi\hbar^2}\right)^2 \left|\int V(\boldsymbol{r})\exp(\mathrm{i}\boldsymbol{\kappa}\,.\,\boldsymbol{r})\,\mathrm{d}\boldsymbol{r}\right|^2. \tag{2.26}$$

Now we know that $V(\boldsymbol{r})$ is short range. Let us make it really short range and put
$$V(\boldsymbol{r}) = a\delta(\boldsymbol{r}), \tag{2.27}$$
where a is a real constant. $\delta(\boldsymbol{r})$ is a three-dimensional Dirac delta function, i.e.
$$\int_{\substack{\text{all}\\ \text{space}}} \delta(\boldsymbol{r})\,\mathrm{d}\boldsymbol{r} = 1. \tag{2.28}$$

Then
$$\int V(\boldsymbol{r})\exp(\mathrm{i}\boldsymbol{\kappa}\,.\,\boldsymbol{r})\,\mathrm{d}\boldsymbol{r} = a\int \delta(\boldsymbol{r})\exp(\mathrm{i}\boldsymbol{\kappa}\,.\,\boldsymbol{r})\,\mathrm{d}\boldsymbol{r} = a. \tag{2.29}$$

Therefore
$$\frac{\mathrm{d}\sigma}{\mathrm{d}\Omega} = \left(\frac{m}{2\pi\hbar^2}\right)^2 a^2. \tag{2.30}$$

But from Section 1.4
$$\frac{\mathrm{d}\sigma}{\mathrm{d}\Omega} = b^2, \tag{2.31}$$

where b is the scattering length. Therefore

$$a = \frac{2\pi\hbar^2}{m} b. \tag{2.32}$$

Inserting this value in (2.27) gives

$$V(\boldsymbol{r}) = \frac{2\pi\hbar^2}{m} b\,\delta(\boldsymbol{r}). \tag{2.33}$$

This potential, known as the *Fermi pseudopotential*, is the one we shall adopt. The positive sign in (2.32) comes from the definition of b in (1.18), which implies a positive scattering length for a repulsive potential.

We recall that our entire derivation of the cross-section is based on Fermi's golden rule, which, for scattering processes, is equivalent to the Born approximation; both are based on first-order perturbation theory. Now the conditions for this theory to apply do not hold for the nuclear scattering of thermal neutrons. The justification for the use of the golden rule in these circumstances is that, when combined with the pseudopotential, it gives the required result of isotropic scattering for a single fixed nucleus.†

It may be noted that the pseudopotential does not correspond even approximately to the actual potential. Equation (2.33) shows that a repulsive pseudopotential gives a positive, and an attractive pseudopotential a negative scattering length. However, a positive scattering length does not imply that the actual potential is repulsive. If we simulate the actual potential by a hypothetical 'square-well' (or square-barrier) of range r_0 and depth or height V, we can solve the Schrödinger equation without approximation. The relation between the scattering length and the parameter $x = (2mV)^{1/2} r_0/\hbar$ is shown in Fig. 2.3. For a repulsive potential b is positive for all values of x. For an attractive potential it may be negative or positive. The actual potential is basically attractive. The details of its shape, depth, and range determine the magnitude and sign of the scattering length.

The scattering length defined in (1.18) relates to a fixed nucleus and is sometimes known as the *bound* scattering length. If the nucleus is free, the scattering must be treated in the centre-of-mass system. The result is the same as if the nucleus were fixed, but the mass m of the neutron must be replaced by the reduced mass μ of the nucleus–

† For further discussion of this point see Fermi (1936) and Breit (1947).

neutron system. This is given by

$$\mu = \frac{mM}{m+M}, \tag{2.34}$$

where M is the mass of the nucleus. The scattering length for this process is called the *free* scattering length. Denote it by b_f. Since the potential is the same whether the nucleus is fixed or free, the expression for the pseudopotential (2.33) shows that

$$\frac{b_f}{\mu} = \frac{b}{m}, \tag{2.35}$$

i.e.

$$b_f = \frac{M}{m+M} b. \tag{2.36}$$

For most nuclei the free scattering length is only slightly less than the bound one. But in the extreme case of light hydrogen

$$b_f = \tfrac{1}{2} b. \tag{2.37}$$

We return to the expression for the cross-section for a general scattering system. If the jth nucleus has scattering length b_j, its potential is

$$V_j(\boldsymbol{x}_j) = \frac{2\pi\hbar^2}{m} b_j \delta(\boldsymbol{x}_j). \tag{2.38}$$

Fig. 2.3 Relation between scattering length b and the parameter $x = (2mV)^{1/2}\, r_0/\hbar$ for a square-well potential. r_0 is the range of the potential and V its depth (or height).

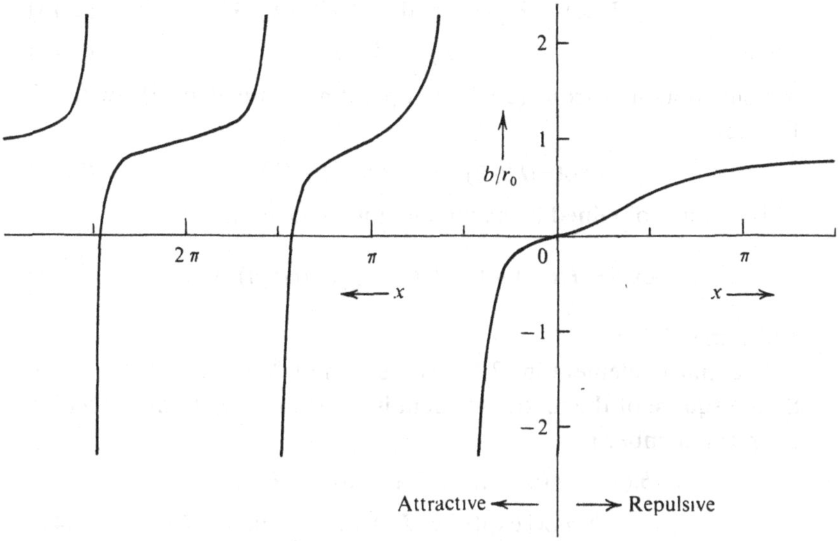

Inserting this in (2.22) gives

$$V_j(\boldsymbol{\kappa}) = \frac{2\pi\hbar^2}{m} b_j. \tag{2.39}$$

From (2.15), (2.21), and (2.39)

$$\left(\frac{d^2\sigma}{d\Omega\, dE'}\right)_{\lambda'\to\lambda} = \frac{k'}{k}\left|\sum_j b_j \langle\lambda'|\exp(i\boldsymbol{\kappa}\,.\,\boldsymbol{R}_j)|\lambda\rangle\right|^2 \delta(E_\lambda - E_{\lambda'} + E - E'). \tag{2.40}$$

Integral representation of the δ-function for energy

We now express the δ-function for energy as an integral with respect to time. The reason for doing this will become apparent later in the section.

$$\delta(E_\lambda - E_{\lambda'} + E - E') = \frac{1}{2\pi\hbar}\int_{-\infty}^{\infty} \exp\{i(E_{\lambda'} - E_\lambda)t/\hbar\}\exp(-i\omega t)\, dt, \tag{2.41}$$

where ω is defined by

$$\hbar\omega = E - E'. \tag{2.42}$$

Eq. (2.41) is derived in Appendix A.2.

We also need the following results. Let H be the Hamiltonian of the scattering system. The states λ and λ' are eigenfunctions of H with eigenvalues E_λ and $E_{\lambda'}$, i.e.

$$H|\lambda\rangle = E_\lambda|\lambda\rangle, \quad \text{and} \quad H|\lambda'\rangle = E_{\lambda'}|\lambda'\rangle. \tag{2.43}$$

Then

$$H^n|\lambda\rangle = E_\lambda^n|\lambda\rangle, \tag{2.44}$$

a result obtained from (2.43) by operating n times on $|\lambda\rangle$ with H. Further

$$\exp(-iHt/\hbar)|\lambda\rangle = \exp(-iE_\lambda t/\hbar)|\lambda\rangle. \tag{2.45}$$

This result is obtained by expanding the exponential

$$\exp(-iHt/\hbar) = 1 - iHt/\hbar + \frac{1}{2!}(-iHt/\hbar)^2 + \ldots \tag{2.46}$$

and using (2.44).

The matrix element in (2.40) is the sum of N terms ($j = 1, \ldots N$). So the square of the matrix element is the sum of N^2 terms, of which a typical member is

$$\begin{aligned} &b_{j'}^* b_j \langle\lambda'|\exp(i\boldsymbol{\kappa}\,.\,\boldsymbol{R}_{j'})|\lambda\rangle^* \langle\lambda'|\exp(i\boldsymbol{\kappa}\,.\,\boldsymbol{R}_j)|\lambda\rangle \\ &\quad = b_{j'} b_j \langle\lambda|\exp(-i\boldsymbol{\kappa}\,.\,\boldsymbol{R}_{j'})|\lambda'\rangle\langle\lambda'|\exp(i\boldsymbol{\kappa}\,.\,\boldsymbol{R}_j)|\lambda\rangle. \end{aligned} \tag{2.47}$$

In this equation we have used the fact that the scattering length is real, and also the relation (C.3). From (2.40), (2.41), and (2.47)

$$\left(\frac{\mathrm{d}^2\sigma}{\mathrm{d}\Omega\,\mathrm{d}E'}\right)_{\lambda\to\lambda'}$$

$$=\frac{k'}{k}\sum_{jj'} b_{j'}b_j\langle\lambda|\exp(-\mathrm{i}\boldsymbol{\kappa}\,.\,\boldsymbol{R}_{j'})|\lambda'\rangle\langle\lambda'|\exp(\mathrm{i}\boldsymbol{\kappa}\,.\,\boldsymbol{R}_j)|\lambda\rangle$$

$$\times\frac{1}{2\pi\hbar}\int_{-\infty}^{\infty}\exp\{\mathrm{i}(E_{\lambda'}-E_\lambda)t/\hbar\}\exp(-\mathrm{i}\omega t)\,\mathrm{d}t \qquad (2.48)$$

$$=\frac{k'}{k}\frac{1}{2\pi\hbar}\sum_{jj'} b_{j'}b_j\int_{-\infty}^{\infty}\langle\lambda|\exp(-\mathrm{i}\boldsymbol{\kappa}\,.\,\boldsymbol{R}_{j'})|\lambda'\rangle$$

$$\times\langle\lambda'|\exp(\mathrm{i}Ht/\hbar)\exp(\mathrm{i}\boldsymbol{\kappa}\,.\,\boldsymbol{R}_j)\exp(-\mathrm{i}Ht/\hbar)|\lambda\rangle\exp(-\mathrm{i}\omega t)\,\mathrm{d}t. \qquad (2.49)$$

The last step follows from (2.45).

Sum over λ', average over λ

In an actual experiment we do not measure the cross-section for a process in which the scattering system goes from a specific state λ to another state λ'. Instead we measure $\mathrm{d}^2\sigma/\mathrm{d}\Omega\,\mathrm{d}E'$, the cross-section defined in (1.11). To obtain this quantity we must first sum $(\mathrm{d}^2\sigma/\mathrm{d}\Omega\,\mathrm{d}E')_{\lambda\to\lambda'}$ over all final states λ', keeping the initial state λ fixed, and then average over all λ.

To carry out the first step we use the closure relation proved in Appendix C. For a pair of operators A and B

$$\sum_{\lambda'}\langle\lambda|A|\lambda'\rangle\langle\lambda'|B|\lambda\rangle=\langle\lambda|AB|\lambda\rangle. \qquad (2.50)$$

To average the last quantity over λ, we multiply by p_λ, the probability that the scattering system is in the state λ, and then sum over λ. p_λ is given by the Boltzmann distribution. If the temperature of the scattering system is T

$$p_\lambda=\frac{1}{Z}\exp(-E_\lambda\beta), \qquad (2.51)$$

where

$$Z=\sum_\lambda\exp(-E_\lambda\beta), \qquad (2.52)$$

and

$$\beta=\frac{1}{k_\mathrm{B}T}. \qquad (2.53)$$

Z is known as the *partition function* and is inserted to ensure that $\sum_\lambda p_\lambda=1$.

From (2.49) and (2.50) we have

$$\frac{d^2\sigma}{d\Omega\, dE'} = \sum_{\lambda\lambda'} p_\lambda \left(\frac{d^2\sigma}{d\Omega\, dE'}\right)_{\lambda\to\lambda'}$$

$$= \frac{k'}{k}\frac{1}{2\pi\hbar}\sum_{jj'} b_j b_{j'} \int_{-\infty}^{\infty} \exp(-i\omega t)\, dt \sum_\lambda p_\lambda$$

$$\times \langle\lambda|\exp(-i\boldsymbol{\kappa}.\boldsymbol{R}_{j'})\exp(iHt/\hbar)\exp(i\boldsymbol{\kappa}.\boldsymbol{R}_j)\exp(-iHt/\hbar)|\lambda\rangle. \tag{2.54}$$

The last expression may be written in terms of the time-dependent Heisenberg operator $\boldsymbol{R}_j(t)$, defined by

$$\boldsymbol{R}_j(t) = \exp(iHt/\hbar)\boldsymbol{R}_j \exp(-iHt/\hbar). \tag{2.55}$$

It follows from the definition that

$$\exp\{i\boldsymbol{\kappa}.\boldsymbol{R}_j(t)\} = \exp(iHt/\hbar)\exp(i\boldsymbol{\kappa}.\boldsymbol{R}_j)\exp(-iHt/\hbar). \tag{2.56}$$

Heisenberg operators are discussed in Appendix D, and (2.56) is justified there. Note that

$$\boldsymbol{R}_j(0) = \boldsymbol{R}_j. \tag{2.57}$$

We denote the thermal average of the operator A at temperature T by $\langle A\rangle$, i.e.

$$\langle A\rangle = \sum_\lambda p_\lambda \langle\lambda|A|\lambda\rangle. \tag{2.58}$$

From (2.54), (2.56), and (2.58) we have

$$\frac{d^2\sigma}{d\Omega\, dE'} = \frac{k'}{k}\frac{1}{2\pi\hbar}\sum_{jj'} b_j b_{j'} \int_{-\infty}^{\infty} \langle\exp\{-i\boldsymbol{\kappa}.\boldsymbol{R}_{j'}(0)\}\exp\{i\,\boldsymbol{\kappa}.\boldsymbol{R}_j(t)\}\rangle$$

$$\times\exp(-i\omega t)\, dt. \tag{2.59}$$

This is our basic expression for the partial differential cross-section for nuclear scattering. It is a compact expression and may appear simple, but its evaluation, except for the most elementary scattering systems, is not a simple matter. The properties of the scattering system are contained in the Hamiltonian H. They are therefore contained in the Heisenberg operators, and the eigenstates $|\lambda\rangle$. In the next three chapters we shall be concerned with evaluating the cross-section for specific physical systems.

We can see at this stage the purpose of expressing the δ-function for conservation of energy as an integral with respect to time in (2.41). The $E_{\lambda'}$ term in the argument of the δ-function prevents us from using (2.50) to sum over λ'. By means of (2.41) we are able to bring the $E_{\lambda'}$ term inside one of the matrix elements, where it reappears as an operator depending on the Hamiltonian. There is now no term in λ' outside the two matrix elements, and the sum over λ' can be carried out immediately.

2.4 Coherent and incoherent scattering

Consider a scattering system consisting of a single element where the scattering length b varies from one nucleus to another owing to nuclear spin or the presence of isotopes or both. Let the value b_i occur with relative frequency f_i, i.e.

$$\sum_i f_i = 1. \tag{2.60}$$

Then the average value of b for the system is

$$\bar{b} = \sum_i f_i b_i, \tag{2.61}$$

and the average value of b^2 is

$$\overline{b^2} = \sum_i f_i b_i^2. \tag{2.62}$$

We assume there is no correlation between the values of b for any two nuclei, that is to say, whatever the value of b for one nucleus, the probability that another nucleus has the value b_i is simply f_i.

Imagine that we have a large number of scattering systems. They are identical in every way as regards the positions and motions of the nuclei. Also the total number of each b_i is the same for all the systems. But each system has a different distribution of the bs among the nuclei, every possible distribution being represented once. Now, provided the system contains a large number of nuclei – a condition usually well satisfied – the cross-section we measure is very close to the cross-section averaged over all the systems. This is given by

$$\frac{d^2\sigma}{d\Omega\, dE'} = \frac{k'}{k}\frac{1}{2\pi\hbar}\sum_{jj'} \overline{b_{j'}b_j} \int \langle j', j\rangle \exp(-i\omega t)\, dt, \tag{2.63}$$

where

$$\langle j', j\rangle = \langle \exp\{-i\boldsymbol{\kappa} \,.\, \boldsymbol{R}_{j'}(0)\} \exp\{i\boldsymbol{\kappa} \,.\, \boldsymbol{R}_j(t)\}\rangle. \tag{2.64}$$

(We use this notation for the moment, because only the j', j values of the matrix element are relevant to the discussion.)

On the assumption of no correlation between the b values of different nuclei

$$\begin{aligned} \overline{b_{j'}b_j} &= (\bar{b})^2, \qquad j' \neq j, \\ \overline{b_{j'}b_j} &= \overline{b^2}, \qquad j' = j. \end{aligned} \tag{2.65}$$

So

$$\frac{d^2\sigma}{d\Omega\, dE'} = \frac{k'}{k}\frac{1}{2\pi\hbar}(\bar{b})^2 \sum_{\substack{jj' \\ j' \neq j}} \int \langle j', j\rangle \exp(-i\omega t)\, dt + \frac{k'}{k}\frac{1}{2\pi\hbar}\overline{b^2} \sum_j \int \langle j, j\rangle \exp(-i\omega t)\, dt \tag{2.66}$$

$$= \frac{k'}{k}\frac{1}{2\pi\hbar}(\bar{b})^2 \sum_{jj'} \int \langle j', j\rangle \exp(-i\omega t)\, dt + \frac{k'}{k}\frac{1}{2\pi\hbar}\{\overline{b^2} - (\bar{b})^2\} \sum_j \int \langle j, j\rangle \exp(-i\omega t)\, dt. \tag{2.67}$$

In going from (2.66) to (2.67) we have added and subtracted $(k'/k)(1/2\pi\hbar)(\bar{b})^2 \int \sum_j \langle j, j\rangle \exp(-i\omega t)\, dt$. The first term in (2.67) is known as the *coherent* and the second term as the *incoherent* scattering cross-section. We write

$$\left(\frac{d^2\sigma}{d\Omega\, dE'}\right)_{coh} = \frac{\sigma_{coh}}{4\pi}\frac{k'}{k}\frac{1}{2\pi\hbar}\sum_{jj'}\int_{-\infty}^{\infty} \langle \exp\{-i\boldsymbol{\kappa}\,.\,\boldsymbol{R}_{j'}(0)\} \exp\{i\boldsymbol{\kappa}\,.\,\boldsymbol{R}_j(t)\}\rangle \times \exp(-i\omega t)\, dt, \tag{2.68}$$

$$\left(\frac{d^2\sigma}{d\Omega\, dE'}\right)_{inc} = \frac{\sigma_{inc}}{4\pi}\frac{k'}{k}\frac{1}{2\pi\hbar}\sum_{j}\int_{-\infty}^{\infty} \langle \exp\{-i\boldsymbol{\kappa}\,.\,\boldsymbol{R}_{j}(0)\} \exp\{i\boldsymbol{\kappa}\,.\,\boldsymbol{R}_j(t)\}\rangle \times \exp(-i\omega t)\, dt, \tag{2.69}$$

where
$$\sigma_{coh} = 4\pi(\bar{b})^2, \qquad \sigma_{inc} = 4\pi\{\overline{b^2} - (\bar{b})^2\}. \tag{2.70}$$

We see from the equations that the coherent scattering depends on the correlation between the positions of the *same* nucleus at different times, and on the correlation between the positions of *different* nuclei at different times. It therefore gives *interference* effects. The incoherent scattering depends only on the correlation between the positions of the *same* nucleus at different times. It does not give interference effects.

The physical interpretation of (2.67) is as follows. The actual scattering system has different scattering lengths associated with different nuclei. The coherent scattering is the scattering the same system (same nuclei with the same positions and motions) would give if all the scattering lengths were equal to $\bar{b}$. The incoherent scattering is the term we must add to this to obtain the scattering due to the actual system. Physically the incoherent scattering arises from the random distribution of the deviations of the scattering lengths from their mean value.

We now derive expressions for the frequencies f_i and for $\bar{b}$ and $\overline{b^2}$. The simplest case is when the scattering system consists of a single isotope with zero nuclear spin. Then all the bs are equal, and the scattering is entirely coherent.

Suppose the system consists of a single isotope with nuclear spin I. The spin of the nucleus–neutron system has the values $I+\frac{1}{2}$ or $I-\frac{1}{2}$. Denote the scattering lengths for the two spin values by b^+ and b^-. The number of states associated with spin $I+\frac{1}{2}$ is

$$2(I+\tfrac{1}{2})+1=2I+2, \tag{2.71}$$

and the number of states associated with spin $I-\frac{1}{2}$ is

$$2(I-\tfrac{1}{2})+1=2I. \tag{2.72}$$

If the neutrons are unpolarised and the nuclear spins are randomly oriented, each spin state has the same *a priori* probability. So the scattering length b^+ occurs with frequency

$$f^+=\frac{2I+2}{4I+2}=\frac{I+1}{2I+1}, \tag{2.73}$$

and the scattering length b^- occurs with frequency

$$f^-=\frac{2I}{4I+2}=\frac{I}{2I+1}. \tag{2.74}$$

Thus

$$\bar{b}=\frac{1}{2I+1}\{(I+1)b^+ + Ib^-\}. \tag{2.75}$$

If the neutrons are polarised or the nuclear spins are aligned, the $4I+2$ spin states of the nucleus–neutron system are not equally probable. But unless *both* conditions apply, f^+ and f^- have the values shown in (2.73) and (2.74).

If there are several isotopes in the scattering system, then for each isotope the quantities f^+ and f^- must be multiplied by the relative abundance of the isotope to obtain the relative frequency of the scattering length. So, in general,

$$\bar{b}=\sum_{\xi}\frac{c_\xi}{2I_\xi+1}\{(I_\xi+1)b_\xi^+ + I_\xi b_\xi^-\}, \tag{2.76}$$

$$\overline{b^2}=\sum_{\xi}\frac{c_\xi}{2I_\xi+1}\{(I_\xi+1)(b_\xi^+)^2 + I_\xi(b_\xi^-)^2\}, \tag{2.77}$$

where c_ξ is the relative abundance of the ξth isotope, I_ξ its nuclear spin, and b_ξ^+ and b_ξ^- its scattering lengths. The quantity $\bar{b}$ is known as the *coherent* scattering length of the element or nuclide. It is conventional to quote the values of $\bar{b}$ and $\overline{b^2}$ in terms of the two quantities σ_{coh} and σ_{inc} defined in (2.70). A list of the values of σ_{coh} and σ_{inc} for the elements, together with a description of the methods of measuring these quantities, has been given by Koester (1977). A few of the values are given in Table 2.1.

Table 2.1 Values of σ_{coh} and σ_{inc}

Element or nuclide	Z	σ_{coh}	σ_{inc}	Element	Z	σ_{coh}	σ_{inc}
^1H	1	1.8	80.2	V	23	0.02	5.0
^2H	1	5.6	2.0	Fe	26	11.5	0.4
C	6	5.6	0.0	Co	27	1.0	5.2
O	8	4.2	0.0	Ni	28	13.4	5.0
Mg	12	3.6	0.1	Cu	29	7.5	0.5
Al	13	1.5	0.0	Zn	30	4.1	0.1

The units of σ_{coh} and σ_{inc} are 10^{-28} m^2. The values are taken from Koester (1977).

The extension of the theory to scattering systems containing more than one element is readily made. If for example the scattering system is a crystal of NaCl, the coherent scattering is that due to a hypothetical crystal in which all the sodium nuclei have scattering lengths equal to $\bar{b}$ for sodium, and all the chlorine nuclei have scattering lengths equal to $\bar{b}$ for chlorine. The incoherent scattering is the sum of the incoherent scattering from the sodium nuclei and the incoherent scattering from the chlorine nuclei.

3

Nuclear scattering by crystals

3.1 Introduction

In the present chapter we evaluate the cross-sections when the scattering system is a single crystal. We start by considering a Bravais crystal, i.e. a crystal with one atom per unit cell. Denote the sides of the unit cell by $\boldsymbol{a}_1$, $\boldsymbol{a}_2$, $\boldsymbol{a}_3$ (see Fig. 3.1). Then a lattice vector is given by

$$\boldsymbol{l} = l_1\boldsymbol{a}_1 + l_2\boldsymbol{a}_2 + l_3\boldsymbol{a}_3, \tag{3.1}$$

where l_1, l_2, l_3 are integers. The volume of the unit cell is

$$v_0 = \boldsymbol{a}_1 \, . \, [\boldsymbol{a}_2 \times \boldsymbol{a}_3]. \tag{3.2}$$

We define the reciprocal lattice to be a lattice with unit-cell vectors $\boldsymbol{\tau}_1$, $\boldsymbol{\tau}_2$, $\boldsymbol{\tau}_3$, where

$$\boldsymbol{\tau}_1 = \frac{2\pi}{v_0}[\boldsymbol{a}_2 \times \boldsymbol{a}_3], \qquad \boldsymbol{\tau}_2 = \frac{2\pi}{v_0}[\boldsymbol{a}_3 \times \boldsymbol{a}_1], \qquad \boldsymbol{\tau}_3 = \frac{2\pi}{v_0}[\boldsymbol{a}_1 \times \boldsymbol{a}_2]. \tag{3.3}$$

The volume of the unit cell in the reciprocal lattice is

$$\boldsymbol{\tau}_1 \, . \, [\boldsymbol{\tau}_2 \times \boldsymbol{\tau}_3] = \frac{(2\pi)^3}{v_0}. \tag{3.4}$$

Fig. 3.1 Unit cell of crystal.

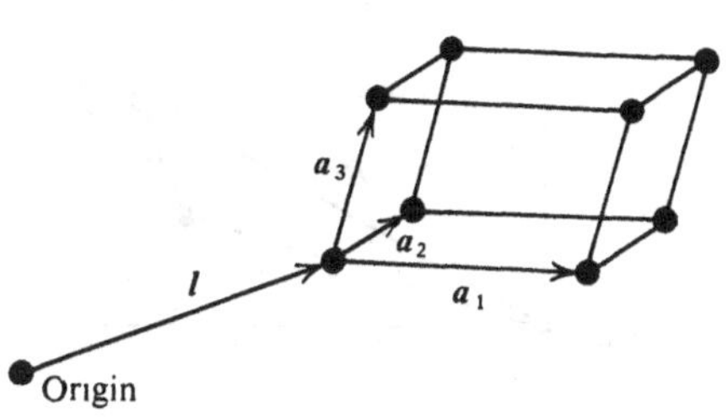

From (3.3)
$$\boldsymbol{a}_i \,.\, \boldsymbol{\tau}_j = 2\pi\delta_{ij}. \tag{3.5}$$

Owing to thermal motion the nucleus l is displaced from its equilibrium position $\boldsymbol{l}$. Its instantaneous position is

$$\boldsymbol{R}_l = \boldsymbol{l} + \boldsymbol{u}_l, \tag{3.6}$$

where $\boldsymbol{u}_l$ is the displacement from the equilibrium position (Fig. 3.2). The index j with which we previously labelled a nucleus now becomes l.

For a Bravais crystal the correlation between the positions of the nuclei l and l' depends only on $\boldsymbol{l}-\boldsymbol{l}'$. So in (2.68), for each value of l', the sum over l is the same. We may thus put $l'=0$. Similarly in (2.69) each term in l is the same and equal to the term $l=0$. Therefore

$$\sum_{ll'} \langle \exp\{-\mathrm{i}\boldsymbol{\kappa}\,.\,\boldsymbol{R}_{l'}(0)\}\exp\{\mathrm{i}\boldsymbol{\kappa}\,.\,\boldsymbol{R}_l(t)\}\rangle$$
$$= N\sum_l \exp(\mathrm{i}\boldsymbol{\kappa}\,.\,\boldsymbol{l})\langle \exp\{-\mathrm{i}\boldsymbol{\kappa}\,.\,\boldsymbol{u}_0(0)\}\exp\{\mathrm{i}\boldsymbol{\kappa}\,.\,\boldsymbol{u}_l(t)\}\rangle, \tag{3.7}$$

$$\sum_{l} \langle \exp\{-\mathrm{i}\boldsymbol{\kappa}\,.\,\boldsymbol{R}_{l}(0)\}\exp\{\mathrm{i}\boldsymbol{\kappa}\,.\,\boldsymbol{R}_l(t)\}\rangle$$
$$= N\langle \exp\{-\mathrm{i}\boldsymbol{\kappa}\,.\,\boldsymbol{u}_0(0)\}\exp\{\mathrm{i}\boldsymbol{\kappa}\,.\,\boldsymbol{u}_0(t)\}\rangle, \tag{3.8}$$

where N is the number of nuclei in the crystal. $\boldsymbol{u}_l(t)$ is the Heisenberg operator for $\boldsymbol{u}_l$. In these equations we have used the relation

$$\boldsymbol{R}_l(t) = \boldsymbol{l} + \boldsymbol{u}_l(t), \tag{3.9}$$

which follows from (3.6), since $\boldsymbol{l}$ is a constant.

3.2 Normal modes

We assume that the interatomic forces in the crystal are harmonic, i.e. that the forces are linear functions of the displacements. For such forces the displacements $\boldsymbol{u}_l$ can be expressed as the sum of displace-

Fig. 3.2 Position of nucleus l: ● equilibrium position, ○ actual (instantaneous) position.

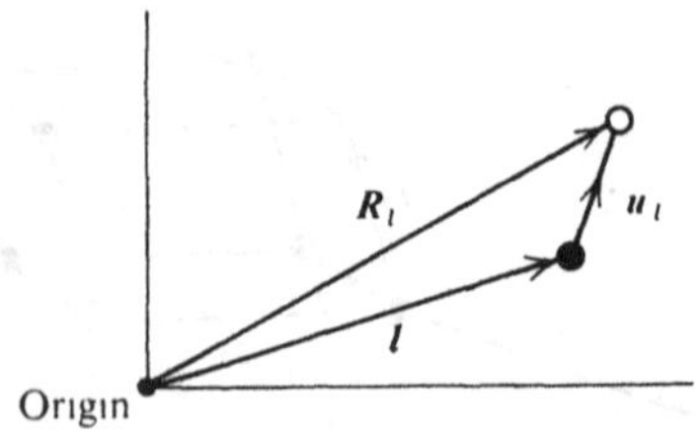

ments due to a set of normal modes. Thus

$$\boldsymbol{u}_l=\left(\frac{\hbar}{2MN}\right)^{1/2}\sum_s\frac{\boldsymbol{e}_s}{\sqrt{\omega_s}}\{a_s\exp(\mathrm{i}\boldsymbol{q}\,.\,\boldsymbol{l})+a_s^+\exp(-\mathrm{i}\boldsymbol{q}\,.\,\boldsymbol{l})\},\qquad(3.10)$$

where $\boldsymbol{q}$ is the wavevector of the mode, and j is its polarisation index ($j=1,2,3$). s stands for the double index $\boldsymbol{q},j$. ω_s is the angular frequency of mode s, and $\boldsymbol{e}_s$ is its polarisation vector. The sum over s is over the N values of $\boldsymbol{q}$ in the 1st Brillouin zone, and over the three values of j. M is the mass of an atom – assumed to be the same for all the atoms. a_s is the annihilation operator for the mode s, and a_s^+, its Hermitian conjugate, is the creation operator. These operators are discussed in Appendix E. Normal modes are discussed in Appendix G, where (3.10) is derived.

$\boldsymbol{u}_l(t)$ is obtained from (3.10) by replacing a_s and a_s^+ by their Heisenberg operators, $a_s(t)$ and $a_s^+(t)$. It is shown in Appendix E.1 that

$$a_s(t)=\exp(\mathrm{i}Ht/\hbar)a_s\exp(-\mathrm{i}Ht/\hbar)=a_s\exp(-\mathrm{i}\omega_st),\qquad(3.11)$$

$$a_s^+(t)=\exp(\mathrm{i}Ht/\hbar)a_s^+\exp(-\mathrm{i}Ht/\hbar)=a_s^+\exp(\mathrm{i}\omega_st).\qquad(3.12)$$

Thus

$$\boldsymbol{\kappa}\,.\,\boldsymbol{u}_l(t)=\left(\frac{\hbar}{2MN}\right)^{1/2}\sum_s\frac{\boldsymbol{\kappa}\,.\,\boldsymbol{e}_s}{\sqrt{\omega_s}}[a_s\exp\{\mathrm{i}(\boldsymbol{q}\,.\,\boldsymbol{l}-\omega_st)\}+a_s^+\exp\{-\mathrm{i}(\boldsymbol{q}\,.\,\boldsymbol{l}-\omega_st)\}].\qquad(3.13)$$

3.3 Probability function for a harmonic oscillator

The theory of the scattering of thermal neutrons by crystals is much simplified by the use of a result, first derived by Bloch (1932), for the probability function of a harmonic oscillator.

We first define the probability function for a single bound particle - not necessarily a harmonic oscillator – moving in one dimension. Denote the displacement or position variable by Q. Denote the Hamiltonian by H, and its normalised eigenfunctions and eigenvalues by ψ_n and E_n. Then

$$H\psi_n=E_n\psi_n.\qquad(3.14)$$

If the particle is in an energy eigenstate n, the probability of finding the displacement between Q and $Q+\mathrm{d}Q$ is $|\psi_n(Q)|^2\,\mathrm{d}Q$. Suppose the particle is not in a single state n, but in an incoherent mixture of states. If the particle is a member of an ensemble of similar particles at temperature T, the probability of its being in the state n is

$$p_n = \frac{1}{Z}\exp(-E_n\beta), \qquad Z = \sum_n \exp(-E_n\beta), \qquad \beta = \frac{1}{k_B T}. \tag{3.15}$$

We define the *probability function* $f(Q)$ by

$$f(Q) = \sum_n p_n |\psi_n(Q)|^2. \tag{3.16}$$

$f(Q)\,dQ$ is the probability of finding the displacement between Q and $Q+dQ$. Since the ψ_n are normalised, and $\sum_n p_n = 1$, it follows from (3.16) that

$$\int_{-\infty}^{\infty} f(Q)\,dQ = 1, \tag{3.17}$$

a necessary result in view of the probability significance of $f(Q)$.

We can express $\langle A(Q)\rangle$, the thermal average of a function $A(Q)$, in terms of $f(Q)$. From (2.58)

$$\begin{aligned}\langle A(Q)\rangle &= \sum_n p_n \int_{-\infty}^{\infty} A(Q)|\psi_n(Q)|^2\,dQ\\ &= \int_{-\infty}^{\infty} A(Q)\sum_n p_n|\psi_n(Q)|^2\,dQ\\ &= \int_{-\infty}^{\infty} A(Q)f(Q)\,dQ.\end{aligned} \tag{3.18}$$

Bloch showed that, for a one-dimensional harmonic oscillator, the probability function is a Gaussian, given by

$$f(Q) = C\exp(-Q^2/2\sigma^2), \tag{3.19}$$

where

$$\sigma^2 = \frac{\hbar}{2M\omega}\coth(\tfrac{1}{2}\hbar\omega\beta). \tag{3.20}$$

M is the mass of the particle, and ω the angular frequency. C is a normalising constant obtained from (3.17). It follows from (3.18) and (3.19) that the thermal averages of $\exp Q$ and Q^2 are related by

$$\langle \exp Q\rangle = \exp\{\tfrac{1}{2}\langle Q^2\rangle\}. \tag{3.21}$$

The results (3.19) to (3.21) are derived in Appendix E.2.

3.4 Development of $\langle \exp U \exp V\rangle$

We first evaluate the coherent cross-section. From (2.68) and (3.7)

$$\left(\frac{d^2\sigma}{d\Omega\,dE'}\right)_{\text{coh}} = \frac{\sigma_{\text{coh}}}{4\pi}\frac{k'}{k}\frac{N}{2\pi\hbar}\sum_l \exp(i\boldsymbol{\kappa}\,.\,\boldsymbol{l})\int_{-\infty}^{\infty}\langle \exp U \exp V\rangle \times\exp(-i\omega t)\,dt, \tag{3.22}$$

where

$$U = -\mathrm{i}\boldsymbol{\kappa} \, . \, \boldsymbol{u}_0(0) = -\mathrm{i} \sum_s g_s a_s + g_s a_s^+, \tag{3.23}$$

$$V = \mathrm{i}\boldsymbol{\kappa} \, . \, \boldsymbol{u}_l(t) = \mathrm{i} \sum_s h_s a_s + h_s^* a_s^+, \tag{3.24}$$

$$g_s = \left(\frac{\hbar}{2MN}\right)^{1/2} \frac{\boldsymbol{\kappa} \, . \, \boldsymbol{e}_s}{\sqrt{\omega_s}}, \tag{3.25}$$

$$h_s = \left(\frac{\hbar}{2MN}\right)^{1/2} \frac{\boldsymbol{\kappa} \, . \, \boldsymbol{e}_s}{\sqrt{\omega_s}} \exp\{\mathrm{i}(\boldsymbol{q} \, . \, \boldsymbol{l} - \omega_s t)\}. \tag{3.26}$$

Eqs. (3.23) to (3.26) follow from (3.13).

We now develop the expression $\langle \exp U \exp V \rangle$. In addition to the result that the probability function for a harmonic oscillator is a Gaussian, we need the following results:

(i) The a and a^+ operators for different oscillators commute. For the same oscillator, the commutation relation for a and a^+ is given by (E.8). The two results are combined in the equation

$$[a_s, a_{s'}^+] = \delta_{ss'}. \tag{3.27}$$

(ii) If A and B are any two operators whose commutator is a c-number (i.e. a number as opposed to an operator), then

$$\exp A \exp B = \exp(A + B) \exp\{\tfrac{1}{2}(AB - BA)\}. \tag{3.28}$$

This result is proved in Appendix I.1.

We first prove that $UV - VU$ is a c-number. From (3.23) and (3.24)

$$UV - VU = \sum_s (g_s a_s + g_s a_s^+) \sum_{s'} (h_{s'} a_{s'} + h_{s'}^* a_{s'}^+)$$
$$- \sum_{s'} (h_{s'} a_{s'} + h_{s'}^* a_{s'}^+) \sum_s (g_s a_s + g_s a_s^+). \tag{3.29}$$

Eq. (3.27) shows that all the terms on the right-hand side of (3.29) give zero, except those with $s' = s$. Thus

$$UV - VU = \sum_s (g_s h_s^* - g_s h_s)(a_s a_s^+ - a_s^+ a_s)$$
$$= \sum_s (g_s h_s^* - g_s h_s), \tag{3.30}$$

which is a c-number.

We next use (3.28) for the operators U and V, and take the thermal average

$$\langle \exp U \exp V \rangle = \langle \exp(U + V) \rangle \exp\{\tfrac{1}{2}(UV - VU)\}. \tag{3.31}$$

Note that the second term on the right-hand side is a number that does not depend on T. The quantity $U + V$ is a linear combination of harmonic displacements. Each displacement has a Gàussian probability function. The probability function for a linear combination of Gaussians is itself a Gaussian. We can therefore apply (3.21) to $U+V$.

$$\langle \exp(U+V)\rangle = \exp\{\tfrac{1}{2}\langle (U+V)^2\rangle\}. \tag{3.32}$$

From (3.31) and (3.32)

$$\begin{aligned}\langle \exp U \exp V\rangle &= \exp\{\tfrac{1}{2}\langle (U+V)^2\rangle\}\exp\{\tfrac{1}{2}\langle UV-VU\rangle\}\\ &= \exp\{\tfrac{1}{2}\langle U^2+V^2+UV+VU+UV-VU\rangle\}\\ &= \exp\{\tfrac{1}{2}\langle U^2+V^2\rangle\}\exp\langle UV\rangle.\end{aligned} \tag{3.33}$$

Now $$\langle U^2\rangle = \langle V^2\rangle. \tag{3.34}$$

This can be proved formally, but it can be seen on physical grounds. U is proportional to the component in the direction of $\boldsymbol{\kappa}$ of the displacement of the origin atom at time zero. V is the corresponding quantity – apart from a change of sign – for the atom $\boldsymbol{l}$ at time t. But the zero of time is arbitrary, and for a Bravais crystal all the atoms are equivalent. So the average values of U^2 and V^2 are equal.

From (3.33) and (3.34)

$$\langle \exp U \exp V\rangle = \exp\langle U^2\rangle \exp\langle UV\rangle. \tag{3.35}$$

This completes the development of $\langle \exp U \exp V\rangle$. Substituting (3.35) in (3.22) gives

$$\left(\frac{d^2\sigma}{d\Omega\, dE'}\right)_{\text{coh}} = \frac{\sigma_{\text{coh}}}{4\pi}\frac{k'}{k}\frac{N}{2\pi\hbar}\exp\langle U^2\rangle \sum_{l}\exp(i\boldsymbol{\kappa}\cdot\boldsymbol{l}) \times \int_{-\infty}^{\infty}\exp\langle UV\rangle \exp(-i\omega t)\,dt. \tag{3.36}$$

3.5 Phonon expansion

A crystal of N atoms has $3N$ normal modes. The initial state λ of the crystal is given by specifying $n_1, n_2 \ldots n_{3N}$, the quantum numbers of the $3N$ oscillators corresponding to the normal modes. In a general scattering process, the state of the crystal changes to λ', which is given by another set of quantum numbers $n'_1, n'_2 \ldots n'_{3N}$. The scattering process may be classified according to the changes in the quantum numbers.

Elastic process. All the quantum numbers remain unchanged, i.e.

$$n_i' = n_i \tag{3.37}$$

for all i from 1 to $3N$.

One-phonon process. All the quantum numbers remain unchanged except for one, that of oscillator α, which changes by unity, i.e.

$$\left.\begin{aligned} n_i' &= n_i, \qquad \text{all } i \text{ except } \alpha. \\ n_\alpha' &= n_\alpha \pm 1. \end{aligned}\right\} \tag{3.38}$$

Two-phonon process. All the quantum numbers remain unchanged except for two, those of oscillators α and β, which change by unity, i.e.

$$\left.\begin{aligned} n_i' &= n_i \qquad \text{all } i \text{ except } \alpha \text{ and } \beta, \\ n_\alpha' &= n_\alpha \pm 1, \qquad n_\beta' = n_\beta \pm 1. \end{aligned}\right\} \tag{3.39}$$

Similarly for three, four, etc. phonon processes.

If we expand the term $\exp\langle UV\rangle$ in (3.36)

$$\exp\langle UV\rangle = 1 + \langle UV\rangle + \frac{1}{2!}\langle UV\rangle^2 + \ldots + \frac{1}{p!}\langle UV\rangle^p + \ldots, \tag{3.40}$$

then the pth term gives the cross-section for all p-phonon processes. Thus the first term, 1, gives the elastic cross-section. The next term $\langle UV\rangle$ gives the cross-section for all one-phonon processes in which α is in turn each of the numbers 1 to $3N$, and, for each α, n_α either increases or decreases by unity. The term $(1/2!)\langle UV\rangle^2$ gives the cross-section for all two-phonon processes in which the combination α, β is in turn each of the $3N(3N-1)/2$ combinations of two oscillators selected from $3N$, and, for each combination, n_α and n_β increase or decrease by unity. And so on.

The statement that the pth term in the expansion of $\exp\langle UV\rangle$ corresponds to a p-phonon process can be justified in two ways. One way is to go back to expression (2.49) for the cross-section for a specific $\lambda \to \lambda'$ transition. Instead of summing over all λ', sum only over those λ' which, for a fixed λ, correspond to a p-phonon process. Then average over λ as before. The result, after a somewhat lengthy calculation, is the expression in (3.36) with $\exp\langle UV\rangle$ replaced by $(1/p!)\langle UV\rangle^p$. The second way is simply to inspect the expressions for each term. They contain δ-functions which show that the process is elastic, one-phonon, two-phonon, and so on. This is the line we shall follow.

3.6 Coherent elastic scattering

Bragg's law

Replace $\exp\langle UV\rangle$ by 1 in (3.36)

$$\left(\frac{d^2\sigma}{d\Omega\, dE'}\right)_{\text{coh el}} = \frac{\sigma_{\text{coh}}}{4\pi}\frac{k'}{k}\frac{N}{2\pi\hbar}\exp\langle U^2\rangle \sum_l \exp(i\boldsymbol{\kappa}\,.\,\boldsymbol{l}) \int_{-\infty}^{\infty} \exp(-i\omega t)\, dt. \tag{3.41}$$

The integral with respect to t is

$$\int_{-\infty}^{\infty} \exp(-i\omega t)\, dt = 2\pi\delta(\omega) = 2\pi\hbar\delta(\hbar\omega) \tag{3.42}$$

(see Appendix A.2). Since $\hbar\omega$ is the change in the energy of the neutron we see that the scattering is elastic. Thus

$$|\boldsymbol{k}'| = |\boldsymbol{k}|, \tag{3.43}$$

and (3.41) becomes

$$\left(\frac{d^2\sigma}{d\Omega\, dE'}\right)_{\text{coh el}} = \frac{\sigma_{\text{coh}}}{4\pi} N \exp\langle U^2\rangle \sum_l \exp(i\boldsymbol{\kappa}\,.\,\boldsymbol{l})\delta(\hbar\omega). \tag{3.44}$$

We can immediately integrate with respect to E' to obtain the differential scattering cross-section $d\sigma/d\Omega$. The energy E of the incident neutrons is fixed. Therefore

$$dE' = -d(\hbar\omega), \tag{3.45}$$

and we have

$$\left(\frac{d\sigma}{d\Omega}\right)_{\text{coh el}} = \int_0^{\infty} \left(\frac{d^2\sigma}{d\Omega\, dE'}\right)_{\text{coh el}} dE'$$

$$= \frac{\sigma_{\text{coh}}}{4\pi} N \exp\langle U^2\rangle \sum_l \exp(i\boldsymbol{\kappa}\,.\,\boldsymbol{l}). \tag{3.46}$$

The lattice sum can be written in the form (see Appendix A.4)

$$\sum_l \exp(i\boldsymbol{\kappa}\,.\,\boldsymbol{l}) = \frac{(2\pi)^3}{v_0} \sum_{\boldsymbol{\tau}} \delta(\boldsymbol{\kappa} - \boldsymbol{\tau}), \tag{3.47}$$

where v_0 is the volume of the unit cell of the crystal, and $\boldsymbol{\tau}$ is a vector in the reciprocal lattice. Thus

$$\left(\frac{d\sigma}{d\Omega}\right)_{\text{coh el}} = \frac{\sigma_{\text{coh}}}{4\pi} N \frac{(2\pi)^3}{v_0} \exp(-2W) \sum_{\boldsymbol{\tau}} \delta(\boldsymbol{\kappa} - \boldsymbol{\tau}), \tag{3.48}$$

where

$$2W = -\langle U^2\rangle = \langle\{\boldsymbol{\kappa}\,.\,\boldsymbol{u}_0(0)\}^2\rangle. \tag{3.49}$$

Eq. (3.48) tells us that the scattering occurs only when

$$\boldsymbol{\kappa} = \boldsymbol{k} - \boldsymbol{k}' = \boldsymbol{\tau}. \tag{3.50}$$

This condition is the same as Bragg's law for X-ray scattering. So coherent elastic scattering of neutrons is simply Bragg scattering.

We can represent (3.50) by a diagram in reciprocal space. Fig. 3.3 shows the reciprocal lattice with the origin at O. AO represents the wavevector of the incident neutrons, and AB that of the scattered neutrons. Since the scattering is elastic, $AO = AB$. In general the point B does not coincide with a reciprocal lattice point, and there is no coherent elastic scattering (Fig. 3.3*a*). But for special orientations of $\boldsymbol{k}$ with respect to the crystal lattice (and hence to the reciprocal lattice), and for special scattering angles θ, B can coincide with a reciprocal lattice point (Fig. 3.3*b*). Coherent scattering then occurs. In Fig. 3.3*b*, OAB is an isosceles triangle with $BO = \boldsymbol{\tau}$. Thus

$$\tau = 2k \sin \tfrac{1}{2}\theta. \tag{3.51}$$

The vector $\boldsymbol{\tau}$ is perpendicular to a set of crystal planes. Its magnitude is

$$\tau = n\frac{2\pi}{d}, \tag{3.52}$$

where d is the spacing of the planes, and n is an integer. Also

$$k = \frac{2\pi}{\lambda}, \tag{3.53}$$

where λ is the wavelength of the neutrons. Substituting these relations in (3.51) gives

$$n\lambda = 2d \sin \tfrac{1}{2}\theta, \tag{3.54}$$

which is the familiar form of Bragg's law.

Fig. 3.3 Bragg's law in reciprocal space. (a) $\boldsymbol{\kappa} \neq \boldsymbol{\tau}$; no coherent elastic scattering. (b) $\boldsymbol{\kappa} = \boldsymbol{\tau}$; coherent elastic scattering occurs.

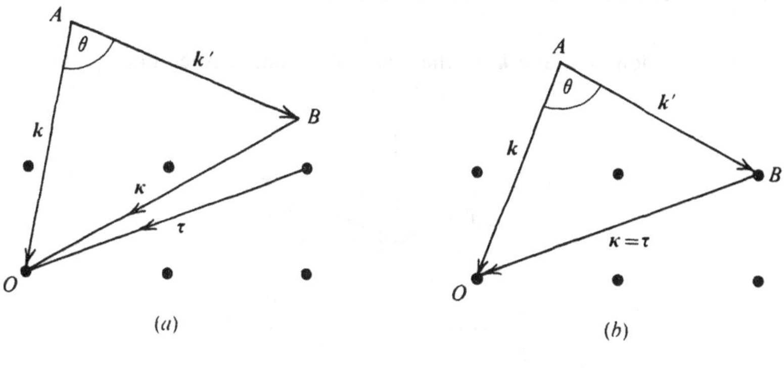

Fig. 3.3*b* is more usually drawn as in Fig. 3.4, which shows the crystal planes acting as a mirror for the incident and reflected neutrons. It can be seen that $\frac{1}{2}\theta$ is the angle between $\boldsymbol{k}$ and the crystal planes.

Debye–Waller factor

The exponential term in (3.48) is known as the *Debye–Waller factor.* From (3.23)

$$2W = -\langle U^2\rangle = \sum_\lambda p_\lambda \sum_{ss'} g_s g_{s'} \langle \lambda|(a_s + a_s^+)(a_{s'} + a_{s'}^+)|\lambda\rangle$$

$$= \sum_\lambda p_\lambda \sum_s g_s^2 \langle \lambda|a_s a_s^+ + a_s^+ a_s|\lambda\rangle. \qquad (3.55)$$

The last step follows because the matrix elements are zero except for terms with $a_s a_s^+$ and $a_s^+ a_s$ (see Appendix E.1). From (E.13)

$$\langle \lambda|a_s a_s^+ + a_s^+ a_s|\lambda\rangle = 2n_s + 1, \qquad (3.56)$$

where n_s is the quantum number of the *s*th oscillator for the state λ. The form of p_λ in (2.51) shows that, in averaging over λ, each term in the sum over s may be averaged independently. Thus from (3.25)

$$2W = \frac{\hbar}{2MN} \sum_s \frac{(\boldsymbol{\kappa}\,.\,\boldsymbol{e}_s)^2}{\omega_s} \langle 2n_s + 1\rangle. \qquad (3.57)$$

It is shown in Appendix E.3 that

$$\langle 2n_s + 1\rangle = \coth(\tfrac{1}{2}\hbar\omega_s\beta), \qquad (3.58)$$

whence

$$2W = \frac{\hbar}{2MN} \sum_s \frac{(\boldsymbol{\kappa}\,.\,\boldsymbol{e}_s)^2}{\omega_s} \coth(\tfrac{1}{2}\hbar\omega_s\beta). \qquad (3.59)$$

Cubic crystal. For a cubic crystal this expression can be developed further. If, for the same polarisation branch, we sum over all the $\boldsymbol{q}$ points related by symmetry, ω_s remains the same, and the mean value

Fig. 3.4 Relation of $\boldsymbol{k}$ and $\boldsymbol{k}'$ to the planes of atoms (dashed lines) in Bragg scattering.

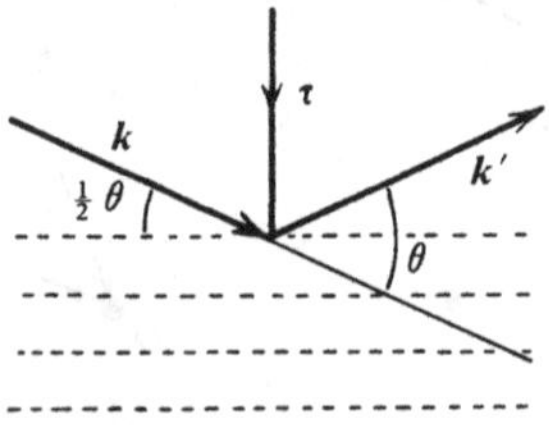

of $(\boldsymbol{\kappa}\,.\,\boldsymbol{e}_s)^2$ is $\frac{1}{3}\kappa^2$. Thus

$$2W=\frac{\hbar}{2MN}\frac{\kappa^2}{3}\sum_s\frac{1}{\omega_s}\coth(\tfrac{1}{2}\hbar\omega_s\beta). \tag{3.60}$$

This result may be expressed another way. For a Bravais crystal the mean-square displacement of an atom is independent of the lattice site. We may therefore write

$$2W=\langle(\boldsymbol{\kappa}\,.\,\boldsymbol{u})^2\rangle=\kappa^2\langle u_\kappa^2\rangle, \tag{3.61}$$

where u_κ is the component of $\boldsymbol{u}$ in the direction of $\boldsymbol{\kappa}$. This equation is true for any crystal. In general the quantity $\langle u_\kappa^2\rangle$ depends on the direction of $\boldsymbol{\kappa}$, but we can see from (3.60) that for a cubic crystal it is independent of direction. If u_x, u_y, u_z are the components of $\boldsymbol{u}$ along three orthogonal directions, it follows that for a cubic crystal

$$\langle u_x^2\rangle=\langle u_y^2\rangle=\langle u_z^2\rangle=\tfrac{1}{3}\langle u^2\rangle, \tag{3.62}$$

where $\langle u^2\rangle$ is the mean value of u^2. Thus

$$2W=\tfrac{1}{3}\kappa^2\langle u^2\rangle. \tag{3.63}$$

Since, for a cubic crystal, the Debye–Waller factor depends only on the frequencies of the normal modes and not on the polarisation vectors, it can be expressed in terms of $Z(\omega)$, the phonon density of states. The definition of this function is that $Z(\omega)\,\mathrm{d}\omega$ is the fraction of the normal modes whose frequencies lie in the range ω to $\omega+\mathrm{d}\omega$. So

$$\int_0^{\omega_m} Z(\omega)\,\mathrm{d}\omega=1, \tag{3.64}$$

where ω_m is the maximum frequency of the normal modes. The total number of normal modes is $3N$. We can therefore put

$$\sum_s\frac{1}{\omega_s}\coth(\tfrac{1}{2}\hbar\omega_s\beta)=3N\int_0^{\omega_m}\frac{1}{\omega}\coth(\tfrac{1}{2}\hbar\omega\beta)Z(\omega)\,\mathrm{d}\omega. \tag{3.65}$$

Thus from (3.60)

$$2W=\frac{\hbar\kappa^2}{2M}\int_0^{\omega_m}\frac{1}{\omega}\coth(\tfrac{1}{2}\hbar\omega\beta)Z(\omega)\,\mathrm{d}\omega. \tag{3.66}$$

and from (3.63)

$$\langle u^2\rangle=\frac{3\hbar}{2M}\int_0^{\omega_m}\frac{1}{\omega}\coth(\tfrac{1}{2}\hbar\omega\beta)Z(\omega)\,\mathrm{d}\omega. \tag{3.67}$$

Even for a non-cubic crystal the value of $(\boldsymbol{\kappa}\,.\,\boldsymbol{e}_s)^2$, averaged over the normal modes with the same value of ω_s, is often close to $\frac{1}{3}\kappa^2$, in which case the above formulae for $2W$ and $\langle u^2\rangle$ are still approximately correct.

Non-Bravais crystals

We consider coherent elastic scattering from a non-Bravais crystal, i.e. a crystal with more than one atom per unit cell. Let the equilibrium position of the dth atom in the unit cell be $\boldsymbol{d}$. The position of atom d in unit cell l is

$$\boldsymbol{R}_{ld} = \boldsymbol{l} + \boldsymbol{d} + \boldsymbol{u}\binom{l}{d}, \tag{3.68}$$

where $\boldsymbol{l}+\boldsymbol{d}$ is the equilibrium position of the atom, and $\boldsymbol{u}\binom{l}{d}$ is the displacement from equilibrium (Fig. 3.5). In general there is a different type of nucleus at each d position. So the mean value of the scattering length (i.e. the average over isotopes and nuclear spin) is different for each d position. Denote it by $\bar{b}_d$. We return to (2.68). The factor

$$\frac{\sigma_{\text{coh}}}{4\pi} = (\bar{b})^2 \tag{3.69}$$

must now be taken inside the double sum over the atoms. With (3.68) the equation becomes

$$\left(\frac{\mathrm{d}^2\sigma}{\mathrm{d}\Omega\,\mathrm{d}E'}\right)_{\text{coh}} = \frac{k'}{k}\frac{1}{2\pi\hbar}\sum_{ld}\sum_{l'd'} \bar{b}_d\bar{b}_{d'} \exp\{\mathrm{i}\boldsymbol{\kappa}\,.\,(\boldsymbol{l}+\boldsymbol{d}-\boldsymbol{l}'-\boldsymbol{d}')\}$$
$$\times\int_{-\infty}^{\infty}\left\langle \exp\left\{-\mathrm{i}\boldsymbol{\kappa}\,.\,\boldsymbol{u}\binom{l'}{d'}\,0\right\}\exp\left\{\mathrm{i}\boldsymbol{\kappa}\,.\,\boldsymbol{u}\binom{l}{d}\,t\right\}\right\rangle \exp(-\mathrm{i}\omega t)\,\mathrm{d}t. \tag{3.70}$$

This is the coherent cross-section for all processes, elastic and inelastic.

Fig. 3.5 Position vectors of the atom l,d for a non-Bravais crystal: ● equilibrium position, ○ actual (instantaneous) position.

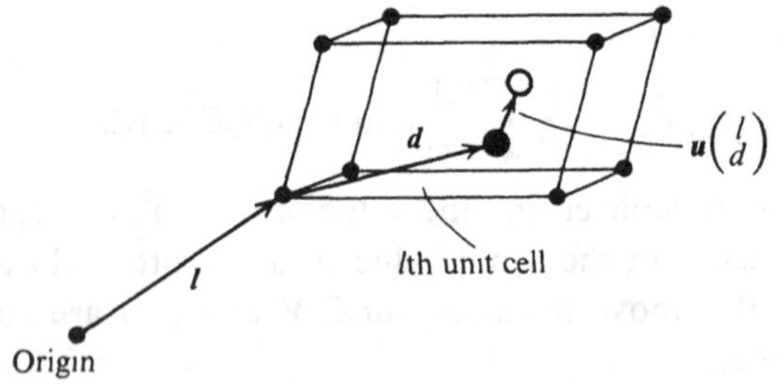

The analysis then proceeds as for the Bravais crystal and gives the result (corresponding to 3.46)

$$\left(\frac{d\sigma}{d\Omega}\right)_{\text{coh el}} = N \sum_{l} \exp(i\boldsymbol{\kappa} \cdot \boldsymbol{l}) \left| \sum_{d} \bar{b}_d \exp(i\boldsymbol{\kappa} \cdot \boldsymbol{d}) \exp(-W_d) \right|^2; \quad (3.71)$$

where

$$W_d = \tfrac{1}{2}\left\langle \left\{ \boldsymbol{\kappa} \cdot \boldsymbol{u}\begin{pmatrix} l \\ d \end{pmatrix} \right\}^2 \right\rangle. \quad (3.72)$$

As before the quantity $\boldsymbol{u}\begin{pmatrix} l \\ d \end{pmatrix}$ is expressed in terms of the displacements due to a set of normal modes. The equation corresponding to (3.10) is†

$$\boldsymbol{u}\begin{pmatrix} l \\ d \end{pmatrix} = \left(\frac{\hbar}{2M_d N}\right)^{1/2} \sum_{s} \frac{1}{\sqrt{\omega_s}} \{\boldsymbol{e}_{ds} a_s \exp(i\boldsymbol{q} \cdot \boldsymbol{l}) + \boldsymbol{e}^*_{ds} a_s^+ \exp(-i\boldsymbol{q} \cdot \boldsymbol{l})\}. \quad (3.73)$$

N is the number of unit cells in the crystal, and M_d is the mass of the atom at position d. As before s stands for the double index $\boldsymbol{q}$, j. $\boldsymbol{e}_{ds}$ is the polarisation vector for the atom at position d for the mode s. The polarisation index j takes values 1 to $3r$, where r is the number of atoms per unit cell. The development of W_d is similar to the Bravais case and leads to

$$W_d = \frac{\hbar}{4M_d N} \sum_{s} \frac{|\boldsymbol{\kappa} \cdot \boldsymbol{e}_{ds}|^2}{\omega_s} \langle 2n_s + 1 \rangle. \quad (3.74)$$

In (3.71) the sum over l is carried out as before and gives

$$\left(\frac{d\sigma}{d\Omega}\right)_{\text{coh el}} = N \frac{(2\pi)^3}{v_0} \sum_{\boldsymbol{\tau}} \delta(\boldsymbol{\kappa} - \boldsymbol{\tau}) |F_N(\boldsymbol{\kappa})|^2, \quad (3.75)$$

where

$$F_N(\boldsymbol{\kappa}) = \sum_{d} \bar{b}_d \exp(i\boldsymbol{\kappa} \cdot \boldsymbol{d}) \exp(-W_d). \quad (3.76)$$

$F_N(\boldsymbol{\kappa})$ is known as the *nuclear unit-cell structure factor.*

Methods of measuring Bragg scattering

The δ-function is an elegant and powerful tool for working with highly peaked functions, but the expressions in (3.48) and (3.75) need some manipulation before they can be compared directly with the results of an actual measurement of coherent elastic scattering. In such a measurement we determine the intensity of a Bragg peak, i.e.

† The normal modes of a non-Bravais crystal are discussed in Appendix G. See also Ghatak and Kothari (1972), Appendix C.

the integrated number of neutrons as, by varying some experimental parameter, we pass through the condition $\boldsymbol{\kappa}=\boldsymbol{\tau}$. Owing to instrumental resolution and mosaic spread in the crystal, the δ-function in the theoretical cross-section is spread out into a peak with finite width. We shall assume that the collimation of the incident and scattered neutrons is always sufficiently relaxed for all the neutrons scattered into the Bragg peak to be counted. The quantity actually measured is therefore the total cross-section σ_{tot} (see Section 1.3).

To calculate σ_{tot} we have to integrate $\mathrm{d}\sigma/\mathrm{d}\Omega$ with respect to Ω over all directions in space, i.e.

$$\sigma_{\text{tot}}=\int_{\substack{\text{all}\\ \text{directions}}}\left(\frac{\mathrm{d}\sigma}{\mathrm{d}\Omega}\right)_{\text{coh el}}\mathrm{d}\Omega. \tag{3.77}$$

Consider the case where neutrons of wavevector $\boldsymbol{k}$ are incident on a crystal, so that the angle ψ between $\boldsymbol{k}$ and a specific reciprocal lattice vector $\boldsymbol{\tau}$ is fixed (Fig. 3.6). Put

$$\boldsymbol{\rho}=\boldsymbol{k}-\boldsymbol{\tau}. \tag{3.78}$$

Then $\boldsymbol{\rho}$ is also fixed, and

$$\rho^2=k^2+\tau^2-2k\tau\cos\psi. \tag{3.79}$$

The integration in (3.77) is over all directions of $\boldsymbol{k}'$, with k' fixed and equal to k, i.e. over the sphere shown in Fig. 3.6. We need consider only the δ-function term in (3.75), i.e.

$$\int\delta(\boldsymbol{\kappa}-\boldsymbol{\tau})\,\mathrm{d}\Omega=\int\delta(\boldsymbol{\rho}-\boldsymbol{k}')\,\mathrm{d}\Omega. \tag{3.80}$$

Fig. 3.6 Integration of the function $\delta(\boldsymbol{\kappa}-\boldsymbol{\tau})$ over all directions of $\boldsymbol{k}'$ in Bragg scattering.

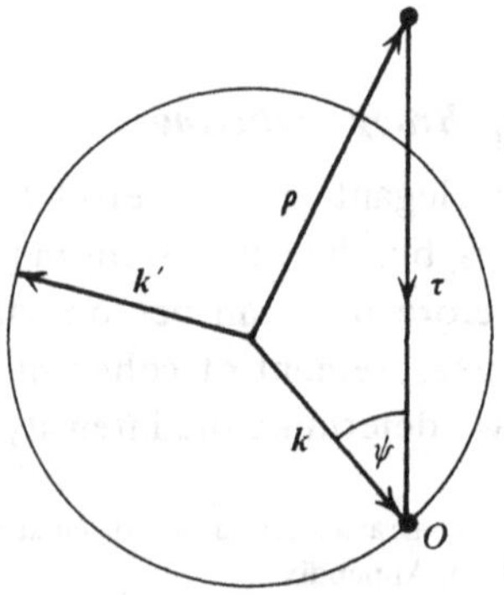

The integral is zero unless $\rho = k'$, so it can be written as

$$\int \delta(\boldsymbol{\rho} - \boldsymbol{k}')\, \mathrm{d}\Omega = c\delta(\rho^2 - k'^2). \tag{3.81}$$

The value of the constant c is obtained as follows. We note that $\delta(\boldsymbol{\rho} - \boldsymbol{k}')$ is a 3-dimensional δ-function, i.e.

$$\int \delta(\boldsymbol{\rho} - \boldsymbol{k}')\, \mathrm{d}\boldsymbol{k}' = 1, \tag{3.82}$$

where the integral is taken over all $\boldsymbol{k}'$ space. Now

$$\begin{aligned}\int \delta(\boldsymbol{\rho} - \boldsymbol{k}')\, \mathrm{d}\boldsymbol{k}' &= \int \delta(\boldsymbol{\rho} - \boldsymbol{k}')\, \mathrm{d}\Omega\, k'^2\, \mathrm{d}k' \\ &= \tfrac{1}{2}c \int \delta(\rho^2 - k'^2) k' d(k'^2) = \tfrac{1}{2}c\rho.\end{aligned} \tag{3.83}$$

Therefore

$$c = \frac{2}{\rho}, \tag{3.84}$$

whence

$$\int \delta(\boldsymbol{\kappa} - \boldsymbol{\tau})\, \mathrm{d}\Omega = \frac{2}{\rho}\delta(\rho^2 - k'^2) = \frac{2}{\rho}\delta(\rho^2 - k^2). \tag{3.85}$$

From (3.75), (3.77), (3.79), and (3.85)

$$\sigma_{\mathrm{tot}\,\boldsymbol{\tau}} = N\frac{(2\pi)^3}{v_0}\frac{2}{\rho}|F_{\mathrm{N}}(\boldsymbol{\tau})|^2\delta(\tau^2 - 2k\tau\cos\psi). \tag{3.86}$$

We now calculate the intensity of the Bragg peak for three standard methods, originally devised for the scattering of X-rays.

Laue method. Neutrons in a fixed direction (*DO* in Fig. 3.7), but with a continuous range of wavelengths, are incident on a crystal of fixed orientation. The angle between the scattered neutrons and the

Fig. 3.7 Diagram for the Laue method. $AB = AO$, $CT = CO$.

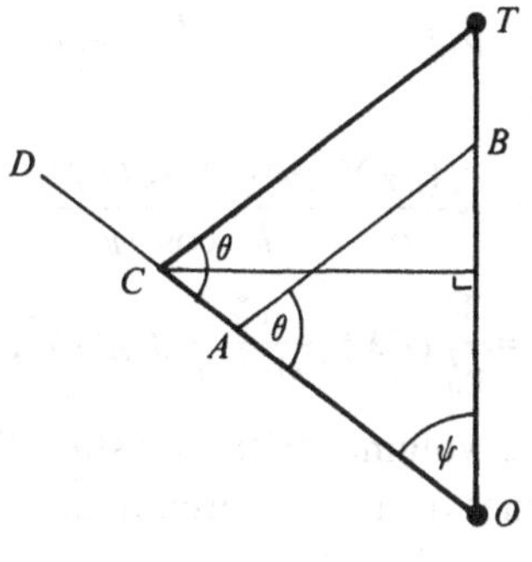

incident direction is set to the value

$$\theta = \pi - 2\psi. \tag{3.87}$$

For a general value of k (AO in Fig. 3.7) there is no Bragg scattering. The wavevector of the scattered neutrons is AB, and BO ($\boldsymbol{\kappa}$) does not equal TO ($\boldsymbol{\tau}$). But for the particular value

$$k = \frac{\tau}{2\cos\psi}, \tag{3.88}$$

represented by CO in Fig. 3.7, $\boldsymbol{\kappa}$ is equal to $\boldsymbol{\tau}$, and the Bragg condition is satisfied. The incident neutrons contain a continuous range of k values, but only the particular value given in (3.88) appears in the scattered beam. This is a standard method of obtaining neutrons with a single k value.

Let the flux of incident neutrons with wavelengths between λ and $\lambda + \mathrm{d}\lambda$ be $\phi(\lambda)\,\mathrm{d}\lambda$. Then the intensity for the peak corresponding to $\boldsymbol{\tau}$, i.e. the number of neutrons per second in the scattered beam, is

$$\begin{aligned} P &= \int \phi(\lambda)\sigma_{\mathrm{tot}\,\boldsymbol{\tau}}\,\mathrm{d}\lambda \\ &= N\frac{(2\pi)^3}{v_0}|F_{\mathrm{N}}(\boldsymbol{\tau})|^2 \int \frac{2}{\rho}\delta(\tau^2 - 2k\tau\cos\psi)\phi(\lambda)\,\mathrm{d}\lambda. \end{aligned} \tag{3.89}$$

Change the variable of integration to

$$x = 2k\tau\cos\psi. \tag{3.90}$$

Then, since

$$\lambda = \frac{2\pi}{k}, \tag{3.91}$$

$$\mathrm{d}\lambda = -\frac{2\pi}{k^2}\,\mathrm{d}k = -\frac{2\pi}{k^2}\frac{\mathrm{d}x}{2\tau\cos\psi}. \tag{3.92}$$

As usual, when an integrand contains a δ-function the remaining terms are evaluated at values for which the argument of the δ-function is zero. We therefore put

$$\rho = k, \qquad \tau = 2k\cos\psi, \tag{3.93}$$

and obtain the result

$$\begin{aligned} P &= N\frac{(2\pi)^3}{v_0}\frac{\pi\phi(\lambda)|F_{\mathrm{N}}(\boldsymbol{\tau})|^2}{k^4\cos^2\psi} \\ &= \frac{V}{v_0^2}\phi(\lambda)\frac{\lambda^4}{2\sin^2\frac{1}{2}\theta}|F_{\mathrm{N}}(\boldsymbol{\tau})|^2, \end{aligned} \tag{3.94}$$

where $V = Nv_0$ is the volume of the crystal. The value of λ to be inserted in (3.94) is that of the scattered neutrons.

Rotation of crystal. In this method of measuring a Bragg peak, a monochromatic beam of neutrons of wavevector $\boldsymbol{k}$ is incident on a crystal that can be rotated. The scattering angle θ is set to satisfy the relation

$$\tau = 2k \sin \tfrac{1}{2}\theta. \tag{3.95}$$

The crystal is rotated about an axis perpendicular to the plane containing $\boldsymbol{k}$ and $\boldsymbol{k}'$, so that the reciprocal lattice vector $\boldsymbol{\tau}$ remains in that plane.

Let $\boldsymbol{k}$, $\boldsymbol{k}'$ and $\boldsymbol{\tau}$ be represented by AO, AB and TO respectively (Fig. 3.8). As the crystal is rotated, i.e. as the angle ψ between $\boldsymbol{k}$ and $\boldsymbol{\tau}$ is varied, the point T traces out a circle, and Bragg scattering occurs as T passes through B. We again assume that, although θ is nominally fixed, the instrumental range of θ is sufficiently large for all the scattered neutrons to be counted as ψ is varied. The counting rate as a function of ψ is known as a *rocking curve.*

If Φ is the flux of the incident neutrons, the integrated number of scattered neutrons per unit time in the Bragg peak is

$$\begin{aligned} P &= \Phi \int_0^\pi \sigma_{\mathrm{tot}\,\boldsymbol{\tau}}\, \mathrm{d}\psi \\ &= N\frac{(2\pi)^3}{v_0}\Phi|F_{\mathrm{N}}(\boldsymbol{\tau})|^2 \int_0^\pi \frac{2}{\rho}\delta(\tau^2 - 2k\tau\cos\psi)\,\mathrm{d}\psi. \end{aligned} \tag{3.96}$$

Put $\quad x = 2k\tau \cos\psi, \quad$ with $\mathrm{d}x = -2k\tau \sin\psi \,\mathrm{d}\psi. \quad$ (3.97)

The integral is evaluated, again with the relations in (3.93), giving the result

$$P = N\frac{(2\pi)^3}{v_0}\Phi|F_{\mathrm{N}}(\boldsymbol{\tau})|^2/Q, \tag{3.98}$$

Fig. 3.8 Diagram for the method of crystal rotation.

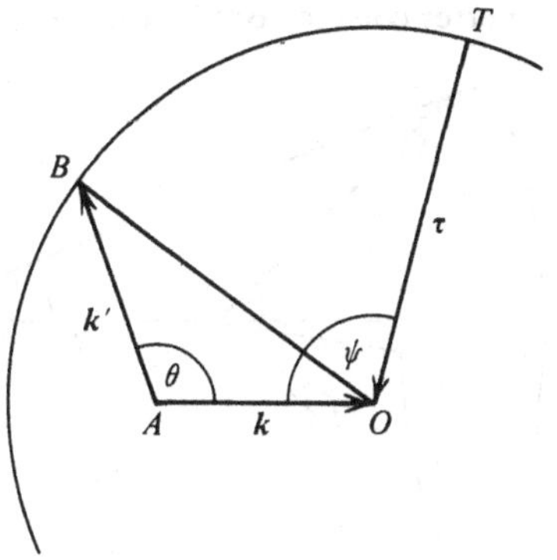

where $$Q = k^2\tau \sin\psi = k^3 \sin 2\psi = k^3 \sin\theta. \tag{3.99}$$

Thus $$P = \frac{V}{v_0^2}\Phi\frac{\lambda^3}{\sin\theta}|F_N(\boldsymbol{\tau})|^2, \tag{3.100}$$

where λ is the wavelength of the incident neutrons.

We are often interested, not in the absolute intensity of a Bragg peak, but in the relative intensity of two different Bragg peaks in the same crystal. If P_1 and P_2 are the intensities of peaks for reciprocal lattice vectors $\boldsymbol{\tau}_1$ and $\boldsymbol{\tau}_2$, and θ_1 and θ_2 are the corresponding scattering angles, then

$$\frac{P_1}{P_2} = \frac{|F_N(\boldsymbol{\tau}_1)|^2/\sin\theta_1}{|F_N(\boldsymbol{\tau}_2)|^2/\sin\theta_2}. \tag{3.101}$$

It may be noted that to observe a Bragg peak corresponding to a particular $\boldsymbol{\tau}$, the value of k must be greater than $\frac{1}{2}\tau$, otherwise (3.95) cannot be satisfied.

Powder method. A monochromatic beam of neutrons with wavevector $\boldsymbol{k}$ is incident on a powder sample, i.e. a sample of many small single crystals with random orientations. For a specified value of $|\boldsymbol{\tau}|$ $(<2k)$, the wavevector $\boldsymbol{k}'$ of the scattered neutrons lies on a cone, known as a *Debye–Scherrer cone* (see Fig. 3.9). The axis of the cone is along $\boldsymbol{k}$ and its semi-angle θ is given by (3.95). Only those microcrystals whose $\boldsymbol{\tau}$ vectors lie on a cone with axis along $\boldsymbol{k}$ and semi-angle

$$\psi = \tfrac{1}{2}\pi - \tfrac{1}{2}\theta \tag{3.102}$$

contribute to the scattering.

The direction of $\boldsymbol{k}$ is fixed. For each microcrystal, the vector $\boldsymbol{\tau}$ points in any direction in space with equal probability. Thus the

Fig. 3.9 Debye–Scherrer cone for Bragg scattering from a powder.

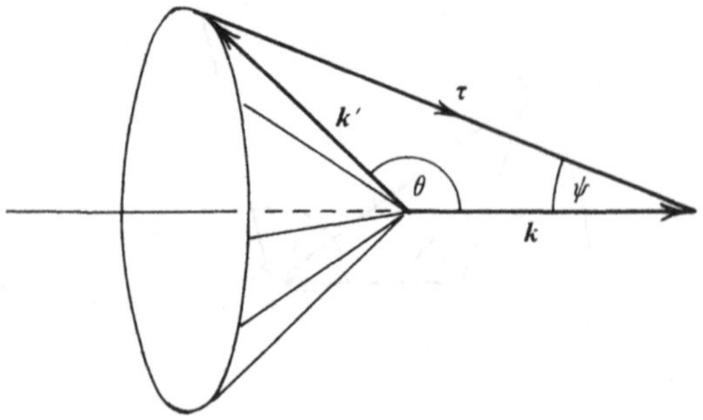

probability that the angle between $\boldsymbol{\tau}$ and $\boldsymbol{k}$ lies between ψ and $\psi + d\psi$ is $2\pi \sin\psi \, d\psi/4\pi$. Again we assume relaxed collimation. Then the total cross-section for each cone is

$$\sigma_{\text{tot}\,\tau}(\text{cone}) = N\frac{(2\pi)^3}{v_0}\frac{2}{k}\sum_{\tau}|F_N(\boldsymbol{\tau})|^2\int_0^{\pi/2}\delta(\tau^2 - 2k\tau\cos\psi)\tfrac{1}{2}\sin\psi\,d\psi$$

$$= \frac{V}{v_0^2}\frac{\lambda^3}{4\sin\frac{1}{2}\theta}\sum_{\tau}|F_N(\boldsymbol{\tau})|^2. \tag{3.103}$$

The sum over $\boldsymbol{\tau}$ is the sum over all reciprocal lattice vectors with the same value of $|\boldsymbol{\tau}|$.

If the neutron detector is at a distance r from the target and has an effective diameter d, it intercepts a fraction $d/2\pi r\sin\theta$ of the neutrons in the cone. The counting rate is therefore

$$P = \Phi\frac{d}{2\pi r\sin\theta}\sigma_{\text{tot}\,\tau}(\text{cone}), \tag{3.104}$$

where Φ is the incident flux.

3.7 Coherent one-phonon scattering

Cross-sections

We return to the scattering from a Bravais crystal. The coherent one-phonon cross-section is obtained from (3.36) by taking the term $\langle UV\rangle$ in the expansion of $\exp\{\langle UV\rangle\}$. From (3.23) and (3.24)

$$\langle\lambda|UV|\lambda\rangle = \sum_{ss'}\langle\lambda|(g_s a_s + g_s a_s^+)(h_{s'}a_{s'} + h_{s'}^* a_{s'}^+)|\lambda\rangle. \tag{3.105}$$

The matrix elements on the right-hand side are zero except for terms with $a_s a_s^+$ and $a_s^+ a_s$. For these terms we have from (E.13)

$$\langle\lambda|a_s a_s^+|\lambda\rangle = n_s + 1, \qquad \langle\lambda|a_s^+ a_s|\lambda\rangle = n_s. \tag{3.106}$$

Thus

$$\langle\lambda|UV|\lambda\rangle = \sum_s g_s h_s^*(n_s + 1) + g_s h_s n_s. \tag{3.107}$$

From (3.25), (3.26) and (3.107) we have

$$\langle UV\rangle = \sum_s g_s h_s^*\langle n_s + 1\rangle + g_s h_s\langle n_s\rangle$$

$$= \frac{\hbar}{2MN}\sum_s\frac{(\boldsymbol{\kappa}\cdot\boldsymbol{e}_s)^2}{\omega_s}[\exp\{-i(\boldsymbol{q}\cdot\boldsymbol{l} - \omega_s t)\}\langle n_s + 1\rangle$$

$$+\exp\{i(\boldsymbol{q}\cdot\boldsymbol{l} - \omega_s t)\}\langle n_s\rangle]. \tag{3.108}$$

The expression for the coherent one-phonon cross-section is

$$\left(\frac{d^2\sigma}{d\Omega\, dE'}\right)_{\text{coh 1 ph}}$$
$$=\frac{\sigma_{\text{coh}}}{4\pi}\frac{k'}{k}\frac{N}{2\pi\hbar}\exp(-2W)\sum_l \exp(i\boldsymbol{\kappa}\,.\,\boldsymbol{l})\int_{-\infty}^{\infty}\langle UV\rangle \exp(-i\omega t)\, dt$$
$$=\frac{\sigma_{\text{coh}}}{4\pi}\frac{k'}{k}\frac{1}{4\pi M}\exp(-2W)\sum_l \exp(i\boldsymbol{\kappa}\,.\,\boldsymbol{l})\sum_s \frac{(\boldsymbol{\kappa}\,.\,\boldsymbol{e}_s)^2}{\omega_s}$$
$$\times\int_{-\infty}^{\infty}[\exp\{-i(\boldsymbol{q}\,.\,\boldsymbol{l}-\omega_s t)\}\langle n_s+1\rangle$$
$$+\exp\{i(\boldsymbol{q}\,.\,\boldsymbol{l}-\omega_s t)\}\langle n_s\rangle]\exp(-i\omega t)\, dt. \tag{3.109}$$

The cross-section is the sum of two terms which arise from the two terms in the square brackets. Consider the first term. The integration with respect to t is

$$\int_{-\infty}^{\infty}\exp\{i(\omega_s-\omega)t\}\, dt = 2\pi\delta(\omega-\omega_s). \tag{3.110}$$

The summation with respect to l is

$$\sum_l \exp\{i(\boldsymbol{\kappa}-\boldsymbol{q})\,.\,\boldsymbol{l}\} = \frac{(2\pi)^3}{v_0}\sum_{\boldsymbol{\tau}}\delta(\boldsymbol{\kappa}-\boldsymbol{q}-\boldsymbol{\tau}). \tag{3.111}$$

Thus the cross-section for the first term is

$$\left(\frac{d^2\sigma}{d\Omega\, dE'}\right)_{\text{coh}+1}$$
$$=\frac{\sigma_{\text{coh}}}{4\pi}\frac{k'}{k}\frac{(2\pi)^3}{v_0}\frac{1}{2M}\exp(-2W)\sum_s\sum_{\boldsymbol{\tau}}\frac{(\boldsymbol{\kappa}\,.\,\boldsymbol{e}_s)^2}{\omega_s}\langle n_s+1\rangle$$
$$\times\,\delta(\omega-\omega_s)\delta(\boldsymbol{\kappa}-\boldsymbol{q}-\boldsymbol{\tau}). \tag{3.112}$$

Similarly the cross-section for the second term is

$$\left(\frac{d^2\sigma}{d\Omega\, dE'}\right)_{\text{coh}-1}$$
$$=\frac{\sigma_{\text{coh}}}{4\pi}\frac{k'}{k}\frac{(2\pi)^3}{v_0}\frac{1}{2M}\exp(-2W)\sum_s\sum_{\boldsymbol{\tau}}\frac{(\boldsymbol{\kappa}\,.\,\boldsymbol{e}_s)^2}{\omega_s}\langle n_s\rangle$$
$$\times\,\delta(\omega+\omega_s)\delta(\boldsymbol{\kappa}+\boldsymbol{q}-\boldsymbol{\tau}). \tag{3.113}$$

The cross-section (3.112) contains the factors $\delta(\omega-\omega_s)$ and $\delta(\boldsymbol{\kappa}-\boldsymbol{q}-\boldsymbol{\tau})$. So for scattering to occur two conditions must be satisfied:

$$\omega=\omega_s, \qquad \boldsymbol{\kappa}=\boldsymbol{\tau}+\boldsymbol{q}. \tag{3.114}$$

From the definition of ω (2.42), the first condition is

$$E - E' = \hbar\omega_s, \tag{3.115}$$

i.e. the energy of the neutron decreases by an amount equal to the energy of a phonon for the sth normal mode. So the scattering process is one in which the neutron creates a phonon. It is known as *phonon emission.* The energy for the phonon comes from the kinetic energy of the neutron. Eq. (3.115) can be written in the form

$$\frac{\hbar^2}{2m}(k^2 - k'^2) = \hbar\omega_s. \tag{3.116}$$

The second condition in (3.114) is

$$\boldsymbol{k} - \boldsymbol{k}' = \boldsymbol{\tau} + \boldsymbol{q}, \tag{3.117}$$

$\boldsymbol{q}$ is the wavevector of the normal mode s. This equation may be regarded as an expression of conservation of momentum. If we multiply (3.117) by $\hbar$, the quantity $\hbar(\boldsymbol{k} - \boldsymbol{k}')$ is the change in the momentum of the neutron, while $\hbar(\boldsymbol{\tau} + \boldsymbol{q})$ is the momentum imparted to the crystal. However there is no physical significance to the separate terms $\hbar\boldsymbol{\tau}$ and $\hbar\boldsymbol{q}$.

The cross-section (3.113) contains the term $\delta(\omega + \omega_s)$ and $\delta(\boldsymbol{\kappa} + \boldsymbol{q} - \boldsymbol{\tau})$. The conditions that must be satisfied here are thus

$$\frac{\hbar^2}{2m}(k'^2 - k^2) = \hbar\omega_s, \tag{3.118}$$

$$\boldsymbol{k} - \boldsymbol{k}' = \boldsymbol{\tau} - \boldsymbol{q}. \tag{3.119}$$

In this process the neutron annihilates a phonon in the sth normal mode. The energy of the phonon goes into an increase in the kinetic energy of the neutron. The process is known as *phonon absorption.*

Coherent one-phonon scattering may be regarded as elastic scattering in the frame of a crystal, whose atoms are displaced from their equilibrium positions with a sinusoidal variation given by the wavevector $\boldsymbol{q}$, and which is moving with the wave velocity of the phonon, i.e. ω_s/q, in the direction of $\boldsymbol{q}$. The condition for constructive interference for waves scattered by a sinusoidally modulated lattice gives (3.117) and (3.119), while transforming the velocities of the incident and scattered neutrons in the crystal frame to their values in the laboratory frame gives the energy equations (3.116) and (3.118) – see Example 3.6. An optical analogue of the interference condition is provided by Fraunhofer diffraction from a grating which is ruled incorrectly, so that the spacing of the lines, instead of being constant, has a sinusoidal variation. The main spectra are flanked by faint spectra known as *ghosts*. The wavevectors of the main spectra satisfy $\boldsymbol{\kappa} = \boldsymbol{\tau}$, while those of the ghosts satisfy $\boldsymbol{\kappa} = \boldsymbol{\tau} \pm \boldsymbol{q}$.

We may note the factors $\langle n_s + 1\rangle$ and $\langle n_s\rangle$ in the cross-sections (3.112) and (3.113). As $T \to 0$, $\langle n_s + 1\rangle \to 1$ and $\langle n_s\rangle \to 0$. So the cross-section for phonon absorption tends to zero as the temperature tends to zero. This must be the case, because when the crystal is at zero temperature all the normal-mode oscillators are in their ground states. Thus there are no phonons to be absorbed.

It is straightforward to generalise the expressions in (3.112) and (3.113) to non-Bravais crystals. The cross-section for coherent one-phonon emission becomes

$$\left(\frac{\mathrm{d}^2\sigma}{\mathrm{d}\Omega\,\mathrm{d}E'}\right)_{\mathrm{coh}+1} = \frac{k'}{k}\frac{(2\pi)^3}{2v_0}\sum_s\sum_{\tau}\frac{1}{\omega_s}\left|\sum_d \frac{\overline{b_d}}{\sqrt{M_d}}\exp(-W_d)\exp(i\boldsymbol{\kappa}\,.\,\boldsymbol{d})(\boldsymbol{\kappa}\,.\,\boldsymbol{e}_{ds})\right|^2 \times\langle n_s+1\rangle\delta(\omega-\omega_s)\delta(\boldsymbol{\kappa}-\boldsymbol{q}-\boldsymbol{\tau}), \qquad (3.120)$$

with a similar expression for the absorption cross-section. The notation is the same as on p. 37.

Measurement of phonon dispersion relations

One of the most important applications of the coherent one-phonon scattering process is to measure the *phonon dispersion relations* for the crystal, that is, the frequency ω_s as a function of wavevector $\boldsymbol{q}$ and polarisation index j. Before the advent of neutron scattering techniques these relations were largely unknown. Measurements of quantities like the specific heat give some average value of ω_s for all the normal modes, but not the detailed function itself.

Suppose we do the following experiment. We allow a beam of monoenergetic neutrons to fall on a single crystal and measure the velocity distribution of the neutrons scattered in a fixed direction. The experiment is a straightforward one and can be done on a time-of-flight apparatus (see Brugger, 1965). Consider the measurements in reciprocal space (Fig. 3.10). The vector $\boldsymbol{k}$ (AO in the figure) is fixed relative to the reciprocal lattice of the crystal. Since the scattering angle is fixed, the vector $\boldsymbol{k}'$ lies along the line AD. The velocity of the scattered neutrons is proportional to k'. So analysing the scattered neutrons according to their velocity is equivalent to measuring the cross-section as a function of k'.

Consider the process of phonon absorption. To obtain coherent one-phonon scattering we must satisfy (3.118) and (3.119). Suppose we select an arbitrary value of k', e.g. AB in Fig. 3.10. If (3.119) is to be satisfied that fixes $\boldsymbol{q}$ to be the vector TB. But in general none of the values of ω_s for the normal modes with this wavevector satisfies (3.118). It is only for certain discrete values of k' (AC in the figure) that both conditions are satisfied, and coherent one-phonon scattering occurs.

Fig. 3.11 shows a graphical construction for obtaining these values of k' from the dispersion relations. Since k is fixed, the value of ω_s required to satisfy (3.118) is a quadratic function of k' (curve 1 in the figure). If (3.119) is also to be satisfied, each value of k' fixes $\boldsymbol{q}$, and we may therefore plot the dispersion relations as functions of k'. They are indicated schematically in the figure. Whenever curve 1 crosses one of the dispersion curves, both (3.118) and (3.119) are satisfied. Curve 2 represents ω_s as a function of k' for (3.116), and when it crosses a dispersion curve the two conditions (3.116) and (3.117) for one-phonon emission are satisfied.

It is readily shown that whatever the form of the dispersion curves the two conditions for one-phonon absorption must be satisfied for at least one value of k' for each polarisation branch in every scattering direction. For $k' = k$, the value of ω_s in curve 1 in Fig. 3.11 is zero, and hence less than the values on the dispersion curves at that k' value. As k' becomes large, ω_s becomes large for curve 1 and eventually must be greater than the values on the dispersion curves, because the latter cannot exceed ω_m, the maximum frequency of the normal modes in the crystal. Since all the curves are continuous, curve 1 must cross each of the three dispersion curves at least once. The same reasoning does not apply for one-phonon emission, and this process may not occur in some scattering directions.

If $\boldsymbol{k}$ and the crystal orientation are kept constant, the values of $\boldsymbol{k}'$ that satisfy the pair of conditions for one-phonon scattering define what is known as a *scattering surface*. There is one such surface for each polarisation branch. Sections through the scattering surfaces of aluminium for two values of $\boldsymbol{k}$ are shown in Fig. 3.12.

The curves in Fig. 3.12 are calculated from a set of theoretical dispersion relations for aluminium. But experimentally we proceed in the reverse direction. For fixed $\boldsymbol{k}$, crystal orientation, and scattering direction, we measure a value of k' at which a coherent one-phonon peak occurs. We then substitute into (3.116)–(3.119), and hence

Fig. 3.10 Diagram in reciprocal space for coherent one-phonon scattering; ● reciprocal lattice point.

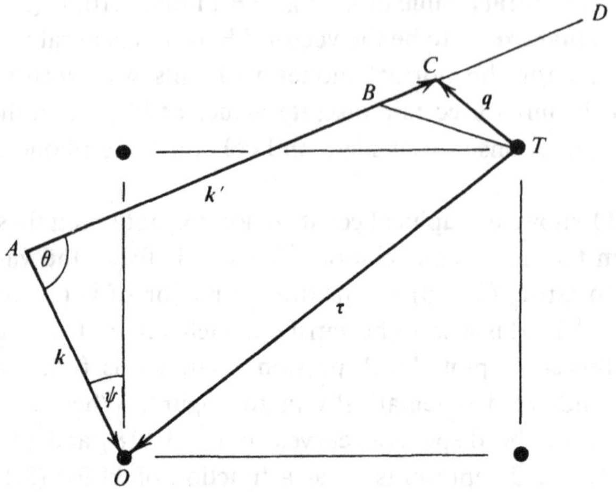

Fig. 3.11 Coherent one-phonon scattering: diagram for determining values of k' for fixed k, ψ, and θ.

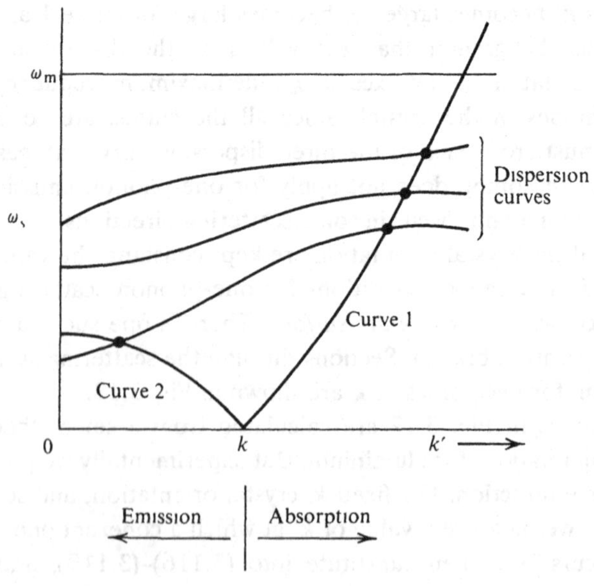

Fig. 3.12 Sections through the scattering surfaces of aluminium in the (001) plane (*a*) for incident neutrons of wavelength 6.74 Å, and (*b*) for incident neutrons of wavelength 1.08 Å. (Squires, 1956.)

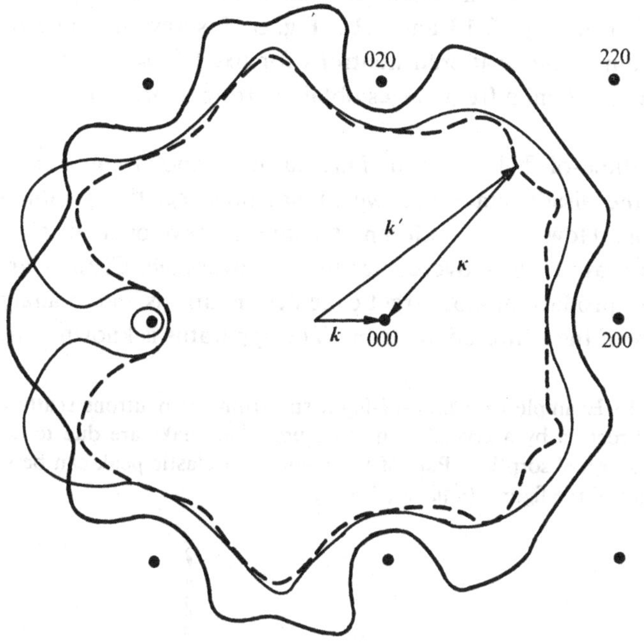

(*a*)

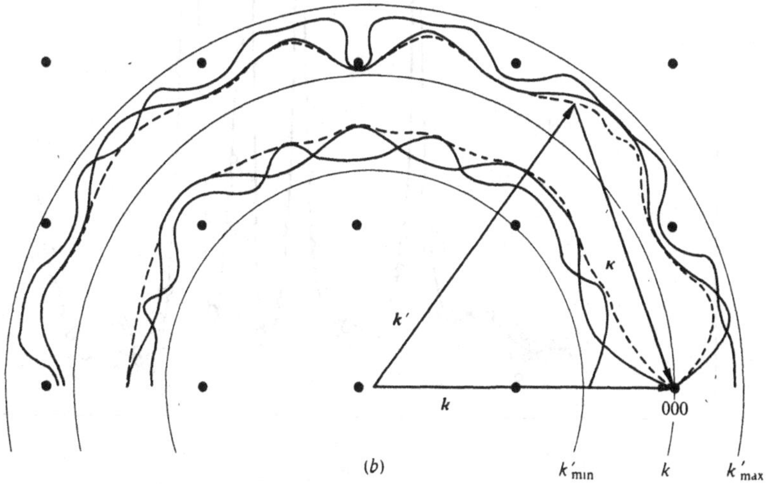

(*b*)

obtain ω_s and $\boldsymbol{q}$ for a particular phonon. By varying the values of $|\boldsymbol{k}|$, the angle between $\boldsymbol{k}$ and the crystal lattice, and the scattering angle, we may determine the ω_s and $\boldsymbol{q}$ values for many phonons. Some results for magnesium, obtained with a time-of-flight spectrometer, are given in Figs. 3.13 and 3.14. Fig. 3.13 shows the time-of-flight spectrum of the scattered neutrons for fixed $\boldsymbol{k}$, ψ, and θ. Fig. 3.14 shows the phonon frequencies obtained from a large number of such spectra.

The time-of-flight method of measuring phonon frequencies suffers from the disadvantage that we cannot preselect the $\boldsymbol{q}$ value of the phonon. However, a different technique, developed originally by Brockhouse (1960), overcomes this disadvantage. Crystals are used both to produce monoenergetic incident neutrons and to analyse the energy of the scattered neutrons. The apparatus is known as a *triple*-

Fig. 3.13 Example of a time-of-flight spectrum for neutrons scattered in a fixed direction by a crystal of magnesium. The peaks are due to coherent one-phonon absorption. Part of the incoherent elastic peak can be seen on the right of the figure. (Squires, 1966.)

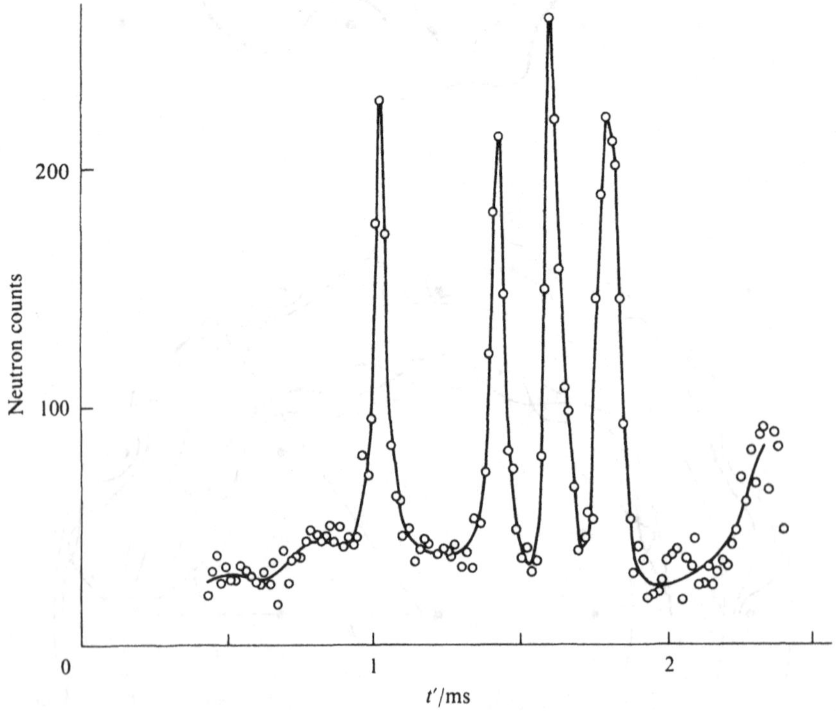

axis spectrometer. It is possible to control the experimental variables – the orientation of the various crystals, and the scattering angle – in such a way that the $\boldsymbol{q}$ value of the phonon to be measured is fixed at any required value. The method is particularly useful when we want to know how the frequency of a particular mode varies with, say, the temperature of the crystal. For a description of the triple-axis spectrometer, and its operation in the constant-$\boldsymbol{q}$ mode, see Iyengar (1965) and Squires (1976).

By means of time-of-flight and crystal spectrometers the phonon dispersion relations have been determined for a large number of crystalline materials, including metallic, ionic, covalent, and rare-gas

Fig. 3.14 Phonon frequencies of magnesium at 290 K along the direction Γ*KM*. The curves correspond to an eight-neighbour axial-force model. Inset: symmetry plane in reciprocal space perpendicular to the hexad axis showing the direction Γ*KM*. (Pynn and Squires, 1972.)

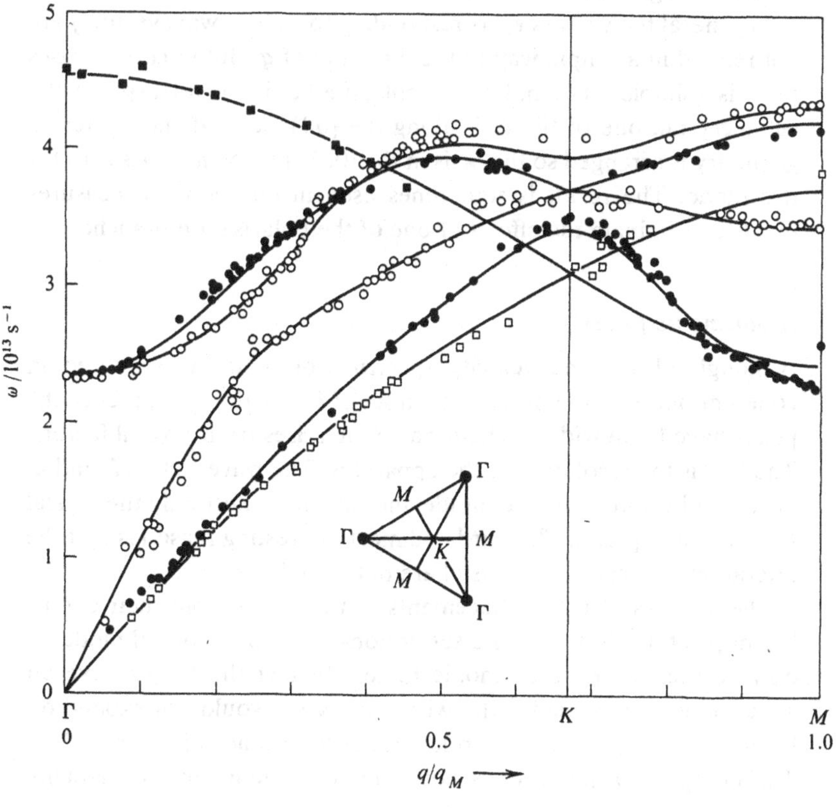

crystals. The object of the measurements is to obtain information about the interatomic forces. A variety of methods have been proposed for interpreting the results. Some employ empirical models with force constants treated as adjustable parameters; others are based on more physical models. A systematic survey of the experimental results, together with a review of the theories, will be found in Venkataraman, Feldkamp, and Sahni (1975). For a briefer account see Cochran (1973).

Polarisation vectors

We may note the term $(\boldsymbol{\kappa} \cdot \boldsymbol{e}_s)^2$ in the cross-sections (3.112) and (3.113). From measurements of the intensities of the peaks for one-phonon scattering, it is possible, in principle, to deduce the polarization vectors $\boldsymbol{e}_s$. There have not been many systematic measurements of this kind, but Brockhouse *et al.* (1963) have reported some on sodium and germanium.

In general the vectors $\boldsymbol{e}_s$ corresponding to a given wavevector $\boldsymbol{q}$ are not related in a simple way to the direction of $\boldsymbol{q}$. But in certain cases there is a simple relation. For example, if $\boldsymbol{q}$ lies in the (001) plane of a cubic crystal, one of the $\boldsymbol{e}_s$ is along the [001] axis. If the scattering geometry is arranged so that $\boldsymbol{\kappa}$ is in the (001) plane, $(\boldsymbol{\kappa} \cdot \boldsymbol{e}_s)$ is zero for this mode. The result is sometimes used in one-phonon measurements to eliminate the effects of one of the polarisation branches.

Anharmonic forces

Although ideally the velocity spectrum of neutrons scattered in coherent one-phonon processes is a set of δ-functions, in practice the peaks have finite widths. The broadening arises from several factors. The first is the resolution of the apparatus. The wavevectors $\boldsymbol{k}$ and $\boldsymbol{k}'$ have a finite spread in magnitude and direction. Secondly the crystal has a mosaic spread. The third and most interesting reason is that the interatomic forces in the crystal are not truly harmonic.

The analysis of the displacements of the atoms from their equilibrium positions in terms of a set of non-interacting normal modes is only correct for pure harmonic forces. In fact the forces have an anharmonic component (otherwise the crystal would not expand on heating) which causes the normal modes to interact with each other. The energy in a mode does not remain constant, but is gradually transferred to other modes. Thus the mode is represented in time by a damped sinusoidal wave, which means that its frequency is no longer sharp. It is spread over a finite range; the greater the anharmonic component, the greater the spread. Hence the scattered neutron groups occur, not at a sharp value of $\boldsymbol{k}'$, but over a range of values.

As the temperature of the crystal is raised, the anharmonic component of the forces increases, and the widths of the neutron peaks increase. By measuring the widths we can study the anharmonic contribution to the forces. In addition to broadening the frequencies of the normal modes, anharmonic forces also produce changes in the mean values of the frequencies. Both effects have been measured. A discussion of the theory of anharmonic forces in crystals, together with references to experimental work, will be found in Cowley (1968).

3.8 Coherent multiphonon scattering

The coherent two-phonon cross-section is obtained from (3.36) by taking the term $(1/2!)\langle UV\rangle^2$ in the expansion of $\exp\langle UV\rangle$. It can readily be shown that the cross-section contains two δ-function terms which give rise to the equations

$$\frac{\hbar^2}{2m}(k^2-k'^2)=\hbar(\pm\omega_{s_1}\pm\omega_{s_2}), \tag{3.121}$$

$$\boldsymbol{k}-\boldsymbol{k}'=\boldsymbol{\tau}\pm\boldsymbol{q}_1\pm\boldsymbol{q}_2. \tag{3.122}$$

Both conditions must be satisfied. The neutron is scattered having simultaneously created or annihilated a single phonon in two different normal modes.

In the one-phonon process we saw that for fixed $\boldsymbol{k}$, scattering angle, and crystal orientation, scattering occurs only for discrete values of k'. However, in the two-phonon process, if we select an arbitrary k' within a certain range, we can always find combinations of two normal modes whose $\boldsymbol{q}$ and ω_s values satisfy (3.121) and (3.122). So two-phonon scattering does not give rise to peaks in the velocity spectrum of the scattered neutrons. It gives a continuous spectrum; in other words it adds to the background (Fig. 3.13). This is fortunate, as it enables us to separate the effects of one- and two-phonon scattering.

For higher phonon processes, we get two equations like (3.121) and (3.122) with additional terms – ω_{s_3}, $\boldsymbol{q}_3$ and so on. It is true, *a fortiori*, that for an arbitrary value of k' we can find combinations of normal modes to satisfy the two equations.

3.9 Incoherent scattering

The basic expression for the incoherent scattering cross-section is given in (2.69). We consider a Bravais crystal and put

$$\boldsymbol{R}_l(t) = \boldsymbol{l} + \boldsymbol{u}_l(t). \tag{3.123}$$

Then

$$\begin{aligned}\sum_l \int \langle \exp\{-\mathrm{i}\boldsymbol{\kappa}\,.\,\boldsymbol{R}_l(0)\} \exp\{\mathrm{i}\boldsymbol{\kappa}\,.\,\boldsymbol{R}_l(t)\}\rangle \exp(-\mathrm{i}\omega t)\,\mathrm{d}t \\ = N\int \langle \exp\{-\mathrm{i}\boldsymbol{\kappa}\,.\,\boldsymbol{u}_0(0)\} \exp\{\mathrm{i}\boldsymbol{\kappa}\,.\,\boldsymbol{u}_0(t)\}\rangle \exp(-\mathrm{i}\omega t)\,\mathrm{d}t \\ = N\int \langle \exp U \exp V_0\rangle \exp(-\mathrm{i}\omega t)\,\mathrm{d}t,\end{aligned} \tag{3.124}$$

where $\quad U = -\mathrm{i}\boldsymbol{\kappa}\,.\,\boldsymbol{u}_0(0), \quad$ and $\quad V_0 = \mathrm{i}\boldsymbol{\kappa}\,.\,\boldsymbol{u}_0(t). \qquad (3.125)$

In summing over l in (3.124) we have used the result that for a Bravais crystal all the terms in the sum are equal. U is the same as the previous U, defined in (3.23). V_0 is the previous V (3.24) evaluated at $l = 0$. From (3.35)

$$\langle \exp U \exp V_0\rangle = \exp\langle U^2\rangle \exp\langle UV_0\rangle. \tag{3.126}$$

Thus

$$\left(\frac{\mathrm{d}^2\sigma}{\mathrm{d}\Omega\,\mathrm{d}E'}\right)_{\text{inc}} = \frac{\sigma_{\text{inc}}}{4\pi}\frac{k'}{k}\frac{N}{2\pi\hbar}\exp\langle U^2\rangle \int \exp\langle UV_0\rangle \exp(-\mathrm{i}\omega t)\,\mathrm{d}t. \tag{3.127}$$

As before $\exp\langle UV_0\rangle$ is expanded in powers of $\langle UV_0\rangle$. The pth term corresponds to a p-phonon process.

To calculate the incoherent elastic scattering we replace $\exp\langle UV_0\rangle$ by unity in (3.127) and use the results (3.42) and (3.43). We then integrate with respect to E', and obtain the result

$$\left(\frac{\mathrm{d}\sigma}{\mathrm{d}\Omega}\right)_{\text{inc el}} = \frac{\sigma_{\text{inc}}}{4\pi} N \exp(-2W). \tag{3.128}$$

The only dependence of this cross-section on the scattering direction is in the Debye–Waller factor, which depends on $\boldsymbol{\kappa}$. At low temperatures the Debye–Waller factor is close to unity, and the scattering is almost isotropic.

The incoherent one-phonon cross-section is obtained from (3.127) by replacing $\exp\langle UV_0\rangle$ by $\langle UV_0\rangle$. From (3.108)

$$\langle UV_0\rangle = \frac{\hbar}{2MN}\sum_s (\boldsymbol{\kappa}\,.\,\boldsymbol{e}_s)^2\frac{1}{\omega_s}\{\langle n_s+1\rangle\exp(\mathrm{i}\omega_s t)+\langle n_s\rangle\exp(-\mathrm{i}\omega_s t)\}, \tag{3.129}$$

$$\left(\frac{\mathrm{d}^2\sigma}{\mathrm{d}\Omega\,\mathrm{d}E'}\right)_{\mathrm{inc\,1\,ph}} = \frac{\sigma_{\mathrm{inc}}}{4\pi}\frac{k'}{k}\frac{1}{2M}\exp(-2W)\sum_s\frac{(\boldsymbol{\kappa}\,.\,\boldsymbol{e}_s)^2}{\omega_s}$$
$$\times\{\langle n_s+1\rangle\delta(\omega-\omega_s)+\langle n_s\rangle\delta(\omega+\omega_s)\}. \tag{3.130}$$

The first term in the curly brackets corresponds to phonon emission and the second to phonon absorption.

Consider the emission cross-section. It contains only one δ-function, $\delta(\omega-\omega_s)$. Thus only the energy condition

$$\omega=\omega_s \quad \text{or} \quad \frac{\hbar^2}{2m}(k^2-k'^2)=\hbar\omega_s \tag{3.131}$$

needs to be satisfied. For incoherent scattering there is no interference condition like (3.117). Therefore, for a given $\boldsymbol{k}$, θ, and crystal orientation, incoherent one-phonon scattering occurs for a continuous range of k' values. For a given k', we get scattering from all normal modes whose ω_s values satisfy (3.131). The cross-section therefore depends on the *number* of modes that have the correct frequency. We can express the cross-section in terms of the phonon density of states $Z(\omega)$.

$$\left(\frac{\mathrm{d}^2\sigma}{\mathrm{d}\Omega\,\mathrm{d}E'}\right)_{\mathrm{inc}+1} = \frac{\sigma_{\mathrm{inc}}}{4\pi}\frac{k'}{k}\frac{3N}{2M}\exp(-2W)\frac{\langle(\boldsymbol{\kappa}\,.\,\boldsymbol{e}_s)^2\rangle_{\mathrm{av}}}{\omega}Z(\omega)\langle n+1\rangle, \tag{3.132}$$

where

$$\omega=(E-E')/\hbar, \tag{3.133}$$

and

$$\langle n+1\rangle=\tfrac{1}{2}\{\coth(\tfrac{1}{2}\hbar\omega\beta)+1\}. \tag{3.134}$$

The quantity $\langle(\boldsymbol{\kappa}\,.\,\boldsymbol{e}_s)^2\rangle_{\mathrm{av}}$ is the value of $(\boldsymbol{\kappa}\,.\,\boldsymbol{e}_s)^2$ averaged over all the modes with frequency ω. For a cubic crystal

$$\langle(\boldsymbol{\kappa}\,.\,\boldsymbol{e}_s)^2\rangle_{\mathrm{av}}=\tfrac{1}{3}\kappa^2, \tag{3.135}$$

and the incoherent one-phonon cross-sections are given by

$$\left(\frac{\mathrm{d}^2\sigma}{\mathrm{d}\Omega\,\mathrm{d}E'}\right)_{\mathrm{inc}\pm1} = \frac{\sigma_{\mathrm{inc}}}{4\pi}\frac{k'}{k}\frac{N}{4M}\kappa^2\exp(-2W)\frac{Z(\omega)}{\omega}\{\coth(\tfrac{1}{2}\hbar\omega\beta)\pm1\}. \tag{3.136}$$

By measuring the incoherent one-phonon scattering as a function of E' for a cubic crystal, the phonon density of states may be determined. The method has been used for vanadium where the scattering is almost entirely incoherent ($\sigma_{coh} = 0.02 \times 10^{-28}\ m^2$, $\sigma_{inc} = 5.0 \times 10^{-28}\ m^2$). Some results are shown in Fig. 3.15. A basic difficulty of the method is that incoherent scattering from multiphonon processes also occurs. Both one-phonon and multiphonon processes give incoherent scattering over a continuous range of k' values, and it is not easy to estimate the contribution of the latter.

To obtain the incoherent cross-sections for non-Bravais crystals we proceed as in Section 3.6. The elastic cross-section is

$$\left(\frac{d\sigma}{d\Omega}\right)_{\text{inc el}} = N \sum_d \{\overline{b_d^2} - (\bar{b}_d)^2\} \exp(-2W_d). \tag{3.137}$$

The incoherent one-phonon emission cross-section is

$$\left(\frac{d^2\sigma}{d\Omega\, dE'}\right)_{\text{inc}+1} = \frac{k'}{k} \sum_d \frac{1}{2M_d} \{\overline{b_d^2} - (\bar{b}_d)^2\} \exp(-2W_d) \times \sum_s \frac{|\boldsymbol{\kappa} \cdot \boldsymbol{e}_{ds}|^2}{\omega_s} \langle n_s + 1 \rangle \delta(\omega - \omega_s). \tag{3.138}$$

Fig. 3.15 Phonon density of states of vanadium from measurements of incoherent scattering. (After Gläser *et al.*, 1965.)

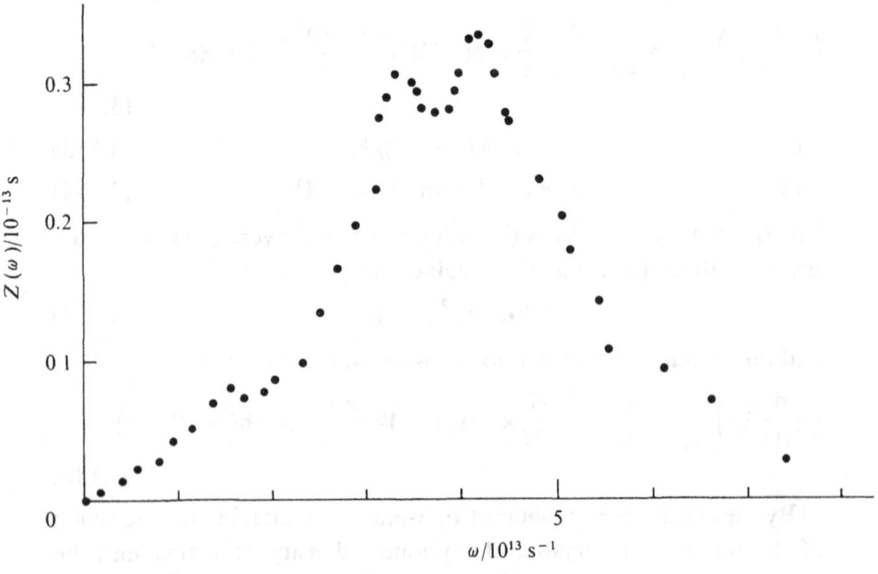

3.10 Multiphonon cross-sections – approximation methods

We have seen that the measurement of one-phonon cross-sections provides useful information. From coherent one-phonon scattering we may obtain dispersion relations. From incoherent one-phonon scattering we may determine the phonon density of states – at any rate for a cubic crystal. Multiphonon processes do not in general give useful information. They just add to the background scattering. Nevertheless we sometimes need to estimate their cross-sections, if only to subtract their contribution to the total scattering. There are two basic difficulties. The first is that as the energy E of the incident neutron increases and as the temperature T of the crystal increases, the convergence of the phonon expansion becomes progressively worse. The second is that, as the number of phonons in the scattering process increases, the cross-section – especially the coherent one – becomes more and more laborious to calculate. Several approximation methods have been devised to overcome these difficulties. We shall mention two.

Incoherent approximation

We write the cross-sections for a process in which p phonons are exchanged between the neutron and the crystal in the form

$$\left(\frac{\mathrm{d}^2\sigma}{\mathrm{d}\Omega\,\mathrm{d}E'}\right)_{\mathrm{coh}\,p}=\sigma_{\mathrm{coh}}F_{\mathrm{coh}\,p}, \qquad (3.139)$$

$$\left(\frac{\mathrm{d}^2\sigma}{\mathrm{d}\Omega\,\mathrm{d}E'}\right)_{\mathrm{inc}\,p}=\sigma_{\mathrm{inc}}F_{\mathrm{inc}\,p}. \qquad (3.140)$$

Then the incoherent approximation is to assume that for $p\geqslant 2$

$$F_{\mathrm{coh}\,p}=F_{\mathrm{inc}\,p}. \qquad (3.141)$$

$F_{\mathrm{coh}\,p}$ is made up of terms from atoms l and l' with both $l=l'$ and $l\neq l'$, while $F_{\mathrm{inc}\,p}$ is made up only of terms $l=l'$. The justification for the incoherent approximation is that for processes involving two or more phonons the cross-sections for both coherent and incoherent scattering are smoothly varying functions of the scattered energy, the incident energy, and the scattering angle. It therefore appears that the interference terms $l\neq l'$ in the coherent cross-section cancel each other to a large extent.

The usefulness of the approximation is that it is much easier to calculate $F_{\mathrm{inc}\,p}$ than $F_{\mathrm{coh}\,p}$. So to calculate $(\mathrm{d}^2\sigma/\mathrm{d}\Omega\,\mathrm{d}E')_{\mathrm{coh}\,p}$ we calculate $F_{\mathrm{inc}\,p}$ and multiply by σ_{coh}. For a discussion of the validity of the approximation see Placzek and Van Hove (1955).

Mass expansion

The mass expansion is a method of calculating the total cross-section – defined in (1.13) – for incoherent scattering by a powder sample or polycrystal. This cross-section is the sum of the elastic and inelastic terms. We have seen that the phonon expansion corresponds to expanding the term $\exp\langle UV_0\rangle$ in the expression for the cross-section in (3.127). Placzek (1954) pointed out that provided M, the mass of a nucleus in the crystal, is large compared to m, the mass of the neutron, a more rapidly converging series is obtained by taking the Debye–Waller factor $\exp\langle U^2\rangle$ together with $\exp\langle UV_0\rangle$ and expanding $\exp\gamma$, where $\gamma=\langle U^2\rangle+\langle UV_0\rangle$. This is the *mass expansion*, so called because γ is proportional to m/M. As the neutron energy E and the crystal temperature T rise, the total elastic cross-section falls, because $\langle U^2\rangle$ (which is a negative quantity) becomes numerically large, and the Debye–Waller factor becomes small. On the other hand, the inelastic cross-section rises, because $\langle UV_0\rangle$ (a positive quantity) also becomes large.† However the quantity γ remains relatively small.

The evaluation of the terms in the mass expansion is usually made on the basis of the Debye approximation, i.e. all the normal modes are assumed to have the same velocity, irrespective of wavevector and polarisation. The cross-section, in units of σ_{inc}, is then a function of the three parameters m/M, $x=(E/k_{\mathrm{B}}\theta_{\mathrm{D}})^{1/2}$, and $\theta=T/\theta_{\mathrm{D}}$, where θ_{D} is the Debye temperature of the crystal. Algebraic expressions for the first few terms are given in Placzek (1957), and a table of numerical values for a range of x and θ is given in Marshall and Lovesey (1971).

The mass expansion may be used in conjunction with the incoherent approximation to obtain an estimate of the total coherent cross-section. However, care is needed at low incident energies, when the incoherent approximation is a poor one for elastic scattering. The total incoherent elastic cross-section is a smoothly varying function of the neutron energy, whereas its coherent counterpart shows dis-

† It may be noted that as E tends to zero the elastic cross-section tends to a constant, but the inelastic cross-section does not tend to zero. As E becomes small compared to $\hbar\omega_{\mathrm{m}}$ the inelastic cross-section starts to rise, being proportional to $1/k$.

continuities as the various crystal planes start to contribute to the scattering. In that case the total incoherent elastic cross-section, calculated from (3.128), is subtracted from the mass expansion estimate to give the inelastic term. The incoherent approximation is then applied to the latter. The total coherent elastic cross-section is obtained separately by summing the expression in (3.103) over all τ (other than zero) for which $\tau < 2k$.

Examples

3.1 (*a*) Derive the dispersion relation (G.4) for a one-dimensional crystal of identical atoms with harmonic forces.

(*b*) Derive the expressions (G.12) and (G.13) for the kinetic and potential energies of the crystal.

(*c*) Show that the time-averaged values of the kinetic and potential energies are equal.

3.2 Show that as the temperature becomes large ($k_B T \gg \hbar\omega$) the expression given in (3.19) and (3.20) for the probability function of a harmonic oscillator tends to the classical expression.

3.3 Show that, for a cubic Bravais crystal, the Debye–Waller factor $\exp(-2W)$ depends on $\overline{1/\omega}$ for $k_B T \ll \hbar\omega_m$, and $\overline{1/\omega^2}$ for $k_B T \gg \hbar\omega_m$, where the average is taken over all normal modes.

3.4 Prove that, for a cubic Bravais crystal having a Debye frequency spectrum with Debye temperature θ_D, the mean square displacement of an atom is given by

$$\langle u^2 \rangle = \frac{9}{4}\frac{\hbar^2}{M}\frac{1}{k_B\theta_D}, \qquad T \ll \theta_D,$$

$$\langle u^2 \rangle = 9\frac{\hbar^2}{M}\frac{T}{k_B\theta_D^2}, \qquad T \gg \theta_D.$$

3.5 (*a*) Estimate the rms displacement of an atom in copper at $T = 20$ K, and at $T = 1000$ K.

(*b*) Neutrons are Bragg scattered by the (111) plane in a copper crystal. By what factor does the intensity of the peak diminish when the temperature of the crystal is changed from 20 K to 1000 K? (Assume a Debye spectrum with $\theta_D = 320$ K. Copper has an atomic weight of 64 and a face-centred cubic structure with cube side = 3.61 Å.)

3.6 Coherent one-phonon scattering may be regarded as elastic scattering in the rest-frame of a moving crystal whose atoms are displaced with a sinusoidal variation. Show that, if the position of atom l in a Bravais crystal is given by

$$\boldsymbol{R}_l = \boldsymbol{l} + \boldsymbol{A}\cos(\boldsymbol{q}\,.\,\boldsymbol{l}), \qquad \text{where } \boldsymbol{A}\,.\,\boldsymbol{\kappa} \ll 1,$$

and the velocity of the crystal is ω_s/q in the direction of $\boldsymbol{q}$, this picture leads to the relations

$$\boldsymbol{k} - \boldsymbol{k}' = \boldsymbol{\tau} \pm \boldsymbol{q}, \qquad \frac{\hbar^2}{2m}(k^2 - k'^2) = \pm\hbar\omega_s.$$

3.7 The point $\boldsymbol{\tau} = 0$ is always a point on the coherent one-phonon scattering surface. (It corresponds to the undisturbed incident beam.) Show that whether it is an isolated point, or whether the scattering surface for a given polarisation branch exists in the neighbourhood of $\boldsymbol{\tau} = 0$, depends on whether the velocity of the incident neutrons is less or greater than the velocity of sound for the branch. (Assume for simplicity that the velocity of sound c is independent of the direction of propagation.)

3.8 Show from the form of the coherent one-phonon absorption cross-section that for fixed $\boldsymbol{k}'$ this cross-section is proportional to $1/k$ (i.e. to $1/v$) as k tends to zero. Interpret this result physically.

4

Correlation functions in nuclear scattering

In the present chapter we relate the cross-sections for neutron scattering to thermal averages of operators belonging to the scattering system. The thermal averages can be expressed in terms of what are known as *correlation functions.* These functions are not only useful for calculating various properties of the scattering system, but they also provide insight into the physical significance of the terms that occur in the scattering cross-sections. This formulation of the subject is due primarily to Van Hove (1954). The calculations are quite general and refer to any system, solid, liquid, or gas, for which the scattering can be divided into coherent and incoherent parts.

4.1 Definitions of $I(\boldsymbol{\kappa}, t)$, $G(\boldsymbol{r}, t)$, and $S(\boldsymbol{\kappa}, \omega)$

We start with the expression (2.68) for the coherent scattering cross-section before we particularised it to a crystal target.

$$\left(\frac{d^2\sigma}{d\Omega\, dE'}\right)_{\text{coh}} = \frac{\sigma_{\text{coh}}}{4\pi}\frac{k'}{k}\frac{1}{2\pi\hbar}\int \sum_{jj'} \langle \exp\{-i\boldsymbol{\kappa}\,.\,\boldsymbol{R}_{j'}(0)\}\exp\{i\boldsymbol{\kappa}\,.\,\boldsymbol{R}_j(t)\}\rangle \times \exp(-i\omega t)\, dt. \tag{4.1}$$

We define a function $I(\boldsymbol{\kappa}, t)$, known as the *intermediate function*, by

$$I(\boldsymbol{\kappa}, t) = \frac{1}{N}\sum_{jj'} \langle \exp\{-i\boldsymbol{\kappa}\,.\,\boldsymbol{R}_{j'}(0)\}\exp\{i\boldsymbol{\kappa}\,.\,\boldsymbol{R}_j(t)\}\rangle, \tag{4.2}$$

where N is the number of nuclei in the scattering system.

We next define functions $G(\boldsymbol{r}, t)$ and $S(\boldsymbol{\kappa}, \omega)$ by

$$G(\boldsymbol{r}, t) = \frac{1}{(2\pi)^3}\int I(\boldsymbol{\kappa}, t)\exp(-i\boldsymbol{\kappa}\,.\,\boldsymbol{r})\, d\boldsymbol{\kappa}, \tag{4.3}$$

$$S(\boldsymbol{\kappa}, \omega) = \frac{1}{2\pi\hbar} \int I(\boldsymbol{\kappa}, t) \exp(-\mathrm{i}\omega t)\, \mathrm{d}t. \tag{4.4}$$

From the inverse relations for Fourier transforms (Appendix B.1)

$$I(\boldsymbol{\kappa}, t) = \int G(\boldsymbol{r}, t) \exp(\mathrm{i}\boldsymbol{\kappa} \cdot \boldsymbol{r})\, \mathrm{d}\boldsymbol{r}, \tag{4.5}$$

$$I(\boldsymbol{\kappa}, t) = \hbar \int S(\boldsymbol{\kappa}, \omega) \exp(\mathrm{i}\omega t)\, \mathrm{d}\omega. \tag{4.6}$$

Thus

$$G(\boldsymbol{r}, t) = \frac{\hbar}{(2\pi)^3} \int S(\boldsymbol{\kappa}, \omega) \exp\{-\mathrm{i}(\boldsymbol{\kappa} \cdot \boldsymbol{r} - \omega t)\}\, \mathrm{d}\boldsymbol{\kappa}\, \mathrm{d}\omega, \tag{4.7}$$

$$S(\boldsymbol{\kappa}, \omega) = \frac{1}{2\pi\hbar} \int G(\boldsymbol{r}, t) \exp\{\mathrm{i}(\boldsymbol{\kappa} \cdot \boldsymbol{r} - \omega t)\}\, \mathrm{d}\boldsymbol{r}\, \mathrm{d}t. \tag{4.8}$$

$G(\boldsymbol{r}, t)$ is known as the *time-dependent pair-correlation function* of the scattering system. $S(\boldsymbol{\kappa}, \omega)$ is known as the *scattering function* of the system.† It can be seen that, apart from a constant factor, $S(\boldsymbol{\kappa}, \omega)$ is the Fourier transform of $G(\boldsymbol{r}, t)$ in space and time. The intermediate function $I(\boldsymbol{\kappa}, t)$ is the Fourier transform of $G(\boldsymbol{r}, t)$ in space, and $S(\boldsymbol{\kappa}, \omega)$ is the Fourier transform of $I(\boldsymbol{\kappa}, t)$ in time. Note that $I(\boldsymbol{\kappa}, t)$ is dimensionless, $G(\boldsymbol{r}, t)$ has dimensions $[\text{volume}]^{-1}$, and $S(\boldsymbol{\kappa}, \omega)$ has dimensions $[\text{energy}]^{-1}$.

We define the self intermediate function by

$$I_{\mathrm{s}}(\boldsymbol{\kappa}, t) = \frac{1}{N} \sum_j \langle \exp\{-\mathrm{i}\boldsymbol{\kappa} \cdot \boldsymbol{R}_j(0)\} \exp\{\mathrm{i}\boldsymbol{\kappa} \cdot \boldsymbol{R}_j(t)\} \rangle. \tag{4.9}$$

Similarly we define

$$G_{\mathrm{s}}(\boldsymbol{r}, t) = \frac{1}{(2\pi)^3} \int I_{\mathrm{s}}(\boldsymbol{\kappa}, t) \exp(-\mathrm{i}\boldsymbol{\kappa} \cdot \boldsymbol{r})\, \mathrm{d}\boldsymbol{\kappa}, \tag{4.10}$$

$$S_{\mathrm{i}}(\boldsymbol{\kappa}, \omega) = \frac{1}{2\pi\hbar} \int I_{\mathrm{s}}(\boldsymbol{\kappa}, t) \exp(-\mathrm{i}\omega t)\, \mathrm{d}t. \tag{4.11}$$

$G_{\mathrm{s}}(\boldsymbol{r}, t)$ is known as the *self time-dependent pair-correlation function*, and $S_{\mathrm{i}}(\boldsymbol{\kappa}, \omega)$ is known as the *incoherent scattering function*.

Eqs. (4.1), (4.2), and (4.4) give

$$\left(\frac{\mathrm{d}^2\sigma}{\mathrm{d}\Omega\, \mathrm{d}E'}\right)_{\mathrm{coh}} = \frac{\sigma_{\mathrm{coh}}}{4\pi} \frac{k'}{k} N S(\boldsymbol{\kappa}, \omega). \tag{4.12}$$

† It is also known as the *scattering law*, though why a function should be called a law is a mystery to the author.

Similarly (2.69), (4.9), and (4.11) give

$$\left(\frac{\mathrm{d}^2\sigma}{\mathrm{d}\Omega\,\mathrm{d}E'}\right)_{\mathrm{inc}} = \frac{\sigma_{\mathrm{inc}}}{4\pi}\frac{k'}{k} N S_{\mathrm{i}}(\boldsymbol{\kappa}, \omega). \tag{4.13}$$

The functions $S(\boldsymbol{\kappa}, \omega)$ and $S_{\mathrm{i}}(\boldsymbol{\kappa}, \omega)$ are thus closely related to the coherent and incoherent scattering cross-sections.

4.2 Expressions for $G(\boldsymbol{r}, t)$ and $G_s(\boldsymbol{r}, t)$

We develop the expression for $G(\boldsymbol{r}, t)$. The algebra is straightforward though rather formal. From (4.2) and (4.3)

$$G(\boldsymbol{r}, t) = \frac{1}{(2\pi)^3}\frac{1}{N}\int \exp(-\mathrm{i}\boldsymbol{\kappa}\,.\,\boldsymbol{r})\,\mathrm{d}\boldsymbol{\kappa} \times \sum_{jj'} \langle \exp\{-\mathrm{i}\boldsymbol{\kappa}\,.\,\boldsymbol{R}_{j'}(0)\} \exp\{\mathrm{i}\boldsymbol{\kappa}\,.\,\boldsymbol{R}_j(t)\}\rangle. \tag{4.14}$$

Put $\langle \exp\{-\mathrm{i}\boldsymbol{\kappa}\,.\,\boldsymbol{R}_{j'}(0)\} \exp\{\mathrm{i}\boldsymbol{\kappa}\,.\,\boldsymbol{R}_j(t)\}\rangle$

$$= \int \langle \delta\{\boldsymbol{r}' - \boldsymbol{R}_{j'}(0)\} \exp(-\mathrm{i}\boldsymbol{\kappa}\,.\,\boldsymbol{r}') \exp\{\mathrm{i}\boldsymbol{\kappa}\,.\,\boldsymbol{R}_j(t)\}\rangle\,\mathrm{d}\boldsymbol{r}'. \tag{4.15}$$

Then

$$G(\boldsymbol{r}, t) = \frac{1}{(2\pi)^3}\frac{1}{N}\sum_{jj'}\int \Big\langle \delta\{\boldsymbol{r}' - \boldsymbol{R}_{j'}(0)\} \times \Big[\int \exp\{-\mathrm{i}\boldsymbol{\kappa}\,.\,\boldsymbol{r} - \mathrm{i}\boldsymbol{\kappa}\,.\,\boldsymbol{r}' + \mathrm{i}\boldsymbol{\kappa}\,.\,\boldsymbol{R}_j(t)\}\,\mathrm{d}\boldsymbol{\kappa}\Big]\Big\rangle\,\mathrm{d}\boldsymbol{r}'. \tag{4.16}$$

From (A.13) the integral in the square brackets has the value

$$(2\pi)^3\delta\{\boldsymbol{r}' + \boldsymbol{r} - \boldsymbol{R}_j(t)\}.$$

Thus

$$G(\boldsymbol{r}, t) = \frac{1}{N}\sum_{jj'}\int \langle \delta\{\boldsymbol{r}' - \boldsymbol{R}_{j'}(0)\}\delta\{\boldsymbol{r}' + \boldsymbol{r} - \boldsymbol{R}_j(t)\}\rangle\,\mathrm{d}\boldsymbol{r}'. \tag{4.17}$$

By similar reasoning we arrive at the result

$$G_s(\boldsymbol{r}, t) = \frac{1}{N}\sum_{j}\int \langle \delta\{\boldsymbol{r}' - \boldsymbol{R}_j(0)\}\delta\{\boldsymbol{r}' + \boldsymbol{r} - \boldsymbol{R}_j(t)\}\rangle\,\mathrm{d}\boldsymbol{r}'. \tag{4.18}$$

The operators $\boldsymbol{R}_{j'}(0)$ and $\boldsymbol{R}_j(t)$ do not commute, except when $t = 0$. This follows from the definitions

$$\boldsymbol{R}_{j'}(0) = \boldsymbol{R}_{j'}, \qquad \boldsymbol{R}_j(t) = \exp(\mathrm{i}Ht/\hbar)\boldsymbol{R}_j \exp(-\mathrm{i}Ht/\hbar). \tag{4.19}$$

The Hamiltonian H, which is a function of the position and momentum operators of all the nuclei in the scattering system, contains the operator $\boldsymbol{p}_{j'}$ (momentum of nucleus j'). $\boldsymbol{R}_{j'}$ does not commute with $\boldsymbol{p}_{j'}$, and hence it does not commute with $\boldsymbol{R}_j(t)$. It is therefore necessary to preserve the order of the operators and keep $\boldsymbol{R}_{j'}(0)$ on the left of $\boldsymbol{R}_j(t)$. We have done this in the preceding algebra.

If we ignore the fact that $\boldsymbol{R}_{j'}(0)$ and $\boldsymbol{R}_j(t)$ do not commute we can carry out the integration in (4.17). The result, known as the *classical* form of $G(\boldsymbol{r}, t)$, is

$$G^{\text{cl}}(\boldsymbol{r}, t)=\frac{1}{N}\sum_{jj'}\langle\delta\{\boldsymbol{r}-\boldsymbol{R}_j(t)+\boldsymbol{R}_{j'}(0)\}\rangle. \tag{4.20}$$

We could have obtained this result directly from (4.14). If $\boldsymbol{R}_{j'}(0)$ and $\boldsymbol{R}_j(t)$ are allowed to commute, we can put

$$\exp\{-\mathrm{i}\boldsymbol{\kappa}\,.\,\boldsymbol{R}_{j'}(0)\}\exp\{\mathrm{i}\boldsymbol{\kappa}\,.\,\boldsymbol{R}_j(t)\}=\exp[-\mathrm{i}\boldsymbol{\kappa}\,.\,\{\boldsymbol{R}_{j'}(0)-\boldsymbol{R}_j(t)\}] \tag{4.21}$$

in (4.14), and (4.20) follows immediately.

Assume for simplicity that all the nuclei are equivalent. Then, for fixed j', the sum over j in (4.20) gives the same value whatever the value of j'. So the sum over j' is N times the term with $j'=0$. Thus

$$G^{\text{cl}}(\boldsymbol{r}, t)=\sum_j\langle\delta\{\boldsymbol{r}-\boldsymbol{R}_j(t)+\boldsymbol{R}_0(0)\}\rangle. \tag{4.22}$$

We conclude from this equation that $G^{\text{cl}}(\boldsymbol{r}, t)\,\mathrm{d}\boldsymbol{r}$ is the probability that, given a particle at the origin at time $t=0$, any particle (including the origin particle) is in the volume $\mathrm{d}\boldsymbol{r}$ at position $\boldsymbol{r}$ at time t. Similarly

$$G_{\mathrm{s}}^{\text{cl}}(\boldsymbol{r}, t)=\langle\delta\{\boldsymbol{r}-\boldsymbol{R}_0(t)+\boldsymbol{R}_0(0)\}\rangle. \tag{4.23}$$

Thus $G_{\mathrm{s}}^{\text{cl}}(\boldsymbol{r}, t)\,\mathrm{d}\boldsymbol{r}$ is the probability that, given a particle at the origin at time $t=0$, the same particle is in the volume $\mathrm{d}\boldsymbol{r}$ at position $\boldsymbol{r}$ at time t.

From these interpretations of $G^{\text{cl}}(\boldsymbol{r}, t)$ and $G_{\mathrm{s}}^{\text{cl}}(\boldsymbol{r}, t)$ it must be the case that

$$\int G^{\text{cl}}(\boldsymbol{r}, t)\,\mathrm{d}\boldsymbol{r}=N, \qquad \int G_{\mathrm{s}}^{\text{cl}}(\boldsymbol{r}, t)\,\mathrm{d}\boldsymbol{r}=1. \tag{4.24}$$

These results may be verified from the expressions in (4.20) and (4.23). It follows from (4.17) and (4.18) that $G(\boldsymbol{r}, t)$ and $G_{\mathrm{s}}(\boldsymbol{r}, t)$ also satisfy the relations

$$\int G(\boldsymbol{r}, t)\,\mathrm{d}\boldsymbol{r}=N, \qquad \int G_{\mathrm{s}}(\boldsymbol{r}, t)\,\mathrm{d}\boldsymbol{r}=1. \tag{4.25}$$

At $t=0$

$$\boldsymbol{R}_{j'}(0)=\boldsymbol{R}_{j'} \quad\text{and}\quad \boldsymbol{R}_j(0)=\boldsymbol{R}_j. \tag{4.26}$$

The two operators commute, and we have

$$G(\boldsymbol{r}, 0) = \sum_j \langle \delta(\boldsymbol{r} - \boldsymbol{R}_j + \boldsymbol{R}_0) \rangle$$
$$= \delta(\boldsymbol{r}) + g(\boldsymbol{r}), \tag{4.27}$$

where

$$g(\boldsymbol{r}) = \sum_{j \neq 0} \langle \delta(\boldsymbol{r} - \boldsymbol{R}_j + \boldsymbol{R}_0) \rangle. \tag{4.28}$$

$g(\boldsymbol{r})$ is known as the *static pair-distribution function.* It gives the average particle-density with respect to any particle as the origin. Similarly

$$G_s(\boldsymbol{r}, 0) = \delta(\boldsymbol{r}). \tag{4.29}$$

It should be noted that the expression for the cross-section in (4.1) is valid only when there is no correlation between the scattering length and the nuclear site. When there is such a correlation, for example for a crystal composed of more than one element, the scattering lengths have to be included in the expressions for $I(\boldsymbol{\kappa}, t)$ and the other functions. The resulting expression for $G^{\mathrm{cl}}(\boldsymbol{r}, t)$ does not then have the simple probability interpretation of the expression in (4.22), owing to the presence of the scattering lengths, which act as weighting factors. We shall confine the discussion to simple systems where the mean value of the scattering length is the same for all the nuclear sites.

4.3 Analytic properties of the correlation functions

It is not possible to derive exact expressions for the functions $I(\boldsymbol{\kappa}, t)$, $G(\boldsymbol{r}, t)$, $S(\boldsymbol{\kappa}, \omega)$, except for the very simplest scattering systems. Approximations have to be made. It is therefore important to establish the basic analytic properties of the functions in order to check that the approximate functions have these properties.

We define an operator

$$\rho(\boldsymbol{r}, t) = \sum_j \delta\{\boldsymbol{r} - \boldsymbol{R}_j(t)\}. \tag{4.30}$$

It gives the number density of particles at position $\boldsymbol{r}$ at time t and is known as the *particle-density operator.* From (A.13) we can express the operator in terms of its Fourier transform

$$\rho(\boldsymbol{r}, t) = \frac{1}{(2\pi)^3} \int \rho_{\boldsymbol{\kappa}}(t) \exp(\mathrm{i}\boldsymbol{\kappa} \,.\, \boldsymbol{r})\, \mathrm{d}\boldsymbol{\kappa}, \tag{4.31}$$

where

$$\rho_{\boldsymbol{\kappa}}(t) = \sum_j \exp\{-\mathrm{i}\boldsymbol{\kappa} \,.\, \boldsymbol{R}_j(t)\}. \tag{4.32}$$

The integral in (4.31) is over all reciprocal space. Since $\boldsymbol{R}_j(t)$ is a Hermitian operator

$$\rho^+(\boldsymbol{r}, t) = \rho(\boldsymbol{r}, t), \tag{4.33}$$

$$\rho^+_{\boldsymbol{\kappa}}(t) = \rho_{-\boldsymbol{\kappa}}(t). \tag{4.34}$$

From (4.2) and (4.32)

$$I(\boldsymbol{\kappa}, t) = \frac{1}{N}\langle \rho_{\boldsymbol{\kappa}}(0)\rho_{-\boldsymbol{\kappa}}(t)\rangle, \tag{4.35}$$

and from (4.17) and (4.30)

$$G(\boldsymbol{r}, t) = \frac{1}{N}\int \langle \rho(\boldsymbol{r}', 0)\rho(\boldsymbol{r}'+\boldsymbol{r}, t)\rangle\, \mathrm{d}\boldsymbol{r}'. \tag{4.36}$$

We may use the properties of the particle-density operator $\rho(\boldsymbol{r}, t)$ to prove the following results:

$$I(\boldsymbol{\kappa}, t) = I^*(\boldsymbol{\kappa}, -t), \tag{4.37}$$

$$G(\boldsymbol{r}, t) = G^*(-\boldsymbol{r}, -t), \tag{4.38}$$

$$S(\boldsymbol{\kappa}, \omega) = S^*(\boldsymbol{\kappa}, \omega), \tag{4.39}$$

$$I(\boldsymbol{\kappa}, t) = I(-\boldsymbol{\kappa}, -t + \mathrm{i}\hbar\beta), \tag{4.40}$$

$$G(\boldsymbol{r}, t) = G(-\boldsymbol{r}, -t + \mathrm{i}\hbar\beta), \tag{4.41}$$

$$S(\boldsymbol{\kappa}, \omega) = \exp(\hbar\omega\beta)S(-\boldsymbol{\kappa}, -\omega). \tag{4.42}$$

From (C.21) the complex conjugate of $I(\boldsymbol{\kappa}, t)$ is obtained by reversing the two operators in (4.35) and taking their Hermitian conjugates. Eqs. (4.34) and (D.13) then give

$$I^*(\boldsymbol{\kappa}, -t) = \frac{1}{N}\langle \rho_{\boldsymbol{\kappa}}(0)\rho_{-\boldsymbol{\kappa}}(t)\rangle = I(\boldsymbol{\kappa}, t).$$

Eq. (4.38) follows from (4.3) and (4.37), and (4.39) from (4.4) and (4.37).

From (4.35), (D.13), and (D.14)

$$I(\boldsymbol{\kappa}, t) = \frac{1}{N}\langle \rho_{\boldsymbol{\kappa}}(0)\rho_{-\boldsymbol{\kappa}}(t)\rangle = \frac{1}{N}\langle \rho_{-\boldsymbol{\kappa}}(t)\rho_{\boldsymbol{\kappa}}(\mathrm{i}\hbar\beta)\rangle$$

$$= \frac{1}{N}\langle \rho_{-\boldsymbol{\kappa}}(0)\rho_{\boldsymbol{\kappa}}(-t + \mathrm{i}\hbar\beta)\rangle = I(-\boldsymbol{\kappa}, -t + \mathrm{i}\hbar\beta).$$

Eq. (4.41) follows from (4.3) and (4.40). From (4.4) and (4.40)

$$S(\boldsymbol{\kappa}, \omega) = \frac{1}{2\pi\hbar}\int I(-\boldsymbol{\kappa}, -t + \mathrm{i}\hbar\beta)\exp(-\mathrm{i}\omega t)\,\mathrm{d}t$$

$$= \exp(\hbar\omega\beta)\frac{1}{2\pi\hbar}\int I(-\boldsymbol{\kappa}, t')\exp(\mathrm{i}\omega t')\,\mathrm{d}t'$$

$$= \exp(\hbar\omega\beta)S(-\boldsymbol{\kappa}, -\omega).$$

It may be noted that it is the quantum properties of the scattering system that cause $G(\boldsymbol{r}, t)$ to be complex. To obtain $G^*(\boldsymbol{r}, t)$ we simply reverse the order of the two Hermitian operators in (4.36). For a classical system these two operators commute, and $G(\boldsymbol{r}, t)$ is real.

We next define three new functions $\tilde{I}(\boldsymbol{\kappa}, t)$, $\tilde{G}(\boldsymbol{r}, t)$, $\tilde{S}(\boldsymbol{\kappa}, \omega)$ as follows

$$\tilde{I}(\boldsymbol{\kappa}, t) = I(\boldsymbol{\kappa}, t + \tfrac{1}{2}\mathrm{i}\hbar\beta), \tag{4.43}$$

$$\tilde{G}(\boldsymbol{r}, t) = \frac{1}{(2\pi)^3} \int \tilde{I}(\boldsymbol{\kappa}, t) \exp(-\mathrm{i}\boldsymbol{\kappa} \cdot \boldsymbol{r})\, \mathrm{d}\boldsymbol{\kappa}, \tag{4.44}$$

$$\tilde{S}(\boldsymbol{\kappa}, \omega) = \frac{1}{2\pi\hbar} \int \tilde{I}(\boldsymbol{\kappa}, t) \exp(-\mathrm{i}\omega t)\, \mathrm{d}t. \tag{4.45}$$

Note that the functions $\tilde{G}(\boldsymbol{r}, t)$ and $\tilde{S}(\boldsymbol{\kappa}, \omega)$ are defined in terms of $\tilde{I}(\boldsymbol{\kappa}, t)$ in exactly the same way as $G(\boldsymbol{r}, t)$ and $S(\boldsymbol{\kappa}, \omega)$ are defined in terms of $I(\boldsymbol{\kappa}, t)$. The object in defining the new functions is that, as will be shown, $\tilde{G}(\boldsymbol{r}, t)$ is a real function and serves as a link between $G^{\mathrm{cl}}(\boldsymbol{r}, t)$, which is also real and which may be calculated from some physical model of the scattering system, and the complex $G(\boldsymbol{r}, t)$.

The following results are readily proved:

$$\tilde{G}(\boldsymbol{r}, t) = G(\boldsymbol{r}, t + \tfrac{1}{2}\mathrm{i}\hbar\beta), \tag{4.46}$$

$$\tilde{S}(\boldsymbol{\kappa}, \omega) = \exp(-\tfrac{1}{2}\hbar\omega\beta) S(\boldsymbol{\kappa}, \omega), \tag{4.47}$$

$$\tilde{I}(\boldsymbol{\kappa}, t) = \tilde{I}(-\boldsymbol{\kappa}, -t) = \tilde{I}^*(\boldsymbol{\kappa}, -t), \tag{4.48}$$

$$\tilde{G}(\boldsymbol{r}, t) = \tilde{G}(-\boldsymbol{r}, -t) = \tilde{G}^*(\boldsymbol{r}, t), \tag{4.49}$$

$$\tilde{S}(\boldsymbol{\kappa}, \omega) = \tilde{S}(-\boldsymbol{\kappa}, -\omega) = \tilde{S}^*(\boldsymbol{\kappa}, \omega). \tag{4.50}$$

Eqs. (4.46) and (4.47) follow from the definitions of $\tilde{I}(\boldsymbol{\kappa}, t)$, $\tilde{G}(\boldsymbol{r}, t)$, and $\tilde{S}(\boldsymbol{\kappa}, \omega)$. In deriving (4.47) the variable for the time integration is changed as in the derivation of (4.42). From (4.37), (4.40), and (4.43)

$$\begin{aligned}\tilde{I}(\boldsymbol{\kappa}, t) &= I(\boldsymbol{\kappa}, t + \tfrac{1}{2}\mathrm{i}\hbar\beta) = I(-\boldsymbol{\kappa}, -t + \tfrac{1}{2}\mathrm{i}\hbar\beta) = \tilde{I}(-\boldsymbol{\kappa}, -t) \\ &= I^*(\boldsymbol{\kappa}, -t + \tfrac{1}{2}\mathrm{i}\hbar\beta) = \tilde{I}^*(\boldsymbol{\kappa}, -t).\end{aligned}$$

Eqs. (4.49) and (4.50) then follow from (4.44) and (4.45).

For many scattering systems

$$G(\boldsymbol{r}, t) = G(-\boldsymbol{r}, t). \tag{4.51}$$

This relation is true for all disordered systems – gases and liquids. It is also true for polycrystals, and single crystals with a centre of symmetry. If we assume (4.51), the previous equations, together with

(4.4) and (4.5) give

$$I(\boldsymbol{\kappa}, t) = I(-\boldsymbol{\kappa}, t), \tag{4.52}$$

$$S(\boldsymbol{\kappa}, \omega) = S(-\boldsymbol{\kappa}, \omega), \tag{4.53}$$

$$\tilde{I}(\boldsymbol{\kappa}, t) = \tilde{I}(-\boldsymbol{\kappa}, t) = \tilde{I}(\boldsymbol{\kappa}, -t), \tag{4.54}$$

$$\tilde{G}(\boldsymbol{r}, t) = \tilde{G}(-\boldsymbol{r}, t) = \tilde{G}(\boldsymbol{r}, -t), \tag{4.55}$$

$$\tilde{S}(\boldsymbol{\kappa}, \omega) = \tilde{S}(-\boldsymbol{\kappa}, \omega) = \tilde{S}(\boldsymbol{\kappa}, -\omega). \tag{4.56}$$

Eq. (4.53) is the complement of (4.51). If the correlation function for the particles in the scattering system is the same when $\boldsymbol{r}$ is reversed, then the scattering properties of the system must be the same when $\boldsymbol{\kappa}$ is reversed.

For all scattering systems, the functions $\tilde{G}(\boldsymbol{r}, t)$ and $\tilde{S}(\boldsymbol{\kappa}, \omega)$ are real. If the relation (4.51) holds, then $\tilde{I}(\boldsymbol{\kappa}, t)$ is also real, and the three functions $\tilde{I}(\boldsymbol{\kappa}, t)$, $\tilde{G}(\boldsymbol{r}, t)$, and $\tilde{S}(\boldsymbol{\kappa}, \omega)$ are even in each of their two variables. Further, from (4.38) and (4.51)

$$G(\boldsymbol{r}, t) = G^*(\boldsymbol{r}, -t). \tag{4.57}$$

Thus the real part of $G(\boldsymbol{r}, t)$ is even in t, while the imaginary part is odd in t.

4.4 Principle of detailed balance

Eq. (4.42) is known as the *principle of detailed balance.* It is an important result, and we give an alternative derivation, which shows more clearly its physical significance.

If the δ-function expressing conservation of energy in (2.40) is retained and not expressed as a time integral, the coherent scattering cross-section has the form

$$\left(\frac{\mathrm{d}^2\sigma}{\mathrm{d}\Omega\,\mathrm{d}E'}\right)_{\mathrm{coh}} = \frac{\sigma_{\mathrm{coh}}}{4\pi}\frac{k'}{k}\sum_{\lambda} p_\lambda \sum_{\lambda'} \left|\sum_j \langle\lambda'|\exp(\mathrm{i}\boldsymbol{\kappa}\,.\,\boldsymbol{R}_j)|\lambda\rangle\right|^2 \delta(E_\lambda - E_{\lambda'} + \hbar\omega). \tag{4.58}$$

Comparing this equation with (4.12) and using the explicit form of p_λ in (2.51) we obtain

$$S(\boldsymbol{\kappa}, \omega) = \frac{1}{NZ}\sum_{\lambda\lambda'} \exp(-E_\lambda\beta)\left|\sum_j \langle\lambda'|\exp(\mathrm{i}\boldsymbol{\kappa}\,.\,\boldsymbol{R}_j)|\lambda\rangle\right|^2 \delta(E_\lambda - E_{\lambda'} + \hbar\omega). \tag{4.59}$$

To make the discussion definite let ω be a positive quantity, i.e. let the neutron lose energy in the scattering process. For every transition of the system that contributes to $S(\boldsymbol{\kappa}, \omega)$, the initial state λ has energy $\hbar\omega$ less than the final state λ' (Fig. 4.1*a*).

Consider now the function $S(-\boldsymbol{\kappa}, -\omega)$ where ω is the same positive quantity. This represents a process in which the neutron gains energy. The transitions of the system are between the same pairs of levels as for the previous process, but now λ' is the initial state and λ is the final state (Fig. 4.1*b*). Thus

$$
\begin{aligned}
S(-\boldsymbol{\kappa}, -\omega) \\
&= \frac{1}{NZ}\sum_{\lambda\lambda'} \exp(-E_{\lambda'}\beta)\left|\sum_j \langle\lambda|\exp(-\mathrm{i}\boldsymbol{\kappa}\,.\,\boldsymbol{R}_j)|\lambda'\rangle\right|^2 \delta(E_{\lambda'}-E_\lambda-\hbar\omega) \\
&= \exp\{-(E_{\lambda'}-E_\lambda)\beta\}\frac{1}{NZ}\sum_{\lambda\lambda'} \exp(-E_\lambda\beta)\left|\sum_j \langle\lambda'|\exp(\mathrm{i}\boldsymbol{\kappa}\,.\,\boldsymbol{R}_j)|\lambda\rangle\right|^2 \\
&\qquad \times \delta(E_\lambda - E_{\lambda'}+\hbar\omega) \\
&= \exp(-\hbar\omega\beta)S(\boldsymbol{\kappa}, \omega). \qquad (4.60)
\end{aligned}
$$

In the middle line of the last equation we have used (C.3), which in the present context expresses the physical result that, for a pair of states in the scattering system, the *a priori* probabilities that the neutron will bring about a transition in either direction are the same. The probability of the system being initially in the higher energy state is lower by the factor $\exp(-\hbar\omega\beta)$ than its probability of being in the lower energy state. Hence the function $S(-\boldsymbol{\kappa}, -\omega)$ is less than $S(\boldsymbol{\kappa}, \omega)$ by this factor.

For scattering systems for which reversal of $\boldsymbol{\kappa}$ has no effect, the principle of detailed balance is

$$S(\boldsymbol{\kappa}, -\omega) = \exp(-\hbar\omega\beta)S(\boldsymbol{\kappa}, \omega). \qquad (4.61)$$

Fig. 4.1 Principle of detailed balance: diagram showing transitions for (*a*) $S(\boldsymbol{\kappa}, \omega)$, where ω is positive, and (*b*) $S(-\boldsymbol{\kappa}, -\omega)$, where ω is the same positive quantity.

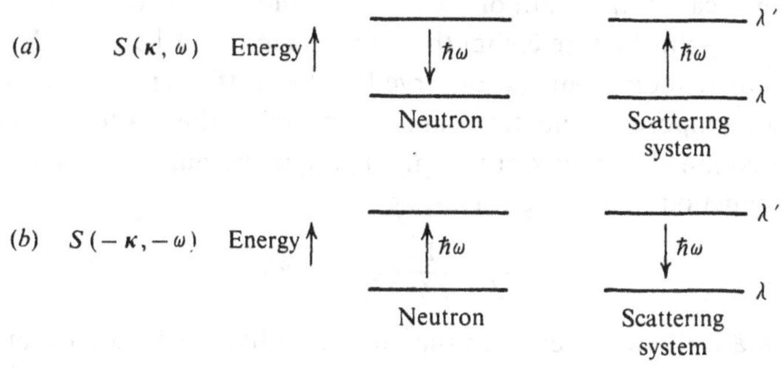

Schofield's prescription for $G(\boldsymbol{r}, t)$

We have mentioned that in general it is not possible to calculate $G(\boldsymbol{r}, t)$ exactly from the atomic properties of the scattering system. Instead, we proceed by setting up a physical model and deriving $G^{\mathrm{cl}}(\boldsymbol{r}, t)$, the classical form of $G(\boldsymbol{r}, t)$, from it. However, this calculated $G^{\mathrm{cl}}(\boldsymbol{r}, t)$ is not a good approximation for $G(\boldsymbol{r}, t)$, because for many systems $G^{\mathrm{cl}}(\boldsymbol{r}, t)$ is real and even in $\boldsymbol{r}$ and t. If we take $G(\boldsymbol{r}, t)$ to be real and even in $\boldsymbol{r}$ and t, then from (4.8)

$$S(\boldsymbol{\kappa}, \omega) = S(-\boldsymbol{\kappa}, -\omega). \qquad (4.62)$$

The result violates the principle of detailed balance.

Schofield (1960) suggested that a better approximation is to assume

$$\tilde{G}(\boldsymbol{r}, t) = G^{\mathrm{cl}}(\boldsymbol{r}, t). \qquad (4.63)$$

We know that the correct $\tilde{G}(\boldsymbol{r}, t)$ is real and even in $\boldsymbol{r}$ and t. So if we put $\tilde{G}(\boldsymbol{r}, t)$ equal to the calculated $G^{\mathrm{cl}}(\boldsymbol{r}, t)$ and then use the result

$$G(\boldsymbol{r}, t) = \tilde{G}(\boldsymbol{r}, t - \tfrac{1}{2}\mathrm{i}\hbar\beta), \qquad (4.64)$$

the resulting $S(\boldsymbol{\kappa}, \omega)$ will satisfy the detailed balance condition.

4.5 Scattering from a single free nucleus

To illustrate the various functions defined in the present chapter we consider the simplest possible case, namely scattering by a single free nucleus of mass M. By this we do not mean an isolated nucleus, but one that is a member of an ensemble at temperature T.

We start with the expression for the cross-section obtained from (2.40)

$$\frac{\mathrm{d}^2\sigma}{\mathrm{d}\Omega\,\mathrm{d}E'} = b^2 \frac{k'}{k} \sum_{\lambda} p_{\lambda} \sum_{\lambda'} |\langle\lambda'|\exp(\mathrm{i}\boldsymbol{\kappa}\,.\,\boldsymbol{R})|\lambda\rangle|^2 \delta(E_{\lambda} - E_{\lambda'} + \hbar\omega). \qquad (4.65)$$

b is the scattering length of the nucleus, and $\boldsymbol{R}$ is its position vector. The first point to note is that the matrix element $\langle\lambda'|\exp(\mathrm{i}\boldsymbol{\kappa}\,.\,\boldsymbol{R})|\lambda\rangle$ is zero unless momentum is conserved in the scattering process. To see this we suppose the neutrons and the nuclei in the scattering system to be confined to a box of volume Y. Since the nuclei are free, their state functions are plane waves, i.e.

$$|\lambda\rangle = \frac{1}{\sqrt{Y}} \exp(\mathrm{i}\boldsymbol{\xi}\,.\,\boldsymbol{R}), \qquad (4.66)$$

where $\boldsymbol{\xi}$ is the wavevector of the nucleus. $\boldsymbol{\xi}$ like $\boldsymbol{k}$ (the wavevector of the neutron) is periodic in the box. Then

$$\langle\lambda'|\exp(\mathrm{i}\boldsymbol{\kappa}.\boldsymbol{R})|\lambda\rangle=\frac{1}{Y}\int_{\text{box}}\exp\{\mathrm{i}(\boldsymbol{\kappa}+\boldsymbol{\xi}-\boldsymbol{\xi}').\boldsymbol{R}\}\,\mathrm{d}\boldsymbol{R}. \quad (4.67)$$

The integral is zero unless

$$\boldsymbol{\kappa}=\boldsymbol{k}-\boldsymbol{k}'=\boldsymbol{\xi}'-\boldsymbol{\xi}. \quad (4.68)$$

(The reasoning is the same as in (A.15).) Multiplication of this equation by $\hbar$ gives conservation of momentum. If the condition is satisfied

$$\langle\lambda'|\exp(\mathrm{i}\boldsymbol{\kappa}.\boldsymbol{R})|\lambda\rangle=1. \quad (4.69)$$

Thus

$$\sum_{\lambda'}|\langle\lambda'|\exp(\mathrm{i}\boldsymbol{\kappa}.\boldsymbol{R})|\lambda\rangle|^2=1, \quad (4.70)$$

and

$$\boldsymbol{\xi}'=\boldsymbol{\kappa}+\boldsymbol{\xi}. \quad (4.71)$$

Then

$$E_\lambda-E_{\lambda'}=\frac{\hbar^2}{2M}(\xi^2-\xi'^2)=-\frac{\hbar^2}{2M}(\kappa^2+2\boldsymbol{\kappa}.\boldsymbol{\xi}). \quad (4.72)$$

The function $S(\boldsymbol{\kappa},\omega)$ is obtained from (4.12) and (4.65).

$$S(\boldsymbol{\kappa},\omega)=\sum_{\lambda\lambda'}p_\lambda|\langle\lambda'|\exp(\mathrm{i}\boldsymbol{\kappa}.\boldsymbol{R})|\lambda\rangle|^2\delta(E_\lambda-E_{\lambda'}+\hbar\omega)$$

$$=\sum_{\boldsymbol{\xi}}p_{\boldsymbol{\xi}}\delta\left\{\hbar\omega-\frac{\hbar^2}{2M}(\kappa^2+2\boldsymbol{\kappa}.\boldsymbol{\xi})\right\}. \quad (4.73)$$

The initial energy of the nucleus is $\hbar^2\xi^2/2M$. The probability $p_{\boldsymbol{\xi}}$ of the nucleus having wavevector $\boldsymbol{\xi}$ is therefore proportional to $\exp(-\hbar^2\xi^2\beta/2M)$. Thus

$$S(\boldsymbol{\kappa},\omega)=\frac{\int\exp(-\hbar^2\xi^2\beta/2M)\delta\{\hbar\omega-(\hbar^2/2M)(\kappa^2+2\boldsymbol{\kappa}.\boldsymbol{\xi})\}\,\mathrm{d}\boldsymbol{\xi}}{\int\exp(-\hbar^2\xi^2\beta/2M)\,\mathrm{d}\boldsymbol{\xi}} \quad (4.74)$$

$$=\left(\frac{\beta}{4\pi E_{\mathrm{r}}}\right)^{1/2}\exp\left\{-\frac{\beta}{4E_{\mathrm{r}}}(\hbar\omega-E_{\mathrm{r}})^2\right\}, \quad (4.75)$$

where

$$E_{\mathrm{r}}=\frac{\hbar^2\kappa^2}{2M}. \quad (4.76)$$

The integrals in (4.74) are evaluated by taking cartesian coordinates in $\boldsymbol{\xi}$ space, with one of the axes in the direction of $\boldsymbol{\kappa}$.

The expressions for $I(\boldsymbol{\kappa},t)$ and $G(\boldsymbol{r},t)$ follow from (4.3), (4.6), (B.6), and (B.7). They are

$$I(\boldsymbol{\kappa},t)=\exp\{-\kappa^2\sigma^2(t)/2\}, \quad (4.77)$$

$$G(\boldsymbol{r},t)=\{2\pi\sigma^2(t)\}^{-3/2}\exp\{-r^2/2\sigma^2(t)\}, \quad (4.78)$$

where

$$\sigma^2(t)=t(t-\mathrm{i}\hbar\beta)/M\beta. \quad (4.79)$$

The functions $\tilde{I}(\boldsymbol{\kappa}, t)$, $\tilde{G}(\boldsymbol{r}, t)$, and $\tilde{S}(\boldsymbol{\kappa}, \omega)$ are obtained from (4.43), (4.46), and (4.47). The results are

$$\tilde{I}(\boldsymbol{\kappa}, t)=\exp\{-\kappa^2\tilde{\sigma}^2(t)/2\}, \tag{4.80}$$

$$\tilde{G}(\boldsymbol{r}, t)=\{2\pi\tilde{\sigma}^2(t)\}^{-3/2}\exp\{-r^2/2\tilde{\sigma}^2(t)\}, \tag{4.81}$$

$$\tilde{S}(\boldsymbol{\kappa}, \omega)=\left(\frac{\beta}{4\pi E_{\mathrm{r}}}\right)^{1/2}\exp\left\{-\frac{\beta}{4E_{\mathrm{r}}}(\hbar^2\omega^2+E_{\mathrm{r}}^2)\right\}, \tag{4.82}$$

where

$$\tilde{\sigma}^2(t)=(t^2+\tfrac{1}{4}\hbar^2\beta^2)/M\beta. \tag{4.83}$$

Eq. (4.75) shows that, for a fixed value of $\boldsymbol{\kappa}$, $S(\boldsymbol{\kappa}, \omega)$ is a Gaussian function of ω, centred on $\hbar\omega=E_{\mathrm{r}}$. From the results in Appendix I.2

$$\int S(\boldsymbol{\kappa}, \omega)\,\mathrm{d}(\hbar\omega)=1, \tag{4.84}$$

$$\int S(\boldsymbol{\kappa}, \omega)\hbar\omega\,\mathrm{d}(\hbar\omega)=E_{\mathrm{r}}, \tag{4.85}$$

$$\int S(\boldsymbol{\kappa}, \omega)(\hbar\omega)^2\,d(\hbar\omega)=\frac{2E_{\mathrm{r}}}{\beta}+E_{\mathrm{r}}^2. \tag{4.86}$$

The following points may be noted:

(i) For a single nucleus the functions $G(\boldsymbol{r}, t)$, $I(\boldsymbol{\kappa}, t)$, and $S(\boldsymbol{\kappa}, \omega)$ are identical with their self counterparts. Although we have been considering scattering from a single nucleus, the results apply also to a perfect gas of identical atoms. For such a system $G_{\mathrm{s}}(\boldsymbol{r}, t)$, $I_{\mathrm{s}}(\boldsymbol{\kappa}, t)$, and $S_{\mathrm{i}}(\boldsymbol{\kappa}, \omega)$ are the same as for a single nucleus, and $G(\boldsymbol{r}, t)$ differs from $G_{\mathrm{s}}(\boldsymbol{r}, t)$ only by a constant independent of $\boldsymbol{r}$ and t. It follows that for a perfect gas the functions $I(\boldsymbol{\kappa}, t)$ and $I_{\mathrm{s}}(\boldsymbol{\kappa}, t)$ are essentially the same. They differ only by a term proportional to $\delta(\boldsymbol{\kappa})$. The same is true for the functions $S(\boldsymbol{\kappa}, \omega)$ and $S_{\mathrm{i}}(\boldsymbol{\kappa}, \omega)$.

(ii) $I(\boldsymbol{\kappa}, t)$ and $G(\boldsymbol{r}, t)$ are complex functions, while $\tilde{I}(\boldsymbol{\kappa}, t)$ and $\tilde{G}(\boldsymbol{r}, t)$ are real and even in both their arguments. $S(\boldsymbol{\kappa}, \omega)$ and $\tilde{S}(\boldsymbol{\kappa}, \omega)$ are both real, and the latter is even in $\boldsymbol{\kappa}$ and ω.

(iii) In Example 4.1 it is shown that for a perfect gas the classical form of $G_{\mathrm{s}}(\boldsymbol{r}, t)$ is

$$G_{\mathrm{s}}^{\mathrm{cl}}(\boldsymbol{r}, t)=\{2\pi\sigma_{\mathrm{cl}}^2(t)\}^{-3/2}\exp\{-r^2/2\sigma_{\mathrm{cl}}^2(t)\}, \tag{4.87}$$

where

$$\sigma_{\mathrm{cl}}^2(t)=t^2/M\beta, \tag{4.88}$$

which we see is the limiting form of $G(\boldsymbol{r}, t)$ and $\tilde{G}(\boldsymbol{r}, t)$ for a single nucleus as $\hbar\to 0$ or $\beta\to 0$ $(T\to\infty)$.

(iv) $S(\boldsymbol{\kappa}, \omega)$ may be calculated from $G(\boldsymbol{r}, t)$ by means of (4.8). If we put $G(\boldsymbol{r}, t) = G_s^{\text{cl}}(\boldsymbol{r}, t)$, the resulting expression for $S(\boldsymbol{\kappa}, \omega)$ is

$$S(\boldsymbol{\kappa}, \omega) = \left(\frac{\beta}{4\pi E_r}\right)^{1/2} \exp\left(-\frac{\beta\hbar^2\omega^2}{4E_r}\right). \tag{4.89}$$

It can be seen that this violates the principle of detailed balance. If we adopt the Schofield prescription and put $\tilde{G}(\boldsymbol{r}, t) = G_s^{\text{cl}}(\boldsymbol{r}, t)$, the resulting expression for $S(\boldsymbol{\kappa}, \omega)$ is

$$S(\boldsymbol{\kappa}, \omega) = \left(\frac{\beta}{4\pi E_r}\right)^{1/2} \exp\left\{-\frac{\beta}{4E_r}(\hbar^2\omega^2 - 2\hbar\omega E_r)\right\}. \tag{4.90}$$

This expression does satisfy the principle of detailed balance and is a good approximation (for $E_r \ll \hbar\omega$) to the correct expression in (4.75).

4.6 Moments of the scattering function

The energy moments of the scattering function $S(\boldsymbol{\kappa}, \omega)$ may be obtained experimentally and often provide a useful check on the measurements. We define the nth moment by

$$S_n(\boldsymbol{\kappa}) = \int_{-\infty}^{\infty} S(\boldsymbol{\kappa}, \omega)(\hbar\omega)^n \,\mathrm{d}(\hbar\omega). \tag{4.91}$$

From (4.2), (4.5), (4.6), and (4.27) we have the following results for the zeroth moment

$$S_0(\boldsymbol{\kappa}) = \int_{-\infty}^{\infty} S(\boldsymbol{\kappa}, \omega)\,\mathrm{d}(\hbar\omega) = I(\boldsymbol{\kappa}, 0) \tag{4.92}$$

$$= \frac{1}{N}\sum_{jj'} \langle \exp\{\mathrm{i}\boldsymbol{\kappa} \,.\, (\boldsymbol{R}_j - \boldsymbol{R}_{j'})\}\rangle \tag{4.93}$$

$$= 1 + \int g(\boldsymbol{r}) \exp(\mathrm{i}\boldsymbol{\kappa} \,.\, \boldsymbol{r})\,\mathrm{d}\boldsymbol{r}. \tag{4.94}$$

For the incoherent scattering function

$$\int_{-\infty}^{\infty} S_{\mathrm{i}}(\boldsymbol{\kappa}, \omega)\,\mathrm{d}(\hbar\omega) = I_{\mathrm{s}}(\boldsymbol{\kappa}, 0) = 1. \tag{4.95}$$

An expression for the higher moments of $S(\boldsymbol{\kappa}, \omega)$ is obtained from (4.6)

$$S_1(\boldsymbol{\kappa}) = \int_{-\infty}^{\infty} S(\boldsymbol{\kappa}, \omega)\hbar\omega \,\mathrm{d}(\hbar\omega) = \frac{\hbar}{\mathrm{i}}\left\{\frac{\partial}{\partial t} I(\boldsymbol{\kappa}, t)\right\}_{t=0}, \tag{4.96}$$

and in general

$$S_n(\boldsymbol{\kappa}) = \left(\frac{\hbar}{\mathrm{i}}\right)^n \left\{\frac{\partial^n}{\partial t^n} I(\boldsymbol{\kappa}, t)\right\}_{t=0}. \tag{4.97}$$

We consider the first moment. From (4.35), (4.96), and (D.6)

$$\int_{-\infty}^{\infty} S(\boldsymbol{\kappa}, \omega)\hbar\omega \, \mathrm{d}(\hbar\omega) = \frac{\hbar}{\mathrm{i}N}\left\langle \rho_{\boldsymbol{\kappa}}(0)\left\{\frac{\partial}{\partial t}\rho_{-\boldsymbol{\kappa}}(t)\right\}_{t=0}\right\rangle$$

$$= -\frac{1}{N}\langle \rho_{\boldsymbol{\kappa}}(0)[\rho_{-\boldsymbol{\kappa}}(0), H]\rangle, \qquad (4.98)$$

where H is the Hamiltonian of the scattering system. In Section 4.5 we showed that for a perfect gas of identical atoms

$$\int S(\boldsymbol{\kappa}, \omega)\hbar\omega \, \mathrm{d}(\hbar\omega) = \frac{\hbar^2\kappa^2}{2M}, \qquad (4.99)$$

where M is the mass of an atom. The usefulness of this result is that it is true, not only for a perfect gas, but for any system, provided the interactions between the atoms depend only on their positions $\boldsymbol{R}_j$ and not on their momenta. This result follows from the form of (4.98). The additional term in the Hamiltonian due to the interatomic forces commutes with $\rho_{-\boldsymbol{\kappa}}(0)$ – since the latter depends only on the positions $\boldsymbol{R}_j$ – and hence contributes nothing to the right-hand side of (4.98).

Eq. (4.99) has a simple physical interpretation for a perfect gas. Consider a neutron scattered with change of wavevector $\boldsymbol{\kappa}$ due to a collision with a single nucleus. Then by conservation of momentum

$$\hbar\boldsymbol{\kappa} = \boldsymbol{p}' - \boldsymbol{p}, \qquad (4.100)$$

where $\boldsymbol{p}$ and $\boldsymbol{p}'$ are the momenta of the nucleus before and after the collision. The gain in the kinetic energy of the nucleus is

$$\frac{1}{2M}(p'^2 - p^2) = \frac{1}{2M}(\hbar^2\kappa^2 + 2\hbar\boldsymbol{p} \, . \, \boldsymbol{\kappa}). \qquad (4.101)$$

For fixed $\boldsymbol{\kappa}$, the average of $\boldsymbol{p} \, . \, \boldsymbol{\kappa}$ is zero. Therefore the average gain of energy of a nucleus, sometimes termed the *recoil energy*, is

$$E_{\mathrm{r}} = \frac{\hbar^2\kappa^2}{2M}. \qquad (4.102)$$

Thus $S_1(\boldsymbol{\kappa})$ is the mean energy transferred from the neutron to the nucleus.

The result (4.99) applies also to the incoherent scattering function $S_{\mathrm{i}}(\boldsymbol{\kappa}, \omega)$. This follows because, for a perfect gas, the coherent and incoherent scattering functions are essentially the same, and by the previous reasoning $\int S_{\mathrm{i}}(\boldsymbol{\kappa}, \omega)\hbar\omega \, \mathrm{d}(\hbar\omega)$ is the same for all systems with velocity-independent interactions.

The expressions for higher moments have been derived by Placzek (1952). He showed that for an isotropic system the second moment of the incoherent scattering function is

$$\int_{-\infty}^{\infty} S_{\mathrm{i}}(\boldsymbol{\kappa}, \omega)(\hbar\omega)^2 \,\mathrm{d}(\hbar\omega) = \tfrac{4}{3}E_{\mathrm{r}}\bar{K} + E_{\mathrm{r}}^2, \tag{4.103}$$

where $\bar{K}$ is the mean kinetic energy of a nucleus. For a classical system

$$\bar{K} = \frac{3}{2\beta}, \tag{4.104}$$

and (4.103) becomes

$$\int_{-\infty}^{\infty} S_{\mathrm{i}}(\boldsymbol{\kappa}, \omega)(\hbar\omega)^2 \,\mathrm{d}(\hbar\omega) = \frac{2E_{\mathrm{r}}}{\beta} + E_{\mathrm{r}}^2, \tag{4.105}$$

which is the same as the result (4.86) for a perfect gas. The general expression for the second moment of the coherent scattering function is more complicated. The leading terms are the same as for the incoherent scattering function, but there are additional quantum terms arising from correlations between the momenta of different nuclei. Expressions for the second and higher moments of the two scattering functions may be found in Placzek (1952) and in Rahman *et al.* (1962). The results for the moments of the scattering functions are known as *sum rules.*

4.7 Relation between elastic scattering and $I(\boldsymbol{\kappa}, \infty)$, $G(\mathbf{r}, \infty)$

We now show that elastic scattering is directly related to the functions $I(\boldsymbol{\kappa}, t)$, $G(\mathbf{r}, t)$ evaluated at $t = \infty$. The relations lead to some useful results. Elastic scattering does not occur for liquids or gases (see Chapter 5) so the present discussion refers to scattering by a solid.

We consider the functions $I(\boldsymbol{\kappa}, t)$ and $S(\boldsymbol{\kappa}, \omega)$ for some fixed $\boldsymbol{\kappa}$ and drop the symbol $\boldsymbol{\kappa}$. The function $I(t)$ tends to the same limit as t tends to $\pm\infty$. Put

$$I(t) = I(\infty) + I'(t), \tag{4.106}$$

where $I(\infty)$ is the limiting value of $I(t)$, and $I'(t)$ is the time-dependent part which tends to zero as $t \to \pm\infty$. This is indicated schematically in Fig. 4.2*a*, where $I(t)$ is taken as real for the purpose of illustration. Then from (4.4)

$$\begin{aligned} S(\omega) &= \frac{1}{2\pi\hbar}\int_{-\infty}^{\infty} \{I(\infty) + I'(t)\} \exp(-\mathrm{i}\omega t)\,\mathrm{d}t \\ &= \frac{1}{\hbar}\delta(\omega) I(\infty) + \frac{1}{2\pi\hbar}\int_{-\infty}^{\infty} I'(t)\exp(-\mathrm{i}\omega t)\,\mathrm{d}t. \end{aligned} \tag{4.107}$$

Since $\hbar\omega$ is the change in energy of the neutron, the first term on the right-hand side of (4.107) represents elastic scattering and the second term inelastic scattering (Fig. 4.2*b*). Note that the second term is not zero at $\omega = 0$, but it is completely swamped by the first term.

From (4.12) and (4.107)

$$\left(\frac{d^2\sigma}{d\Omega\, dE'}\right)_{\text{coh el}} = \frac{\sigma_{\text{coh}}}{4\pi}\frac{N}{\hbar}\delta(\omega)I(\boldsymbol{\kappa}, \infty). \qquad (4.108)$$

Integrating this with respect to E' gives

$$\left(\frac{d\sigma}{d\Omega}\right)_{\text{coh el}} = \frac{\sigma_{\text{coh}}}{4\pi} NI(\boldsymbol{\kappa}, \infty). \qquad (4.109)$$

By definition

$$I(\boldsymbol{\kappa}, \infty) = \frac{1}{N}\sum_{jj'} \langle \exp\{-i\boldsymbol{\kappa} . \boldsymbol{R}_{j'}(0)\} \exp\{i\boldsymbol{\kappa} . \boldsymbol{R}_j(\infty)\}\rangle. \qquad (4.110)$$

Fig. 4.2 (*a*) Schematic representation of $I(t)$. (The function is actually complex.) (*b*) $S(\omega)$, the Fourier transform of $I(t)$.

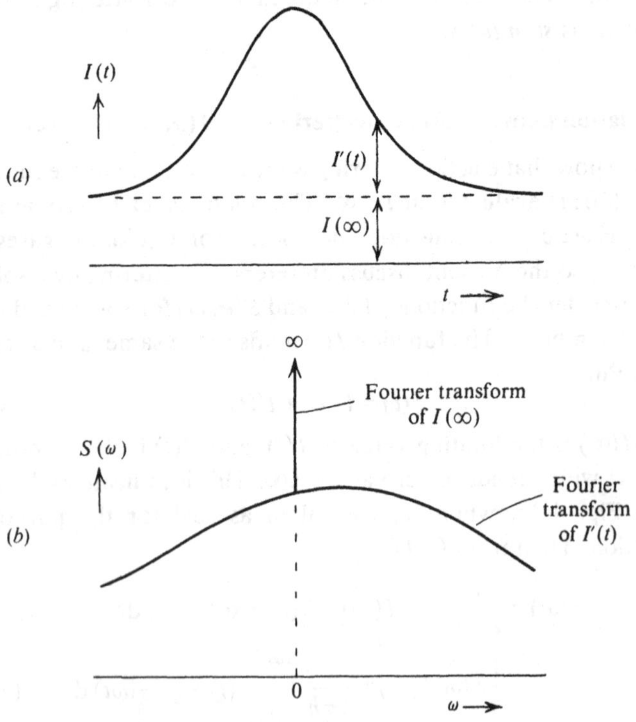

But as $t \to \infty$ the correlation between $\boldsymbol{R}_{j'}(0)$ and $\boldsymbol{R}_j(t)$ becomes independent of t, and we have

$$I(\boldsymbol{\kappa}, \infty) = \frac{1}{N} \sum_{jj'} \langle \exp(-\mathrm{i}\boldsymbol{\kappa} \,.\, \boldsymbol{R}_{j'}) \rangle \langle \exp(\mathrm{i}\boldsymbol{\kappa} \,.\, \boldsymbol{R}_j) \rangle. \qquad (4.111)$$

Note that

$$\langle \exp\{\mathrm{i}\boldsymbol{\kappa} \,.\, \boldsymbol{R}_j(t)\} \rangle = \langle \exp\{\mathrm{i}\boldsymbol{\kappa} \,.\, \boldsymbol{R}_j(0)\} \rangle = \langle \exp(\mathrm{i}\boldsymbol{\kappa} \,.\, \boldsymbol{R}_j) \rangle. \qquad (4.112)$$

This may be verified formally by writing $\langle \exp\{\mathrm{i}\boldsymbol{\kappa} \,.\, \boldsymbol{R}_j(t)\} \rangle$ in terms of the definition of a Heisenberg operator. But it is obvious on physical grounds, since the origin of t is arbitrary. From (4.109) and (4.111)

$$\left(\frac{\mathrm{d}\sigma}{\mathrm{d}\Omega}\right)_{\text{coh el}} = \frac{\sigma_{\text{coh}}}{4\pi} \sum_{jj'} \langle \exp(-\mathrm{i}\boldsymbol{\kappa} \,.\, \boldsymbol{R}_{j'}) \rangle \langle \exp(\mathrm{i}\boldsymbol{\kappa} \,.\, \boldsymbol{R}_j) \rangle. \qquad (4.113)$$

It may be verified that for a Bravais crystal this is the same as (3.48): see Example 4.3.

The same reasoning may be used to obtain an expression for $G(\boldsymbol{r}, \infty)$. From (4.17)

$$G(\boldsymbol{r}, t) = \frac{1}{N} \sum_{jj'} \int \langle \delta\{\boldsymbol{r}' - \boldsymbol{R}_{j'}(0)\} \delta\{\boldsymbol{r}' + \boldsymbol{r} - \boldsymbol{R}_j(t)\} \rangle \, \mathrm{d}\boldsymbol{r}'. \qquad (4.114)$$

Therefore

$$G(\boldsymbol{r}, \infty) = \frac{1}{N} \sum_{jj'} \int \langle \delta(\boldsymbol{r}' - \boldsymbol{R}_{j'}) \rangle \langle \delta(\boldsymbol{r}' + \boldsymbol{r} - \boldsymbol{R}_j) \rangle \, \mathrm{d}\boldsymbol{r}'$$

$$= \frac{1}{N} \int \langle \rho(\boldsymbol{r}') \rangle \langle \rho(\boldsymbol{r}' + \boldsymbol{r}) \rangle \, \mathrm{d}\boldsymbol{r}', \qquad (4.115)$$

where $\rho(\boldsymbol{r})$ is the particle density operator at $t = 0$. The function $G(\boldsymbol{r}, \infty)$ is known as the *Patterson function* and is used as an aid to structure determination in X-ray scattering by crystals.†

The coherent elastic cross-section may be expressed in terms of the Fourier transform of $\langle \rho(\boldsymbol{r}) \rangle$. From (4.5)

$$NI(\boldsymbol{\kappa}, \infty) = N \int G(\boldsymbol{r}, \infty) \exp(\mathrm{i}\boldsymbol{\kappa} \,.\, \boldsymbol{r}) \, \mathrm{d}\boldsymbol{r},$$

$$= \int \langle \rho(\boldsymbol{r}') \rangle \langle \rho(\boldsymbol{r}' + \boldsymbol{r}) \rangle \exp(\mathrm{i}\boldsymbol{\kappa} \,.\, \boldsymbol{r}) \, \mathrm{d}\boldsymbol{r}' \, \mathrm{d}\boldsymbol{r} \qquad (4.116)$$

$$= \int \langle \rho(\boldsymbol{r}') \rangle \langle \rho(\boldsymbol{r}'') \rangle \exp\{\mathrm{i}\boldsymbol{\kappa} \,.\, (\boldsymbol{r}'' - \boldsymbol{r}')\} \, \mathrm{d}\boldsymbol{r}' \, \mathrm{d}\boldsymbol{r}'' \qquad (4.117)$$

$$= \left| \int \langle \rho(\boldsymbol{r}) \rangle \exp(\mathrm{i}\boldsymbol{\kappa} \,.\, \boldsymbol{r}) \, \mathrm{d}\boldsymbol{r} \right|^2. \qquad (4.118)$$

† See for example Woolfson (1970), Section 8.3.

In taking the step from (4.116) to (4.117) we have put $\boldsymbol{r}''=\boldsymbol{r}'+\boldsymbol{r}$. Since the integrals with respect to $\boldsymbol{r}$ and $\boldsymbol{r}''$ are over all space and are taken with $\boldsymbol{r}'$ constant, integrating with respect to $\boldsymbol{r}''$ gives the same result as integrating with respect to $\boldsymbol{r}$. From (4.109) and (4.118)

$$\left(\frac{\mathrm{d}\sigma}{\mathrm{d}\Omega}\right)_{\text{coh el}}=\frac{\sigma_{\text{coh}}}{4\pi}\left|\int\langle\rho(\boldsymbol{r})\rangle\exp(\mathrm{i}\boldsymbol{\kappa}\,.\,\boldsymbol{r})\,\mathrm{d}\boldsymbol{r}\right|^2. \tag{4.119}$$

For the incoherent elastic cross-section, the result corresponding to (4.109) is

$$\begin{aligned}\left(\frac{\mathrm{d}\sigma}{\mathrm{d}\Omega}\right)_{\text{inc el}}&=\frac{\sigma_{\text{inc}}}{4\pi}NI_{\mathrm{s}}(\boldsymbol{\kappa},\infty)\\&=\frac{\sigma_{\text{inc}}}{4\pi}N\int G_{\mathrm{s}}(\boldsymbol{r},\infty)\exp(\mathrm{i}\boldsymbol{\kappa}\,.\,\boldsymbol{r})\,\mathrm{d}\boldsymbol{r},\end{aligned} \tag{4.120}$$

where

$$I_{\mathrm{s}}(\boldsymbol{\kappa},\infty)=\frac{1}{N}\sum_j\langle\exp(-\mathrm{i}\boldsymbol{\kappa}\,.\,\boldsymbol{R}_j)\rangle\langle\exp(\mathrm{i}\boldsymbol{\kappa}\,.\,\boldsymbol{R}_j)\rangle, \tag{4.121}$$

and

$$G_{\mathrm{s}}(\boldsymbol{r},\infty)=\frac{1}{N}\sum_j\int\langle\delta(\boldsymbol{r}'-\boldsymbol{R}_j)\rangle\langle\delta(\boldsymbol{r}'+\boldsymbol{r}-\boldsymbol{R}_j)\rangle\,\mathrm{d}\boldsymbol{r}'. \tag{4.122}$$

Comparison of (3.128) and (4.120) shows that

$$\begin{aligned}\exp(-2W)&=I_{\mathrm{s}}(\boldsymbol{\kappa},\infty)\\&=\int G_{\mathrm{s}}(\boldsymbol{r},\infty)\exp(\mathrm{i}\boldsymbol{\kappa}\,.\,\boldsymbol{r})\,\mathrm{d}\boldsymbol{r},\end{aligned} \tag{4.123}$$

i.e. the Debye–Waller factor is the Fourier transform of $G_{\mathrm{s}}(\boldsymbol{r},\infty)$.

4.8 Static approximation

We return to the discussion of the scattering by a general physical system and consider an important limiting case known as the *static approximation.*

From (2.40)

$$\left(\frac{\mathrm{d}^2\sigma}{\mathrm{d}\Omega\,\mathrm{d}E'}\right)_{\text{coh}}=\frac{\sigma_{\text{coh}}}{4\pi}\frac{k'}{k}\sum_\lambda p_\lambda\sum_{\lambda'}\left|\sum_j\langle\lambda'|\exp(\mathrm{i}\boldsymbol{\kappa}\,.\,\boldsymbol{R}_j)|\lambda\rangle\right|^2$$
$$\times\delta(E_\lambda-E_{\lambda'}+E-E'). \tag{4.124}$$

E is the energy of the incident neutron and E' the energy of the scattered neutron. Suppose we fix $\boldsymbol{k}$ and the direction of the scattered neutrons, and consider the cross-section as a function of $|\boldsymbol{k}'|$, see Fig. 4.3. If k' lies in a small range of values, the value of E' is fixed. For a given initial state λ of the scattering system, the δ-function term in the cross-section picks out certain final states λ', namely, those whose energy is correct for conservation of energy. The sum over λ' is taken only over those states. For each successive interval of k', a different set of λ' states contribute to the scattering. In the development of the theory in Chapter 2 we changed the operator in the matrix element to a time-dependent one, and summed over all λ'. The result was the same as if we had not changed the operator and summed over the correct limited number of λ'.

The static approximation is to ignore the term $E_\lambda - E_{\lambda'}$ in the argument of the δ-function in (4.124). This has two consequences. Firstly, the cross-section becomes a δ-function in $(E-E')$, i.e. the scattering is zero unless $k=k'$. Secondly, in the sum over λ', instead of the matrix element for each λ' being evaluated at the correct value of $\boldsymbol{\kappa}$ (determined by the value of k'), all the matrix elements are evaluated at the same value of $\boldsymbol{\kappa}$, namely $\boldsymbol{\kappa}_0$, the value of $\boldsymbol{\kappa}$ when $k'=k$.

If the term $E_\lambda - E_{\lambda'}$ is ignored in (4.124) the summation over λ' may be carried out by the closure relation.

$$\sum_{\lambda'} \langle\lambda|\exp(-\mathrm{i}\boldsymbol{\kappa}_0 . \boldsymbol{R}_{j'})|\lambda'\rangle\langle\lambda'|\exp(\mathrm{i}\boldsymbol{\kappa}_0 . \boldsymbol{R}_j)|\lambda\rangle$$
$$= \langle\lambda|\exp\{\mathrm{i}\boldsymbol{\kappa}_0 . (\boldsymbol{R}_j - \boldsymbol{R}_{j'})\}|\lambda\rangle. \qquad (4.125)$$

Fig. 4.3 Cross-section $\mathrm{d}^2\sigma/\mathrm{d}\Omega\,\mathrm{d}E'$ as a function of k' for fixed scattering angle. The cross-section is effectively non-zero over a range $\Delta k'$ (shown $\leftrightarrow$). The static approximation is valid for (1) $(\Delta k' \ll k)$, but not for (2).

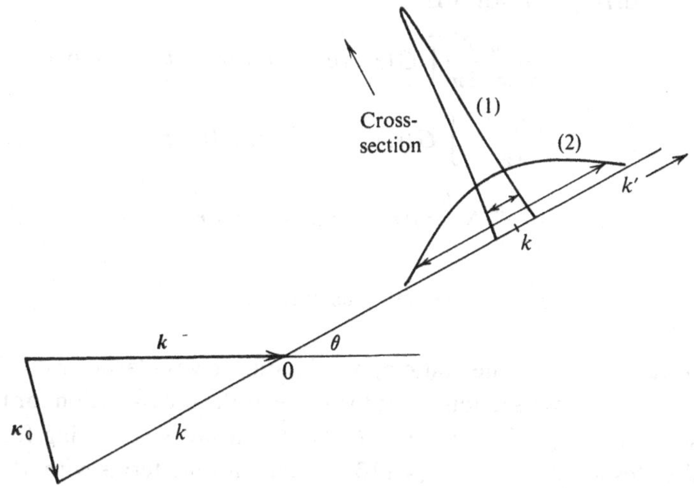

Thus

$$\left(\frac{d^2\sigma}{d\Omega\, dE'}\right)_{coh}^{sa} = \frac{\sigma_{coh}}{4\pi} \sum_{\lambda} p_\lambda \sum_{jj'} \langle\lambda|\exp\{i\boldsymbol{\kappa}_0 . (\boldsymbol{R}_j - \boldsymbol{R}_{j'})\}|\lambda\rangle\delta(E-E'). \tag{4.126}$$

Integration with respect to E' gives

$$\left(\frac{d\sigma}{d\Omega}\right)_{coh}^{sa} = \frac{\sigma_{coh}}{4\pi} \sum_{jj'} \langle\exp\{i\boldsymbol{\kappa}_0 . (\boldsymbol{R}_j - \boldsymbol{R}_{j'})\}\rangle \tag{4.127}$$

$$= \frac{\sigma_{coh}}{4\pi} NI(\boldsymbol{\kappa}_0, 0). \tag{4.128}$$

Alternatively, we may use the formalism of $G(\boldsymbol{r}, t)$ and $S(\boldsymbol{\kappa}, \omega)$. From (4.8) and (4.12)

$$\left(\frac{d^2\sigma}{d\Omega\, dE'}\right)_{coh} = \frac{\sigma_{coh}}{4\pi} \frac{k'}{k} \frac{N}{2\pi\hbar} \int G(\boldsymbol{r}, t) \exp\{i(\boldsymbol{\kappa} . \boldsymbol{r} - \omega t)\}\, d\boldsymbol{r}\, dt. \tag{4.129}$$

Again we keep the scattering angle, i.e. the direction of $\boldsymbol{k}'$, fixed and integrate with respect to E'. In the correct calculation $\boldsymbol{\kappa}$ varies with E'. But in the static approximation the cross-section is a δ-function in $(E'-E)$. So the only contribution to the integral comes from $\boldsymbol{\kappa} = \boldsymbol{\kappa}_0$. Since

$$E - E' = \hbar\omega, \qquad dE' = -\hbar\, d\omega. \tag{4.130}$$

Thus

$$\left(\frac{d\sigma}{d\Omega}\right)_{coh}^{sa} = \int \left(\frac{d^2\sigma}{d\Omega\, dE'}\right)_{coh}^{sa} dE'$$

$$= \frac{\sigma_{coh}}{4\pi} \frac{N}{2\pi} \int G(\boldsymbol{r}, t) \exp\{i(\boldsymbol{\kappa}_0 . \boldsymbol{r} - \omega t)\}\, d\boldsymbol{r}\, dt\, d\omega$$

$$= \frac{\sigma_{coh}}{4\pi} N \int G(\boldsymbol{r}, t) \exp(i\boldsymbol{\kappa}_0 . \boldsymbol{r})\delta(t)\, d\boldsymbol{r}\, dt$$

$$= \frac{\sigma_{coh}}{4\pi} N \int G(\boldsymbol{r}, 0) \exp(i\boldsymbol{\kappa}_0 . \boldsymbol{r})\, d\boldsymbol{r}$$

$$= \frac{\sigma_{coh}}{4\pi} NI(\boldsymbol{\kappa}_0, 0) \quad \text{as before.} \tag{4.131}$$

Note that although the static approximation gives elastic scattering, the resulting cross-section is not the same as the cross-section for true elastic scattering. The former – (4.127) – includes scattering for all final states λ'. The latter – (4.113) – contains only terms with $\lambda' = \lambda$.

Condition for validity of static approximation

Since, in the static approximation, $\boldsymbol{\kappa}$ is replaced by $\boldsymbol{\kappa}_0$ in the matrix elements of (4.124), the approximation will be a good one if, for values of k' not close to k, the states λ' specified by conservation of energy give matrix elements $\sum_j \langle\lambda'|\exp(\mathrm{i}\boldsymbol{\kappa}.\boldsymbol{R}_j)|\lambda\rangle$ which are small.

The matrix elements become small as $|E_{\lambda'} - E_\lambda|$ increases beyond a certain value, because for most scattering systems there is an upper limit to the amount of energy that can be transferred to the neutron. Call this upper limit E_0. For a crystal $E_0 \sim \hbar\omega_\mathrm{m}$, where ω_m is the maximum frequency of a phonon. For a liquid $E_0 \sim \hbar/t_0$, where t_0 is the relaxation time for a disturbance in the liquid. The values of k' therefore lie between a maximum and minimum value given by

$$\frac{\hbar^2}{2m}(k_{\max}^2 - k^2) \sim E_0, \qquad \frac{\hbar^2}{2m}(k^2 - k_{\min}^2) \sim E_0. \tag{4.132}$$

If the maximum and minimum values are not very different from the value of k, (4.132) becomes

$$\frac{\hbar^2}{m}k'\,\Delta k' \sim E_0, \tag{4.133}$$

where $\Delta k'$ is the range of k' (Fig. 4.3). Putting $E = \hbar^2k^2/2m$ gives

$$\frac{\Delta k'}{k} \sim \frac{E_0}{E}. \tag{4.134}$$

The condition for the validity of the static approximation is $\Delta k' \ll k$. From (4.134) this is equivalent to

$$E_0 \ll E, \tag{4.135}$$

i.e. the energy that can be transferred to and from the scattering system is small compared to the energy of the incident neutron.

We may express this condition another way. If v is the velocity and λ the wavelength of the incident neutron,

$$E = \tfrac{1}{2}mv^2 \approx \hbar kv \approx \frac{\hbar v}{\lambda}. \tag{4.136}$$

The condition $E_0 \ll E$ therefore becomes

$$\frac{1}{t_0} \ll \frac{v}{\lambda}. \tag{4.137}$$

Now for scattering to occur other than in the forward direction λ must be of the order of or larger than a, the interatomic spacing in the scattering system. Thus

$$\frac{\lambda}{v} \gtrsim \frac{a}{v} \sim t_1, \tag{4.138}$$

where t_1 is the time taken by the neutron to cross from one atom to the next. So another way of expressing the condition for the validity of the static approximation is

$$t_1 \ll t_0, \tag{4.139}$$

i.e. the time taken for the neutron to cross from one atom to the next is small compared to the characteristic oscillation or relaxation time of the scattering system.

Comments on scattering theory

We bring together the basic results for the coherent cross-sections.

General result:

$$\left(\frac{\mathrm{d}^2\sigma}{\mathrm{d}\Omega\,\mathrm{d}E'}\right)_{\mathrm{coh}} = \frac{\sigma_{\mathrm{coh}}}{4\pi}\frac{k'}{k} NS(\boldsymbol{\kappa}, \omega), \tag{4.140}$$

$$S(\boldsymbol{\kappa}, \omega) = \frac{1}{2\pi\hbar}\int G(\boldsymbol{r}, t)\exp\{\mathrm{i}(\boldsymbol{\kappa}\,.\,\boldsymbol{r} - \omega t)\}\,\mathrm{d}\boldsymbol{r}\,\mathrm{d}t. \tag{4.141}$$

Static approximation:

$$\begin{aligned}\left(\frac{\mathrm{d}\sigma}{\mathrm{d}\Omega}\right)^{\mathrm{sa}}_{\mathrm{coh}} &= \frac{\sigma_{\mathrm{coh}}}{4\pi} N \int G(\boldsymbol{r}, 0)\exp(\mathrm{i}\boldsymbol{\kappa}_0\,.\,\boldsymbol{r})\,\mathrm{d}\boldsymbol{r} \\ &= \frac{\sigma_{\mathrm{coh}}}{4\pi} N\left\{1 + \int g(\boldsymbol{r})\exp(\mathrm{i}\boldsymbol{\kappa}_0\,.\,\boldsymbol{r})\,\mathrm{d}\boldsymbol{r}\right\}.\end{aligned} \tag{4.142}$$

The last line follows from (4.27).

Eq. (4.140) shows that the cross-section is essentially the product of two factors. The first factor σ_{coh} depends on the interaction between the neutron and the individual particles in the scattering system. The second factor $S(\boldsymbol{\kappa}, \omega)$ does not depend on the properties of the neutron at all, neither on its intrinsic properties – mass, energy, etc. – nor on its interaction with the particles in the scattering system. It is a property only of the scattering system and depends on the relative positions and motions of the particles in the system. These depend on the forces between the particles, and on the temperature of the system.

This separation of the cross-section into two factors applies, not only to the scattering of thermal neutrons, but to the scattering of any particle – X-rays, electrons, etc. – provided the Born approximation is valid for the scattering process, and there is no correlation between scattering length (or the equivalent physical parameter for X-rays, etc.) and atomic site.

The scattering function $S(\boldsymbol{\kappa}, \omega)$ is the Fourier transform in space and time of $G(\boldsymbol{r}, t)$, the time-dependent pair-correlation function. When the static approximation applies, the cross-section is proportional to the Fourier transform of $G(\boldsymbol{r}, 0)$, the static pair-correlation function. The relevance of the time variation of the pair-correlation function comes from the interference aspect of scattering. Whatever the nature of the incident particles they may be regarded as waves. The waves scattered by the particles in the scattering system interfere, and their relative phases depend on the relative positions of the particles. Consider the scattering by two particles j' and j. The incident wave reaches j' at time zero and j at time t. The phase difference between the two scattered waves depends on the position of atom j at time t relative to that of j' at time zero. Hence the scattering depends on $G(\boldsymbol{r}, t)$. When the static approximation applies, the waves travel so fast that in the interval between the two scattering events the particle j has not had time to move. Hence the scattering depends on $G(\boldsymbol{r}, 0)$.

For most condensed systems the characteristic time for oscillation or relaxation is

$$t_0 \sim 10^{-13} \text{ to } 10^{-12} \text{ s.}$$

The atomic spacing a is $\sim 10^{-10}$ m. For X-rays or light, travelling with velocity c, the time t_1 for the incident radiation to pass from one particle in the scattering system to the next is

$$t_1 \sim \frac{a}{c} \sim 10^{-18} \text{ s.}$$

Thus $t_1 \ll t_0$ and the static approximation is valid. The scattering of light and X-rays can therefore give information only about $G(\boldsymbol{r}, 0)$, the static pair-correlation function. For thermal neutrons

$$v \sim 10^3 \text{ m s}^{-1}, \quad \text{and} \quad t_1 \sim 10^{-13} \text{ s.}$$

Thus the static approximation does not apply. The scattering of thermal neutrons gives information about $G(\boldsymbol{r}, t)$ for all values of t. It is this property that makes it such a useful tool for studying the properties of condensed matter.

Examples

4.1 Prove that for a perfect gas of atoms of mass M at temperature T the classical form of $G_s(\boldsymbol{r}, t)$ is

$$G_s^{cl}(\boldsymbol{r}, t) = \{2\pi\sigma_{cl}^2(t)\}^{-3/2} \exp\{-r^2/2\sigma_{cl}^2(t)\},$$

where
$$\sigma_{cl}^2(t) = t^2/M\beta.$$

4.2 Prove the following results for an isotropic harmonic oscillator of mass M and frequency ω.

(*a*)
$$I_s(\boldsymbol{\kappa}, t) = \exp\{-\kappa^2\sigma^2(t)/2\},$$
$$G_s(\boldsymbol{r}, t) = \{2\pi\sigma^2(t)\}^{-3/2} \exp\{-r^2/2\sigma^2(t)\},$$

where $\sigma^2(t) = \dfrac{\hbar}{M\omega}\{\coth(\tfrac{1}{2}\hbar\omega\beta)(1-\cos\omega t) - \mathrm{i}\sin\omega t\}.$

(*b*)
$$\tilde{I}_s(\boldsymbol{\kappa}, t) = \exp\{-\kappa^2\tilde{\sigma}^2(t)/2\},$$

where $\tilde{\sigma}^2(t) = \dfrac{\hbar}{M\omega}\{\coth(\tfrac{1}{2}\hbar\omega\beta) - \operatorname{cosech}(\tfrac{1}{2}\hbar\omega\beta)\cos\omega t\}.$

(*c*) The classical form of $I_s(\boldsymbol{\kappa}, t)$ is

$$I_s^{cl}(\boldsymbol{\kappa}, t) = \exp\{-\kappa^2\sigma_{cl}^2(t)/2\},$$

where
$$\sigma_{cl}^2(t) = \frac{2}{M\omega^2\beta}(1-\cos\omega t).$$

(*d*) The classical expression for $\langle r^2(t)\rangle$, the mean-square distance between the positions of the particle at times zero and t, is

$$\langle r_{cl}^2(t)\rangle = \frac{6}{M\omega^2\beta}(1-\cos\omega t).$$

Comment. It follows from the above results that, for a cubic Bravais crystal, $I_s(\boldsymbol{\kappa}, t)$ and $G_s(\boldsymbol{r}, t)$ have the same form as for a single oscillator, with $\sigma^2(t)$ given by

$$\sigma^2(t) = \frac{\hbar}{M}\int_0^{\omega_m} \frac{Z(\omega)}{\omega}\{\coth(\tfrac{1}{2}\hbar\omega\beta)(1-\cos\omega t) - \mathrm{i}\sin\omega t\}\, d\omega. \qquad (4.143)$$

Similarly, the classical expression for the mean-square distance between the positions of an atom at times zero and t is

$$\langle r_{cl}^2(t)\rangle = \frac{6}{M\beta}\int_0^{\omega_m} \frac{Z(\omega)}{\omega^2}(1-\cos\omega t)\, d\omega. \qquad (4.144)$$

4.3 Show that for a Bravais crystal

$$\sum_{ll'} \langle \exp(-\mathrm{i}\boldsymbol{\kappa} \,.\, \boldsymbol{R}_{l'})\rangle \langle \exp(\mathrm{i}\boldsymbol{\kappa} \,.\, \boldsymbol{R}_{l})\rangle$$

$$= N\frac{(2\pi)^3}{v_0}\exp(-2W)\sum_{\boldsymbol{\tau}} \delta(\boldsymbol{\kappa}-\boldsymbol{\tau}).$$

4.4 (*a*) The atoms of a cubic Bravais crystal have mean-square displacement $\langle u^2\rangle$. Show that the thermal average of the particle density operator is

$$\langle \rho(\boldsymbol{r})\rangle = (\pi\sigma^2)^{-3/2}\sum_{l} \exp\{-(\boldsymbol{r}-\boldsymbol{l})^2/\sigma^2\},$$

where $$\sigma^2 = \tfrac{2}{3}\langle u^2\rangle.$$

(*b*) Hence or otherwise show that the self pair-correlation function at $t=\infty$ is

$$G_{\mathrm{s}}(\boldsymbol{r}, \infty) = (2\pi\sigma^2)^{-3/2}\exp(-r^2/2\sigma^2).$$

(*c*) If r is the distance between the positions of an atom at two instants a long time apart, show that

$$\langle r^2\rangle = 2\langle u^2\rangle.$$

4.5 Show that for incoherent one-phonon scattering from a cubic Bravais crystal at high temperature ($\hbar\omega\beta \ll 1$)

$$S_{\mathrm{i}}(\boldsymbol{\kappa}, \omega) = \frac{\kappa^2}{2M\hbar\beta}\exp(-2W)\frac{Z(\omega)}{\omega^2}.$$

5

Scattering by liquids

5.1 Introduction

The theory of the scattering of thermal neutrons by liquids is complicated, mainly because the liquid state itself is complicated. For a crystalline solid we have a relatively simple model, namely a perfect crystal with harmonic forces, which serves as a zero-order approximation for more refined calculations. For a gas we have the perfect gas model – point particles in uncorrelated motion. For a liquid neither of these extreme situations applies.

Coherent neutron scattering gives information about the relative positions and motions of *different* particles in the liquid. From this scattering we may determine what is known as the *structure* of the liquid, which is, in effect, the static pair-correlation function $g(r)$. Measurements of the coherent scattering at low values of momentum transfer also show effects due to excitations of cooperative modes in the liquid. Incoherent scattering depends on the motion of a *single* particle and is therefore easier to interpret.

In the present chapter we give some of the basic results of theory and experiment. We restrict the discussion to classical monatomic liquids, and moreover confine the treatment of coherent scattering to that part that gives information on the *equilibrium*, as opposed to the *dynamic*, properties of liquids. Readers who wish to pursue some of the topics omitted here will find an excellent series of articles in *Reports on Progress in Physics* – by Allen and Higgins (1973) on neutron studies of molecular motion, by Woods and Cowley (1973) on liquid helium, and by Copley and Lovesey (1975) on the dynamic properties of monatomic liquids. See also Egelstaff (1967) and Powles (1973). A readable introduction to theories of the liquid state has been given by Pryde (1966).

5.2 No elastic scattering

The first general result is that there is no truly elastic scattering by a liquid. From (4.108) the coherent elastic scattering is proportional to

$$I(\boldsymbol{\kappa}, \infty)=\int G(\boldsymbol{r}, \infty)\exp(i\boldsymbol{\kappa}\,.\,\boldsymbol{r})\,d\boldsymbol{r}, \tag{5.1}$$

where

$$G(\boldsymbol{r}, \infty)=\frac{1}{N}\int \langle\rho(\boldsymbol{r}')\rangle\langle\rho(\boldsymbol{r}'+\boldsymbol{r})\rangle\,d\boldsymbol{r}'. \tag{5.2}$$

For a liquid the density is uniform, i.e.

$$\langle\rho(\boldsymbol{r})\rangle=\rho=\frac{N}{V}, \tag{5.3}$$

where ρ is the mean number density, i.e. the number of particles per unit volume, N the total number of particles, and V the total volume. Therefore

$$G(\boldsymbol{r}, \infty)=\frac{1}{N}\rho^2 V=\rho. \tag{5.4}$$

Similarly $G_s(\boldsymbol{r}, \infty)$ is constant and given by

$$G_s(\boldsymbol{r}, \infty)=\frac{\rho}{N}, \tag{5.5}$$

which is effectively zero since N is large. Putting $G(\boldsymbol{r}, \infty)$ equal to a constant in (5.1) gives

$$I(\boldsymbol{\kappa}, \infty)\propto\delta(\boldsymbol{\kappa}).$$

So elastic scattering occurs only for $\boldsymbol{\kappa}=0$. But this is not scattering at all; it corresponds to the incident neutrons continuing in the forward direction. The same result applies to the incoherent scattering. We therefore conclude that there is no elastic scattering – coherent or incoherent – from a liquid.

The above reasoning shows that $S(\boldsymbol{\kappa}, \omega)$ has a delta-function behaviour when $\boldsymbol{\kappa}$ and ω are both zero. Since this feature does not correspond to actual scattering we remove it in the following way. The equation relating $S(\boldsymbol{\kappa}, \omega)$ and $G(\boldsymbol{r}, t)$ is

$$S(\boldsymbol{\kappa}, \omega)=\frac{1}{2\pi\hbar}\int G(\boldsymbol{r}, t)\exp\{i(\boldsymbol{\kappa}\,.\,\boldsymbol{r}-\omega t)\}\,d\boldsymbol{r}\,dt. \tag{5.6}$$

We may express $G(\boldsymbol{r}, t)$ as the sum of $G(\boldsymbol{r}, \infty)$, and a time-dependent term $G'(\boldsymbol{r}, t)$ which tends to zero as $t\to\infty$. Thus

$$G(\boldsymbol{r}, t)=G(\boldsymbol{r}, \infty)+G'(\boldsymbol{r}, t)=\rho+G'(\boldsymbol{r}, t). \tag{5.7}$$

If we substitute this expression into (5.6), the ρ part gives rise to the term $\delta(\boldsymbol{\kappa})\,\delta(\omega)$ that we wish to remove. We therefore put

$$S(\boldsymbol{\kappa}, \omega)=\frac{1}{2\pi\hbar}\int\{G(\boldsymbol{r}, t)-\rho\}\exp\{\mathrm{i}(\boldsymbol{\kappa}\,.\,\boldsymbol{r}-\omega t)\}\,\mathrm{d}\boldsymbol{r}\,\mathrm{d}t. \qquad (5.8)$$

It should be emphasised that the change from (5.6) to (5.8) has no effect on the results other than the elimination of the delta function at $\boldsymbol{\kappa}=0$.

5.3 Coherent scattering

The zeroth energy moment of $S(\boldsymbol{\kappa}, \omega)$ is known as the *structure factor* and denoted† by $S(\boldsymbol{\kappa})$. From (4.94), modified according to (5.8), we have

$$S(\boldsymbol{\kappa})=\int_{-\infty}^{\infty} S(\boldsymbol{\kappa}, \omega)\,\mathrm{d}(\hbar\omega)=1+\int\{g(\boldsymbol{r})-\rho\}\exp(\mathrm{i}\boldsymbol{\kappa}\,.\,\boldsymbol{r})\,\mathrm{d}\boldsymbol{r}. \quad (5.9)$$

We are confining the discussion to liquids‡ for which $g(\boldsymbol{r})$ depends only on $|\boldsymbol{r}|$. Eq. (5.9) then becomes

$$\begin{aligned} S(\kappa)&=1+2\pi\int_0^{\infty} r^2\{g(r)-\rho\}\,\mathrm{d}r\int_{-1}^{1}\exp(\mathrm{i}\kappa r\cos\theta)\,\mathrm{d}(\cos\theta)\\ &=1+\frac{4\pi}{\kappa}\int_0^{\infty}\{g(r)-\rho\}\sin(\kappa r)r\,\mathrm{d}r. \qquad (5.10)\end{aligned}$$

Limiting values of the structure factor

We examine the limits of $S(\kappa)$ for large and small κ. For large κ, i.e. $\kappa a \gg 1$ where a is the mean distance between the atoms, the value of the integral on the right-hand side of (5.9) tends to zero. Thus

$$S(\infty)=1. \qquad (5.11)$$

For $\kappa=0$ we have

$$S(0)=1+\int\{g(r)-\rho\}\,\mathrm{d}\boldsymbol{r}. \qquad (5.12)$$

The integral on the right-hand side is related to density fluctuations in the liquid. Consider a fixed volume $\bar{V}$ in the liquid. The number n of

† $S(\boldsymbol{\kappa})$ was denoted by $S_0(\boldsymbol{\kappa})$ in Section 4.6, but the subscript zero is usually omitted.

‡ In the present chapter $g(\boldsymbol{r})$ and $G(\boldsymbol{r}, t)$ depend only on the magnitude of $\boldsymbol{r}$, and $S(\boldsymbol{\kappa}, \omega)$, $S(\boldsymbol{\kappa})$, and $I(\boldsymbol{\kappa}, t)$ depend only on the magnitude of $\boldsymbol{\kappa}$. These functions, therefore appear with plain r and κ in their arguments.

atoms in the volume fluctuates about its mean value $\bar{n}=\rho\bar{V}$. The quantity $\overline{n^2}$ is given by

$$\overline{n^2}=\rho\int_{\bar{V}}\int_{\bar{V}} G(\boldsymbol{r}_2-\boldsymbol{r}_1, 0)\,\mathrm{d}\boldsymbol{r}_1\,\mathrm{d}\boldsymbol{r}_2. \tag{5.13}$$

This follows because $\rho\,\mathrm{d}\boldsymbol{r}_1$ is the *a priori* probability that there is an atom in the element of volume $\mathrm{d}\boldsymbol{r}_1$ at $\boldsymbol{r}_1$, and $G(\boldsymbol{r}_2-\boldsymbol{r}_1, 0)\,\mathrm{d}\boldsymbol{r}_2$ is the probability that, given an atom at $\boldsymbol{r}_1$, there is at the same time an atom in $\mathrm{d}\boldsymbol{r}_2$ at $\boldsymbol{r}_2$. From (4.27)

$$G(\boldsymbol{r}_2-\boldsymbol{r}_1, 0)=\delta(\boldsymbol{r}_2-\boldsymbol{r}_1)+g(\boldsymbol{r}_2-\boldsymbol{r}_1). \tag{5.14}$$

Thus (5.13) becomes

$$\begin{aligned}\overline{n^2}&=\bar{n}\int_{\bar{V}}\{\delta(\boldsymbol{r}_2-\boldsymbol{r}_1)+g(\boldsymbol{r}_2-\boldsymbol{r}_1)\}\,\mathrm{d}\boldsymbol{r}_2\\ &=\bar{n}\left\{1+\int_{\bar{V}} g(\boldsymbol{r})\,\mathrm{d}\boldsymbol{r}\right\}.\end{aligned} \tag{5.15}$$

From (5.12) and (5.15)

$$S(0)=\frac{\overline{n^2}}{\bar{n}}-\bar{n}=\frac{\overline{(\Delta n)^2}}{\bar{n}}, \tag{5.16}$$

where

$$\Delta n=n-\bar{n}. \tag{5.17}$$

Instead of a fixed volume $\bar{V}$ containing a variable number of particles, we can consider a fixed number of particles whose volume V fluctuates about the mean value $\bar{V}$. The previous number fluctuation is then related to the volume fluctuation by

$$\frac{\overline{(\Delta n)^2}}{\bar{n}}=\rho\frac{\overline{(\Delta V)^2}}{\bar{V}}, \tag{5.18}$$

where

$$\Delta V=V-\bar{V}. \tag{5.19}$$

The fixed number of particles form a relatively small subsystem in the liquid, the rest of the liquid being at a fixed temperature T_0 and pressure p_0. The temperature of the subsystem remains constant at T_0 during the fluctuations, but its pressure p varies. If $\bar{E}$ is the mean internal energy of the subsystem and S its entropy, then at equilibrium the quantity $G_0=\bar{E}-T_0S+p_0V$ is a minimum,† and in the neighbourhood of the minimum can be expressed by a Taylor expansion as

$$G_0(V)=G_0(\bar{V})+\left(\frac{\partial G_0}{\partial V}\right)_T\Delta V+\frac{1}{2}\left(\frac{\partial^2 G_0}{\partial V^2}\right)_T(\Delta V)^2+\dots. \tag{5.20}$$

† See Reif (1965), Section 8.3 for proof of this result and of (5.23) below.

The derivatives with respect to V are evaluated at $V=\bar{V}$. At equilibrium

$$\left(\frac{\partial G_0}{\partial V}\right)_T=\left(\frac{\partial \bar{E}}{\partial V}\right)_T-T_0\left(\frac{\partial S}{\partial V}\right)_T+p_0$$
$$=-p+p_0=0, \tag{5.21}$$

which corresponds to the obvious result that at equilibrium the pressure of the subsystem is equal to that of the rest of the liquid. The second derivative of G_0 is

$$\left(\frac{\partial^2 G_0}{\partial V^2}\right)_T=-\left(\frac{\partial p}{\partial V}\right)_T=\frac{1}{\bar{V}\kappa_T}, \tag{5.22}$$

where κ_T is the isothermal compressibility.

The probability that the volume of the system lies between V and $V+\mathrm{d}V$ is

$$f(V)\,\mathrm{d}V\propto\exp(-\beta G_0)\,\mathrm{d}V, \tag{5.23}$$

where $\beta=1/k_{\mathrm{B}}T_0$. Thus from (5.20), (5.21), and (5.22)

$$f(V)\propto\exp\left\{-\frac{\beta}{2\bar{V}\kappa_T}(\Delta V)^2\right\}, \tag{5.24}$$

which is a Gaussian centred at the mean volume $\bar{V}$. The formula for the standard deviation of a Gaussian gives

$$\overline{(\Delta V)^2}=\frac{\bar{V}\kappa_T}{\beta}. \tag{5.25}$$

Therefore, from (5.16) and (5.18)

$$S(0)=\frac{\rho\kappa_T}{\beta}. \tag{5.26}$$

The results in the present section apply to gases as well as liquids. For a perfect gas

$$\kappa_T=\frac{1}{p}, \quad \text{and} \quad \frac{\rho\kappa_T}{\beta}=1. \tag{5.27}$$

These results agree with (4.84), which shows that for a perfect gas $S(\kappa)=1$ for all values of κ. For a liquid, κ_T is much smaller than for a gas, and the value of $S(0)$ is small compared to 1.

Placzek corrections

The most common method of measuring $S(\kappa)$ is to measure the effective differential scattering cross-section $\mathrm{d}\sigma/\mathrm{d}\Omega$. A beam of neutrons with wavevector $\boldsymbol{k}$ and energy E is incident on the liquid specimen and the neutrons scattered into a small solid angle in a direction at an angle θ to the incident direction are counted. The detector thus carries out the integration with respect to the final energy. The effective cross-section measured in this way is

$$\left(\frac{\mathrm{d}\sigma}{\mathrm{d}\Omega}\right)_{\mathrm{eff}} = \int_0^\infty f(E')\frac{\mathrm{d}^2\sigma}{\mathrm{d}\Omega\,\mathrm{d}E'}\,\mathrm{d}E' = \frac{\sigma_{\mathrm{coh}}}{4\pi}\hbar N\int_{-\infty}^{E/\hbar} f(E')\frac{k'}{k}S(\kappa,\omega)\,\mathrm{d}\omega, \tag{5.28}$$

where $f(E')$ is the efficiency of the detector for neutrons of energy E'.

In the static approximation $(\mathrm{d}\sigma/\mathrm{d}\Omega)_{\mathrm{eff}}$ is proportional to $S(\kappa)$. For, in this approximation, $S(\kappa,\omega)$ is effectively a delta function in ω with its peak at $\omega=0$. We then have

$$\begin{aligned}\left(\frac{\mathrm{d}\sigma}{\mathrm{d}\Omega}\right)^{\mathrm{sa}}_{\mathrm{eff}} &= \frac{\sigma_{\mathrm{coh}}}{4\pi}\hbar N f_0\int_{-\infty}^{\infty} S(\kappa_0,\omega)\,\mathrm{d}\omega \\ &= \frac{\sigma_{\mathrm{coh}}}{4\pi}N f_0 S(\kappa_0),\end{aligned} \tag{5.29}$$

where κ_0 is the value for κ for elastic scattering, i.e.

$$\kappa_0^2 = 2k^2(1-\cos\theta), \tag{5.30}$$

and f_0 is the efficiency of the detector for elastically scattered neutrons.

In the correct calculation of (5.28) we have to allow for the variation of $\boldsymbol{\kappa}$, k' and $f(E')$ as ω varies. Placzek (1952) gave a method of calculating the corrections to the static approximation, valid when the mean value of $\hbar\omega$ is small compared to the incident energy E. It consists of expanding the integrand in (5.28) in powers of

$$x = \frac{\hbar\omega}{E} = \frac{E-E'}{E} = \frac{k^2-k'^2}{k^2}. \tag{5.31}$$

The calculation is simplified† if we assume that the efficiency of the counter varies as $1/k'$. Then

$$\left(\frac{\mathrm{d}\sigma}{\mathrm{d}\Omega}\right)_{\mathrm{eff}} = \frac{\sigma_{\mathrm{coh}}}{4\pi}N\hbar f_0\int_{-\infty}^{\infty} S(\kappa,\omega)\,\mathrm{d}\omega. \tag{5.32}$$

† The algebra when $f(E')$ is some general function of E' is straightforward, but longer. It is given by Yarnell *et al.* (1973).

(We have assumed that when $\hbar\omega = E$, $S(\kappa, \omega)$ is sufficiently small for the upper limit of integration in (5.28) to be extended to infinity.) Put

$$S(\kappa, \omega) = S(\kappa_0, \omega) + \Delta\left[\frac{\partial S(\kappa, \omega)}{\partial(\kappa^2)}\right]_{\kappa=\kappa_0} + \frac{\Delta^2}{2!}\left[\frac{\partial^2 S(\kappa, \omega)}{\partial^2(\kappa^2)}\right]_{\kappa=\kappa_0} + \ldots, \tag{5.33}$$

where

$$\Delta = \kappa^2 - \kappa_0^2. \tag{5.34}$$

Now

$$\kappa^2 = k^2 + k'^2 - 2kk' \cos\theta$$
$$= 2k^2 - k^2 x - (1-x)^{1/2}(2k^2 - \kappa_0^2). \tag{5.35}$$

Expanding $(1-x)^{1/2}$ in powers of x and rearranging we have

$$\Delta = -\tfrac{1}{2}\kappa_0^2 x + \tfrac{1}{8}(2k^2 - \kappa_0^2)x^2 + \ldots. \tag{5.36}$$

Inserting (5.33) and (5.36) into (5.32) gives

$$\left(\frac{d\sigma}{d\Omega}\right)_{\text{eff}} = \frac{\sigma_{\text{coh}}}{4\pi} N f_0 \left[S(\kappa_0) - \frac{\kappa_0^2}{2E} S_1'(\kappa_0) + \frac{1}{8E^2}\{(2k^2 - \kappa_0^2) S_2'(\kappa_0) + \kappa_0^4 S_2''(\kappa_0)\} + \ldots \right]. \tag{5.37}$$

$S_n(\kappa)$ is the nth energy moment of $S(\kappa, \omega)$ defined in (4.91), and

$$S_n'(\kappa_0) = \left[\frac{\partial S_n(\kappa)}{\partial(\kappa^2)}\right]_{\kappa=\kappa_0}. \tag{5.38}$$

$S_1(\kappa)$ is given by (4.99), and $S_2(\kappa)$ by (4.103). (We assume the 2nd moments of $S(\kappa, \omega)$ and $S_i(\kappa, \omega)$ are equal.) Eq. (5.37) then gives

$$\left(\frac{d\sigma}{d\Omega}\right)_{\text{eff}} = \frac{\sigma_{\text{coh}}}{4\pi} N f_0 \left[S(\kappa_0) + \frac{m}{M}\left\{\frac{\bar{K}}{3E} - \frac{\kappa_0^2}{2k^2}\left(1 + \frac{\bar{K}}{3E}\right)\right\} + 0\left(\frac{m^2}{M^2}\right)\right], \tag{5.39}$$

where m is the mass of the neutron and M is the mass of a nucleus in the liquid. $\bar{K}$ is the mean kinetic energy of a nucleus; for most liquids its value is close to $3/2\beta$, the classical limit.† Eq. (5.39) shows that the corrections to the static approximation become small for heavy nuclei and high incident energies.

Experimental results and interpretation

The structure factor $S(\kappa)$ has been measured for a number of monatomic liquids – see Page (1973). As an example we consider the results of Yarnell *et al.* (1973) for liquid argon at 85 K. The sample

† See Rahman *et al.* (1962) for estimates of quantum corrections.

consisted of the isotope ^{36}Ar, for which the scattering is entirely coherent. The results are shown in Fig. 5.1. The wavelength of the incident neutrons was 0.9782 Å. Under these conditions the Placzek corrections varied from 0.0012 at $\kappa = 0$ to -0.0426 at $\kappa = 9.08$ Å^{-1}. (The corrections are not given exactly by (5.39) because the detector did not have a $1/k'$ efficiency.) The value of $S(0)$ was taken to be 0.0522, based on the values $\kappa_T = 2.16$ GPa^{-1}, and $\rho = 2.13 \times 10^{28}$ atoms m^{-3}.

The pair distribution function $g(r)$, obtained by Fourier transforming the results for $S(\kappa)$ in Fig. 5.1, is shown in Fig. 5.2. The oscillations at low values of r are spurious. They arise because, for large κ, the quantity $S(\kappa) - 1$ becomes small and hence difficult to measure accurately, but the relevant quantity for the Fourier transform is $\kappa\{S(\kappa) - 1\}$, which may still be appreciable. The spurious oscillations are often inside the atomic diameter and hence do not cause difficulties in interpretation.

The function $g(r)$ is important for calculating the equilibrium properties of a liquid (see Pryde, 1966, Chapter 8). In addition, it is of

Fig. 5.1 The structure factor $S(\kappa)$ for ^{36}Ar at 85 K. The curve through the experimental points is obtained from a molecular dynamics calculation of Verlet based on a Lennard-Jones potential. (After Yarnell *et al.*, 1973.)

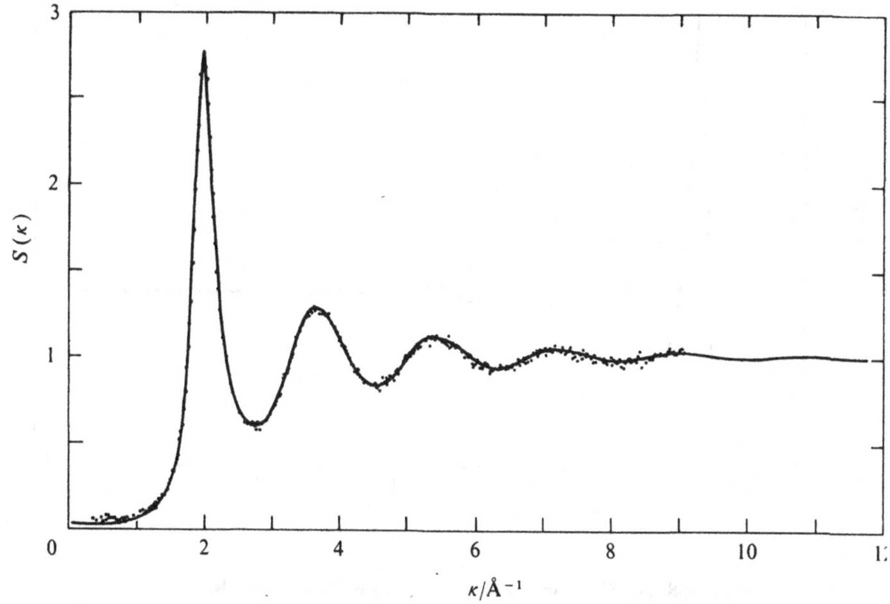

interest to compare the measured values of $g(r)$ with theoretical estimates obtained from a model of the interatomic forces, which are usually taken to be two-body interactions. However, the calculation of $g(r)$ from an assumed two-body potential $\phi(r)$ involves the solutions of complicated integral equations, and sophisticated techniques are necessary. In the last decade alternative methods, based on computer simulation, have been developed for calculating $g(r)$, and other properties of many-body systems, from $\phi(r)$. Systems of a few hundred molecules are considered by two basic methods. In the first, known as the *Monte Carlo method*, a large number of geometrical configurations of the molecules are generated, and the probability of each configuration is calculated from $\phi(r)$ and the Boltzmann factor. In the second method, known as *molecular dynamics*, the molecules are given some initial configuration and the computer calculates the force on each molecule. The molecules are then moved small distances along their trajectories and the calculation is repeated. Both types of method are used to calculate $g(r)$.† The two potentials most

Fig. 5.2 The pair-distribution function $g(r)$ obtained from the experimental results in Fig. 5.1. The mean number density is $\rho = 2.13 \times 10^{28}$ atoms m^{-3}. (After Yarnell *et al.*, 1973.)

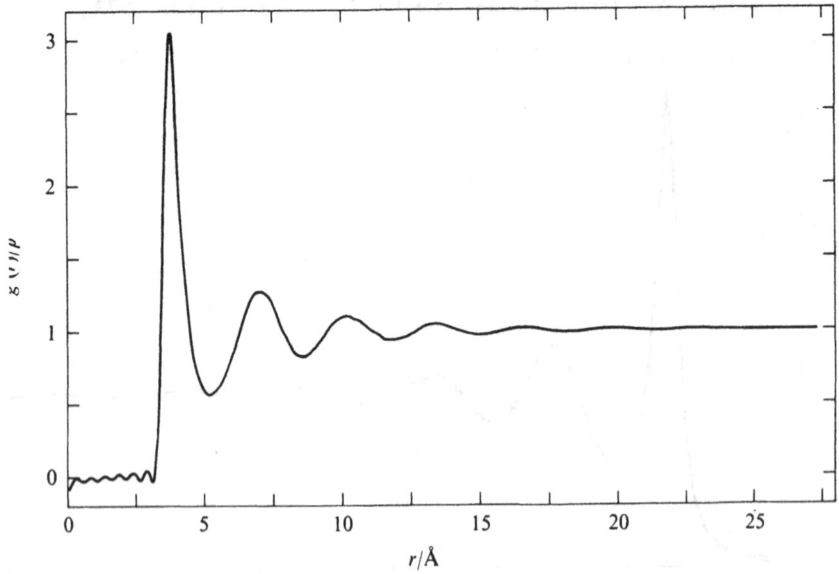

† See Wood (1968) and Croxton (1974) for reviews of these methods.

commonly used in the calculations are the hard-sphere potential, i.e.

$$\begin{aligned} \phi(r) &= \infty \qquad r<r_0 \\ &= 0 \qquad r>r_0, \end{aligned} \tag{5.40}$$

and the Lennard-Jones potential

$$\phi(r) = 4\varepsilon\{(r_0/r)^{12} - (r_0/r)^6\}, \tag{5.41}$$

where ε and r_0 are constants (see Fig. 5.3). It may be noted that these short-range potentials are not appropriate for liquid metals where the potentials have an oscillatory tail extending over several atomic spacings (see Faber, 1972).

Having calculated $g(r)$ from $\phi(r)$ we may proceed to calculate $S(\kappa)$ and compare the result with the measured values. We might hope in this way to determine some of the features of $\phi(r)$. Unfortunately it turns out that $S(\kappa)$ depends mainly on the repulsive core of $\phi(r)$ and is rather insensitive to its long-range behaviour. In Fig. 5.1 a theoretical curve is shown for $S(\kappa)$ obtained from a molecular dynamics calculation based on a Lennard-Jones potential, with ε and

Fig. 5.3 (*a*) Hard-sphere potential, (*b*) Lennard-Jones potential.

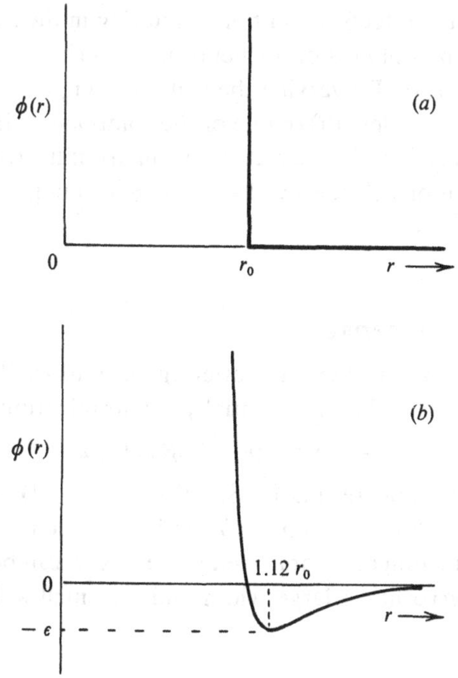

r_0 chosen to fit some thermodynamic data for argon. The fit is clearly very good.

Comparison with X-rays

X-rays can also be used to determine $S(\kappa)$ and have certain advantages over neutrons. They have high intensities, the static approximation holds to a high degree of accuracy obviating the necessity for Placzek corrections, and there is no incoherent scattering. On the other hand, because X-rays are scattered by electrons and not by nuclei, the scattering depends on an atomic form factor. The intensities therefore drop as κ increases and become small for $\kappa \gtrsim 5$ Å^{-1}. X-rays also suffer from large absorption, which means that the scattering must be observed with reflection geometry. This gives problems due to surface contamination for reactive liquids. The advantages of neutrons are that the form factor is independent of κ, and, because the absorption is usually small, the scattering may be observed in transmission. In addition, by comparing the scattering from the liquid with that from a standard vanadium sample the $S(\kappa)$ values may be put on an absolute scale. A major problem with neutrons is correcting for the effects of multiple scattering in the sample.

An interesting application of neutron scattering is in the field of liquid binary alloys. By varying the isotopic composition (and hence the mean scattering length) of one of the components it is possible to measure the partial structure factors arising from the different metal–metal combinations (Enderby, 1968). This technique has no counterpart for X-rays.

5.4 Incoherent scattering

Measurements of incoherent scattering give $(\mathrm{d}^2\sigma/\mathrm{d}\Omega\,\mathrm{d}E')_{\mathrm{inc}}$ as a function of κ and ω. This is essentially a determination of $S_{\mathrm{i}}(\kappa, \omega)$ or

$$\tilde{S}_{\mathrm{i}}(\kappa, \omega) = \exp(-\tfrac{1}{2}\hbar\omega\beta) S_{\mathrm{i}}(\kappa, \omega). \tag{5.42}$$

Fig. 5.4 shows some results by Sköld *et al.* (1972) for incoherent scattering from liquid argon at $T = 85$ K. Each curve represents $\tilde{S}_{\mathrm{i}}(\kappa, \omega)$ as a function of ω for a fixed value of κ. It can be seen that for large κ the curve has a large width and for small κ it has a small width.

These are typical results for incoherent scattering by a liquid, and the two extreme cases may be interpreted as follows. Since $S_i(\kappa, \omega)$ and $G_s(r, t)$ are a Fourier transform pair – see (4.7) and (4.8) – the form of the function $S_i(\kappa, \omega)$ at large κ depends mainly on the behaviour of $G_s(r, t)$ at small values of r. The fact that $S_i(\kappa, \omega)$ extends to large values of ω means that, for small values of r, $G_s(r, t)$ is a highly peaked function of t around $t = 0$. For short times the atom moves as though it were free. So the scattering in this case is the same as for an assembly of free atoms. The liquid behaves as a *perfect gas*. At the other extreme we have $S_i(\kappa, \omega)$ curves corresponding to small values of κ and ω. The scattering function depends mainly on the values of $G_s(r, t)$ at large values of r and t. In long times the atoms make many collisions with each other, so the long-time behaviour of the liquid is governed by *diffusion*. Perfect gas behaviour is found to apply for $t \lesssim 10^{-13}$ s, and diffusion for $t \gtrsim 10^{-12}$ s.

Fig. 5.4 The function $\tilde{S}_i(\kappa,\omega)$ plotted against $\hbar\omega$ for fixed κ for liquid argon at 85 K. The value of κ is shown beside each curve. The experimental points represent smoothed values from positive and negative values of ω. (After Sköld *et al.*, 1972.)

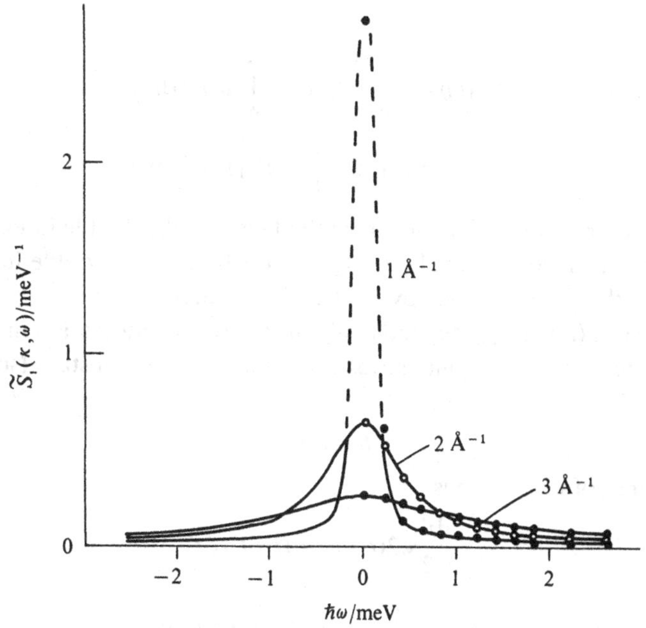

Velocity autocorrelation function

We consider the motion of atoms in a liquid on a classical theory in which each particle has an exact position and velocity at any instant in time. We shall show that the incoherent scattering is related to the *velocity autocorrelation function*, defined as follows. Consider an assembly of atoms in thermal equilibrium. Take the scalar product of the velocity $\boldsymbol{v}(t_0)$ of an atom at time t_0 with the velocity $\boldsymbol{v}(t_0+t)$ of the same atom at time t_0+t. Then the velocity autocorrelation function is this scalar product averaged over all the atoms in the assembly. Because the assembly is in equilibrium, the average is independent of t_0, and we usually write the velocity autocorrelation function in the form $\langle \boldsymbol{v}(0) \, . \, \boldsymbol{v}(t)\rangle$.

We define a quantity $\langle r^2(t)\rangle$ by

$$\langle r^2(t)\rangle = \langle\{\boldsymbol{r}(t)-\boldsymbol{r}(0)\}^2\rangle, \tag{5.43}$$

where $\boldsymbol{r}(0)$ is the position of an atom at some instant, and $\boldsymbol{r}(t)$ is the position of the same atom at a time t later. The relation between $\langle r^2(t)\rangle$ and the velocity autocorrelation function may be derived as follows.

$$\boldsymbol{r}(t)-\boldsymbol{r}(0) = \int_0^t \boldsymbol{v}(t_1)\,\mathrm{d}t_1. \tag{5.44}$$

Therefore

$$\begin{aligned}\langle r^2(t)\rangle &= \left\langle \int_0^t \boldsymbol{v}(t_1)\,\mathrm{d}t_1 \, . \int_0^t \boldsymbol{v}(t_2)\,\mathrm{d}t_2 \right\rangle \\ &= 2\int_0^t \mathrm{d}t_2 \int_0^{t_2} \langle \boldsymbol{v}(t_1) \, . \, \boldsymbol{v}(t_2)\rangle\,\mathrm{d}t_1. \end{aligned} \tag{5.45}$$

The factor 2 in (5.45) arises from the fact that the double integration covers the shaded triangle in Fig. 5.5*a*, whereas the double integration in the previous line covers the whole square.

Since $\langle \boldsymbol{v}(t_1) \, . \, \boldsymbol{v}(t_2)\rangle$ depends only on t_2-t_1, the integration in (5.45) is carried out by diagonal strips for which t_2-t_1 is constant (see Fig. 5.5*b*). Put

$$t' = t_2 - t_1. \tag{5.46}$$

The area of each strip is

$$\frac{\mathrm{d}t'}{\sqrt{2}}\sqrt{2}(t-t') = (t-t')\,\mathrm{d}t'. \tag{5.47}$$

Thus

$$\langle r^2(t)\rangle = 2\int_0^t \langle \boldsymbol{v}(0) \, . \, \boldsymbol{v}(t')\rangle(t-t')\,\mathrm{d}t'. \tag{5.48}$$

Further useful results are obtained by differentiating this equation twice with respect to t. Put

$$\langle \boldsymbol{v}(0) \,.\, \boldsymbol{v}(t') \rangle = g(t'). \tag{5.49}$$

Then

$$\langle r^2(t) \rangle = 2t \int_0^t g(t')\,\mathrm{d}t' - 2 \int_0^t g(t')t'\,\mathrm{d}t'. \tag{5.50}$$

$$\frac{\mathrm{d}}{\mathrm{d}t}\langle r^2(t) \rangle = 2 \int_0^t g(t')\,\mathrm{d}t' + 2tg(t) - 2tg(t)$$

$$= 2 \int_0^t \langle \boldsymbol{v}(0) \,.\, \boldsymbol{v}(t') \rangle\,\mathrm{d}t'. \tag{5.51}$$

$$\frac{\mathrm{d}^2}{\mathrm{d}t^2}\langle r^2(t) \rangle = 2\langle \boldsymbol{v}(0) \,.\, \boldsymbol{v}(t) \rangle. \tag{5.52}$$

Note that since $\langle r^2(t) \rangle$ is real and even in t, $\langle \boldsymbol{v}(0) \,.\, \boldsymbol{v}(t) \rangle$ is also real and even in t.

Velocity frequency function

The Fourier transform of $\langle \boldsymbol{v}(0) \,.\, \boldsymbol{v}(t) \rangle$ is known as the *velocity frequency function*. It is defined by

$$p(\omega) = \frac{M\beta}{3\pi} \int_{-\infty}^{\infty} \langle \boldsymbol{v}(0) \,.\, \boldsymbol{v}(t) \rangle \exp(-\mathrm{i}\omega t)\,\mathrm{d}t, \tag{5.53}$$

where M is the mass of an atom in the assembly. The function is real and even in ω. The constant $M\beta/3\pi$ is chosen to make

$$\int_0^{\infty} p(\omega)\,\mathrm{d}\omega = 1. \tag{5.54}$$

Fig. 5.5 Diagrams showing region of integration for evaluation of $\langle r^2(t) \rangle$. The value of $t' = t_2 - t_1$ is constant for the diagonal strip in (*b*).

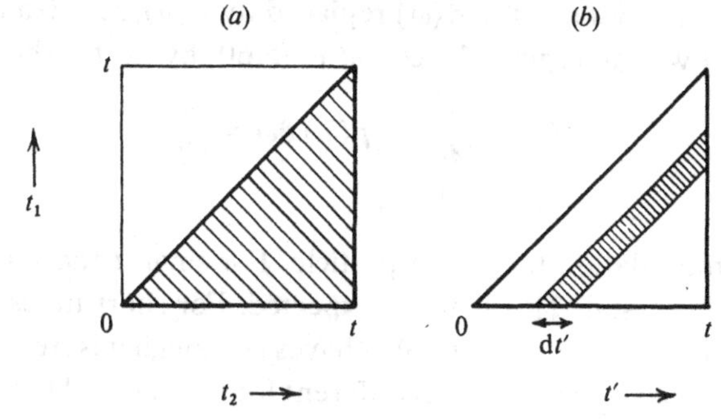

To verify this equation we note that by the equipartition theorem

$$\langle \boldsymbol{v}^2(0)\rangle = \frac{3}{M\beta}. \tag{5.55}$$

Then

$$\begin{aligned}\int_0^{\infty} p(\omega)\,\mathrm{d}\omega &= \frac{M\beta}{3\pi}\int_0^{\infty}\int_{-\infty}^{\infty}\langle \boldsymbol{v}(0)\,.\,\boldsymbol{v}(t)\rangle \exp(-\mathrm{i}\omega t)\ \mathrm{d}t\ \ \mathrm{d}\omega\\ &= \frac{M\beta}{3}\int_{-\infty}^{\infty}\langle \boldsymbol{v}(0)\,.\,\boldsymbol{v}(t)\rangle\delta(t)\,\mathrm{d}t\\ &= \frac{M\beta}{3}\langle \boldsymbol{v}^2(0)\rangle = 1.\end{aligned} \tag{5.56}$$

We now derive a relation between $\langle r^2(t)\rangle$ and $p(\omega)$. From (5.53) and (B.2)

$$\begin{aligned}\langle \boldsymbol{v}(0)\,.\,\boldsymbol{v}(t)\rangle &= \frac{3}{2M\beta}\int_{-\infty}^{\infty} p(\omega)\exp(\mathrm{i}\omega t)\,\mathrm{d}\omega\\ &= \frac{3}{M\beta}\int_0^{\infty} p(\omega)\cos\omega t\ \mathrm{d}\omega,\end{aligned} \tag{5.57}$$

since $p(\omega)$ is real and even. From (5.52)

$$\frac{\mathrm{d}^2}{\mathrm{d}t^2}\langle r^2(t)\rangle = \frac{6}{M\beta}\int_0^{\infty} p(\omega)\cos\omega t\ \mathrm{d}\omega. \tag{5.58}$$

Integrating this equation twice, and using the results

$$\langle r^2(t)\rangle = \frac{\mathrm{d}}{\mathrm{d}t}\langle r^2(t)\rangle = 0 \quad \text{at } t=0, \tag{5.59}$$

we obtain

$$\langle r^2(t)\rangle = \frac{6}{M\beta}\int_0^{\infty} p(\omega)\frac{1-\cos\omega t}{\omega^2}\,\mathrm{d}\omega. \tag{5.60}$$

This expression is the same as the classical result for $\langle r^2(t)\rangle$ for a cubic Bravais crystal, with $Z(\omega)$ replaced by $p(\omega)$; see Example 4.2. For small t we can replace $1-\cos\omega t$ in (5.60) by $\frac{1}{2}\omega^2 t^2$. Then

$$\begin{aligned}\langle r^2(t)\rangle &= \frac{3t^2}{M\beta}\int_0^{\infty} p(\omega)\,\mathrm{d}\omega = \frac{3}{M\beta}t^2\\ &= \langle \boldsymbol{v}^2(0)\rangle t^2,\end{aligned} \tag{5.61}$$

which is the relation for a free particle. For small t the same result applies in a crystal. This is to be expected. For short times an atom whether in a liquid or in a crystal behaves as though it is free. For long times the behaviour of $\langle r^2(t)\rangle$ is different for a liquid and for a crystal.

For a liquid it increases linearly with time, while for a crystal it tends to a constant. The algebraic reason why the two similar equations (5.60) and (4.144) in Example 4.2 lead to these different results is that, as ω tends to zero, $p(\omega)$ tends to a constant, while $Z(\omega)$ tends to zero as ω^2. For large t the integral in (5.60) is dominated by the value of the integrand at small ω. Changing the variable of integration from ω to ωt shows that the integral is proportional to t.

Gaussian approximation

For short times the particles in the liquid behave as though they are free, and we saw in Section 4.5 that for free nuclei (perfect gas) the classical form of $G_s(r, t)$ is

$$G_s^{cl}(r, t) = \{2\pi\sigma^2(t)\}^{-3/2} \exp\{-r^2/2\sigma^2(t)\}, \tag{5.62}$$

where

$$\sigma^2(t) = \frac{t^2}{M\beta}. \tag{5.63}$$

We shall see in the next section that for long times, when the motion is governed by diffusion, $G_s^{cl}(r, t)$ also has the Gaussian form of (5.62), the only difference being in the way $\sigma^2(t)$ varies with time. We now assume that for all values of t the spatial variation of $G_s^{cl}(r, t)$ is given by (5.62). This is known as the *Gaussian approximation*. We make no assumption about the way $\sigma^2(t)$ varies with time. Then

$$I_s^{cl}(\kappa, t) = \int G_s^{cl}(r, t) \exp(i\boldsymbol{\kappa} . \boldsymbol{r})\, d\boldsymbol{r} = \exp\{-\tfrac{1}{2}\kappa^2\sigma^2(t)\}, \tag{5.64}$$

$$\langle r^2(t)\rangle = 4\pi \int_0^\infty r^4 G_s^{cl}(r, t)\, dr = 3\sigma^2(t). \tag{5.65}$$

There is no theoretical justification for the Gaussian approximation other than the fact that it holds for the extremes of short and long times.† The experiments of Sköld *et al.* (1972) show that for liquid argon there are small but significant departures from the Gaussian form for t in the range 5×10^{-13} to 5×10^{-12} s; see (5.83) and Fig. 5.10.

In Fig. 5.6 the time variation of $\sigma^2(t)$ is drawn schematically for various models. The results of Example 4.2 show that $G_s^{cl}(r, t)$ is a Gaussian for a cubic crystal with harmonic forces, and this case is included in the figure.

† For a discussion of the Gaussian approximation see Nijboer and Rahman (1966) and Boutin and Yip (1968).

We now derive a relation between $p(\omega)$ and $S_i(\kappa, \omega)$. Schofield's suggestion for $\tilde{I}_s(\kappa, t)$, together with the Gaussian approximation, give

$$\tilde{I}_s(\kappa, t) = I_s^{cl}(\kappa, t) = \exp\{-\tfrac{1}{2}\kappa^2\sigma^2(t)\} \tag{5.66}$$
$$= 1 - \tfrac{1}{2}\kappa^2\sigma^2(t) + 0(\kappa^4). \tag{5.67}$$

Therefore, from (5.52), (5.57), and (5.65)

$$\lim_{|\kappa|\to 0}\left\{-\frac{2}{\kappa^2}\frac{\partial^2}{\partial t^2}\tilde{I}_s(\kappa, t)\right\}$$
$$= \frac{\partial^2}{\partial t^2}\sigma^2(t)$$
$$= \frac{1}{3}\frac{\partial^2}{\partial t^2}\langle r^2(t)\rangle = \tfrac{2}{3}\langle \boldsymbol{v}(0) \,.\, \boldsymbol{v}(t)\rangle$$
$$= \frac{1}{M\beta}\int_{-\infty}^{\infty} p(\omega)\exp(\mathrm{i}\omega t)\,\mathrm{d}\omega. \tag{5.68}$$

But from (4.45)

$$\tilde{I}_s(\kappa, t) = \hbar\int_{-\infty}^{\infty} \tilde{S}_i(\kappa, \omega)\exp(\mathrm{i}\omega t)\,\mathrm{d}\omega. \tag{5.69}$$

Differentiating this equation twice with respect to t and comparing the result with (5.68) we obtain

$$p(\omega) = 2M\hbar\beta\omega^2 \lim_{|\kappa|\to 0}\frac{\tilde{S}_i(\kappa, \omega)}{\kappa^2}. \tag{5.70}$$

Fig. 5.6 The function $\sigma^2(t)$ for various models in which $G_s^{cl}(r,t)$ is a Gaussian function of r.

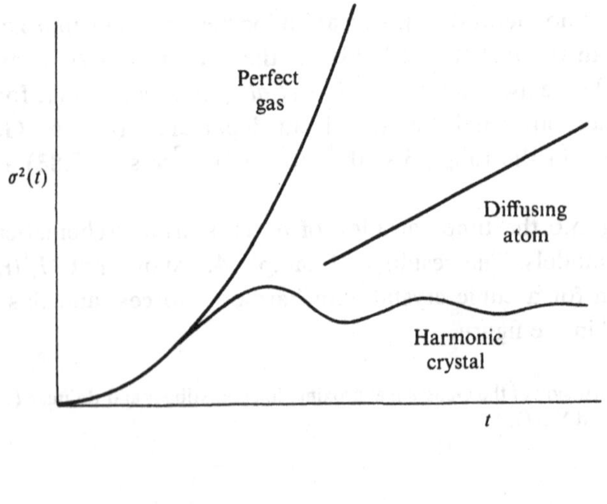

Thus, measuring the incoherent cross-section as a function of κ for a fixed value of ω and extrapolating $\tilde{S}_i(\kappa, \omega)/\kappa^2$ to $\kappa = 0$ gives $p(\omega)$. Carneiro (1976) has calculated $p(\omega)$ in this way from measurements on liquid argon and liquid orthohydrogen. The results for argon are shown in Fig. 5.7.

Diffusion

The form of $G_s(\boldsymbol{r}, t)$ for times long compared to the mean time between collisions of the atoms is governed by the diffusion process. The basic equation for diffusion, known as *Fick's law*, is

$$\frac{\partial n(\boldsymbol{r}, t)}{\partial t} = D\nabla^2 n(\boldsymbol{r}, t), \tag{5.71}$$

where $n(\boldsymbol{r}, t)$ is the number of atoms per unit volume at the point $\boldsymbol{r}$ at time t, and D is the *diffusion constant.*

Fig. 5.7 The velocity frequency function $p(\omega)$ for liquid argon at 85 K. The open circle at $\omega = 0$ is based on an independently measured value of the diffusion constant D. The dip in the values of $p(\omega)$ at small ω is consistent with a $t^{-3/2}$ tail in the velocity autocorrelation function. (After Carneiro, 1976.)

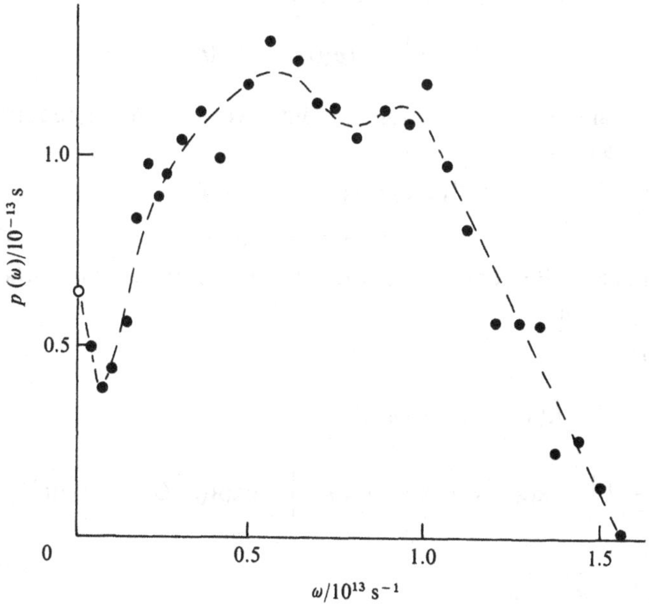

By its probability significance – see (4.23) – the function $G_s^{cl}(\boldsymbol{r}, t)$ must satisfy (5.71). For isotropic diffusion

$$\nabla^2 = \frac{\partial^2}{\partial r^2} + \frac{2}{r}\frac{\partial}{\partial r}. \tag{5.72}$$

It is readily verified that an isotropic solution of (5.71) is

$$G_s^{cl}(r, t) = \{2\pi\sigma^2(t)\}^{-3/2} \exp(-r^2/2\sigma^2(t)\}, \tag{5.73}$$

provided

$$\frac{\mathrm{d}}{\mathrm{d}t}\sigma^2(t) = 2D, \tag{5.74}$$

i.e.

$$\sigma^2(t) = 2D|t| + c, \tag{5.75}$$

where c is a constant whose value depends on the form of the velocity autocorrelation function. This expression for $\sigma^2(t)$ is valid only for times long enough for diffusion theory to apply, and for such times c is usually small compared to $2Dt$.

From (5.51) and (5.65)

$$\frac{\mathrm{d}}{\mathrm{d}t}\sigma^2(t) = \frac{1}{3}\frac{\mathrm{d}}{\mathrm{d}t}\langle r^2(t)\rangle = \frac{2}{3}\int_0^t \langle \boldsymbol{v}(0) \,.\, \boldsymbol{v}(t')\rangle \,\mathrm{d}t'. \tag{5.76}$$

For large t', $\langle \boldsymbol{v}(0) \,.\, \boldsymbol{v}(t')\rangle$ tends to zero, and

$$\frac{\mathrm{d}}{\mathrm{d}t}\sigma^2(t) = \frac{2}{3}\int_0^\infty \langle \boldsymbol{v}(0) \,.\, \boldsymbol{v}(t)\rangle \,\mathrm{d}t. \tag{5.77}$$

Comparison with (5.74) gives the result

$$D = \frac{1}{3}\int_0^\infty \langle \boldsymbol{v}(0) \,.\, \boldsymbol{v}(t)\rangle \,\mathrm{d}t. \tag{5.78}$$

To obtain the cross-section for diffusive motion we require the function $\tilde{S}_i(\kappa, \omega)$. From (5.64)

$$\begin{aligned} I_s^{cl}(\kappa, t) &= \exp\{-\tfrac{1}{2}\kappa^2\sigma^2(t)\} \\ &= \exp\{-\kappa^2 D|t|\}, \end{aligned} \tag{5.79}$$

if we neglect the constant c in (5.75). As before we take $\tilde{I}_s(\kappa, t) = I_s^{cl}(\kappa, t)$. Then

$$\begin{aligned} \tilde{S}_i(\kappa, \omega) &= \frac{1}{2\pi\hbar}\int_{-\infty}^{\infty} \tilde{I}_s(\kappa, t)\exp(-\mathrm{i}\omega t)\,\mathrm{d}t \\ &= \frac{1}{2\pi\hbar}\left[\int_0^\infty \exp\{-(\kappa^2 D + \mathrm{i}\omega)t\}\,\mathrm{d}t + \int_{-\infty}^0 \exp\{(\kappa^2 D - \mathrm{i}\omega)t\}\,\mathrm{d}t\right] \\ &= \frac{1}{\pi\hbar}\frac{D\kappa^2}{(D\kappa^2)^2 + \omega^2}. \end{aligned} \tag{5.80}$$

This function, known as a *Lorentzian*, is plotted in Fig. 5.8 for a fixed value of κ. It has a maximum when $\omega = 0$, and its full-width at half-maximum is

$$\Delta E = \hbar\,\Delta\omega = 2\hbar D\kappa^2. \tag{5.81}$$

In Fig. 5.9*a* the experimental values of ΔE, obtained by Sköld *et al.* (1972) for liquid argon, are shown as a function of κ^2. In Fig. 5.9*b* the experimental values of $\tilde{S}_i(\kappa, 0)$ are shown. From (5.80) the theoretical value of this function for diffusion is

$$\tilde{S}_i(\kappa, 0) = 1/\pi\hbar D\kappa^2. \tag{5.82}$$

The curves given by (5.81) and (5.82), with an independently measured value of the diffusion constant, are shown in both figures, and it can be seen that in the region of small energy transfer the experimental results agree fairly well with simple diffusion theory. The departure from the theory is due to vibrational motion of the atoms and is consistent with the results obtained by Levesque and Verlet (1970) by computer simulation studies.

Sköld *et al.* determined the function $\tilde{I}_s(\kappa, t)$ from their results. The incoherent scattering function $S_i(\kappa, \omega)$ was measured over the range $1.0\,\text{Å}^{-1} \leqslant \kappa \leqslant 4.4\,\text{Å}^{-1}$ and $0 \leqslant \hbar\omega \leqslant 10.6$ meV. The Fourier transform of $\tilde{S}_i(\kappa, \omega)$ for a fixed κ and varying ω gives $\tilde{I}_s(\kappa, t)$. This was

Fig. 5.8 The function $\tilde{S}_i(\kappa,\omega)$ at constant κ for diffusive motion.

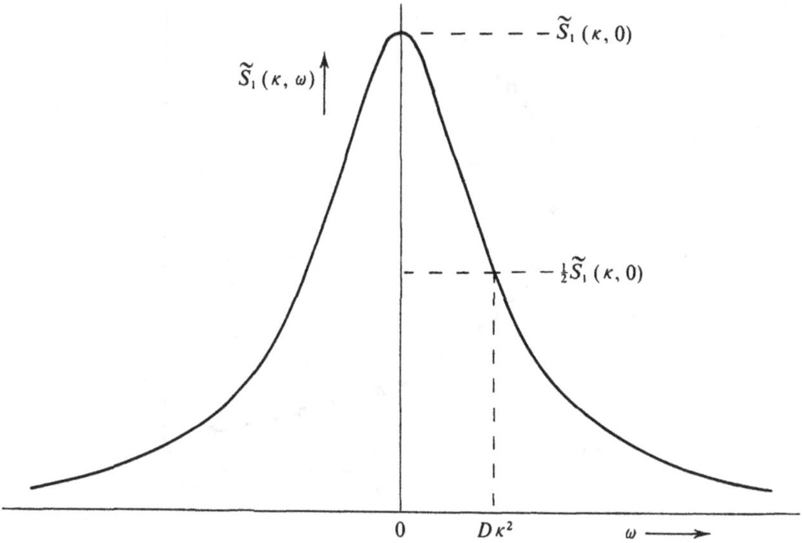

fitted to a mathematical function of the form

$$\tilde{I}_s(\kappa, t) = \exp\{-\tfrac{1}{2}\kappa^2\sigma^2(t)\}\{1 + \alpha(t)\kappa^4\sigma^4(t)\}. \qquad (5.83)$$

The coefficient $\alpha(t)$ is a measure of the departure of $\tilde{I}_s(\kappa, t)$ from a Gaussian function.

Fig. 5.9 Results for liquid argon at 85 K. (*a*) Full-width at half-maximum of $\tilde{S}_i(\kappa, \omega)$ at constant κ, plotted against κ^2. The straight line is the simple diffusion relation (5.81) with $D = 1.94 \times 10^{-9}\ \mathrm{m^2 s^{-1}}$. (*b*) The function $\tilde{S}_i(\kappa, 0)$. The curve is the relation (5.82) with the same value of D. (After Sköld *et al.*, 1972.)

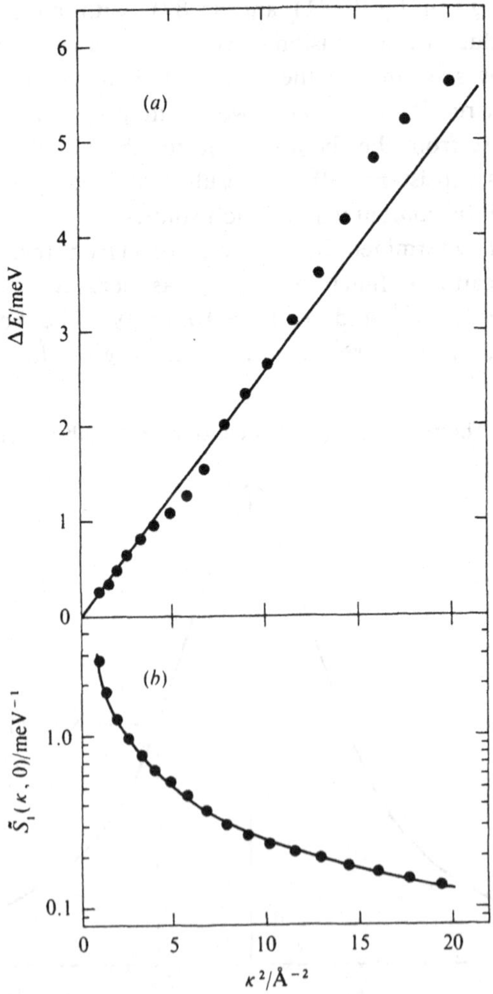

For perfect gas behaviour

$$\sigma^2(t)=\frac{t^2}{M\beta}+\frac{\hbar^2\beta}{4M}. \tag{5.84}$$

This result was derived in Section 4.5 and differs from (5.63) by the addition of a quantum term, which, at the temperature of the measurements, is significant only for $t \leqslant 5\times10^{-14}$ s. For diffusion behaviour

$$\sigma^2(t)=2Dt. \tag{5.85}$$

The functions (5.84) and (5.85) are shown in Fig. 5.10, together with the experimental values of $\sigma^2(t)$, and it can be seen that there is a good agreement. Note the smooth transition from the perfect gas to the diffusion form of $\sigma^2(t)$. The values of the coefficient $\alpha(t)$ given by the measurements are also included in the figure.

Fig. 5.10 Parameters of the function $\tilde{I}_s(\kappa, t)$ for argon at 85 K: • experimental values of $\sigma^2(t)$ (scale on left-hand side). The curve for the free particle is given by (5.84), and the line for the diffusing particle by (5.85). ○ experimental values of $\alpha(t)$ (scale on right-hand side). The curve for $\alpha(t)$ is based on a calculation by Levesque and Verlet. (After Sköld *et al.*, 1972.)

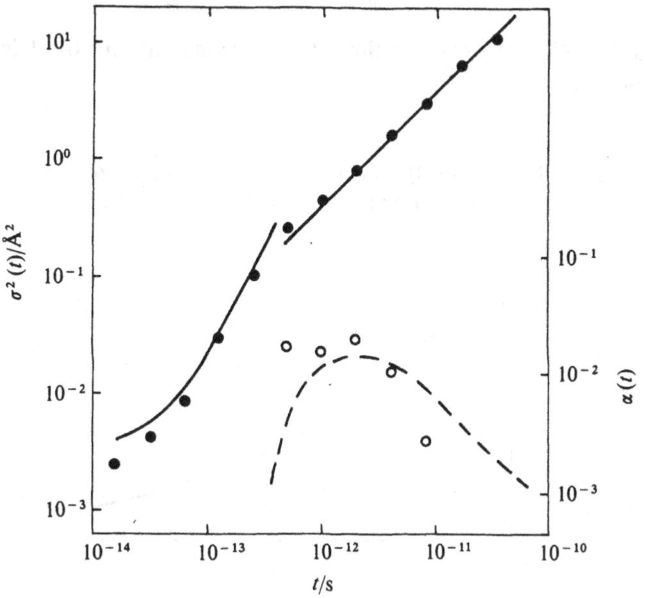

Brownian motion

An approximate expression for the velocity autocorrelation function for diffusion may be obtained from a simple equation of motion, known as the *Langevin equation*,

$$M\frac{d\boldsymbol{v}}{dt} = -\gamma M\boldsymbol{v} + \boldsymbol{R}. \tag{5.86}$$

$-\gamma M\boldsymbol{v}$ is a frictional or viscous force (γ is a constant), and $\boldsymbol{R}$ is a rapidly varying random force due to collisions with other atoms. This equation was originally proposed to describe Brownian motion, i.e. the motion of a heavy particle suspended in a fluid and subjected to bombardment by the molecules of the fluid. A solution of (5.86) is

$$\boldsymbol{v}(t) = \boldsymbol{v}(0)\exp(-\gamma|t|) + \boldsymbol{V}(t). \tag{5.87}$$

This equation is not valid for small t. The $\boldsymbol{V}(t)$ term depends on $\boldsymbol{R}$. Since the latter is randomly fluctuating, the scalar product of $\boldsymbol{v}(0)$ and $\boldsymbol{V}(t)$ is zero when averaged over all the atoms. Thus

$$\langle \boldsymbol{v}(0)\,.\,\boldsymbol{v}(t)\rangle = \langle \boldsymbol{v}^2(0)\rangle \exp(-\gamma|t|). \tag{5.88}$$

Substituting (5.88) in (5.78) gives

$$D = \tfrac{1}{3}\langle \boldsymbol{v}^2(0)\rangle \int_0^\infty \exp(-\gamma t)\,dt = \frac{1}{\gamma M\beta}. \tag{5.89}$$

This is the relation between the diffusion constant and the frictional coefficient γ.

Fig. 5.11 The velocity frequency function $p(\omega)$ given by the Langevin equation.

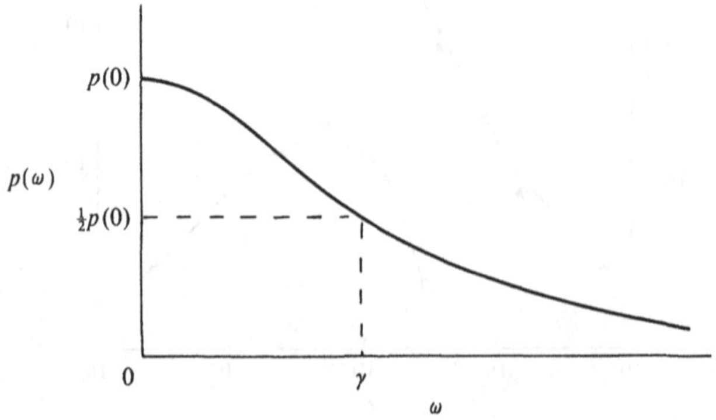

The form of $p(\omega)$ for diffusion is obtained from (5.53) and (5.88).

$$p(\omega)=\frac{M\beta}{3\pi}\int_{-\infty}^{\infty}\langle \boldsymbol{v}(0)\,.\,\boldsymbol{v}(t)\rangle \exp(-\mathrm{i}\omega t)\,\mathrm{d}t$$

$$=\frac{1}{\pi}\left[\int_{0}^{\infty}\exp\{-(\gamma+\mathrm{i}\omega)t\}\,\mathrm{d}t+\int_{-\infty}^{0}\exp\{(\gamma-\mathrm{i}\omega)t\}\,\mathrm{d}t\right]$$

$$=\frac{1}{\pi}\frac{2\gamma}{\gamma^2+\omega^2}. \tag{5.90}$$

The function is plotted in Fig. 5.11. From (5.89) and (5.90)

$$p(0)=\frac{2}{\pi}DM\beta. \tag{5.91}$$

For large values of t the velocity autocorrelation function does not in fact decay exponentially but goes as $t^{-3/2}$ (Ernst *et al.*, 1970). This asymptotic behaviour results in $p(\omega)$ having, for small ω, the form

$$p(\omega)=\frac{2}{\pi}DM\beta-c\omega^{1/2}+\ldots, \tag{5.92}$$

where c is a positive constant that depends on the diffusion constant and the viscosity of the liquid. The results obtained by Carneiro (1976) for $p(\omega)$, see Fig. 5.7, are consistent, at small ω, with the form of (5.92).

6

Neutron optics

Many of the phenomena in optics have been demonstrated with neutrons. These include total reflection, refraction by a prism, diffraction by a slit, and diffraction by a ruled grating. In addition neutron interferometers have been built.† An interesting application of the latter is a demonstration that when the spin of the neutron precesses through an angle of 2π in a magnetic field the wavefunction of the neutron changes sign in accordance with the fermion nature of the neutron (Rauch *et al.*, 1975, Werner *et al.*, 1975).

The discussion in the previous chapters has been concerned with interference between the neutron waves scattered by the nuclei in the scattering system. Optical phenomena arise from interference between the scattered waves and the waves of the incident beam, and we consider this in the present chapter.

6.1 Refractive index

When the scattered wave is small compared to the incident wave, the interference effects can be described in terms of a *refractive index* of the scattering system for thermal neutrons. We first prove that the refractive index n is given by

$$n = 1 - \frac{1}{2\pi}\rho\lambda^2\bar{b}, \tag{6.1}$$

where ρ is the number of nuclei per unit volume, λ the wavelength of the incident neutrons, and $\bar{b}$ the mean value of the scattering length of the nuclei.

† For descriptions and references see Bauspeiss *et al.* (1974) and Bonse and Graeff (1977). The former paper gives references to some of the other optical neutron experiments.

Consider a thin slab of the scattering material of thickness t, perpendicular to the direction of the incident beam (Fig. 6.1). If the incident beam is represented by $\exp(ikz)$, each scattered wave is represented on average by $-(\bar{b}/r)\exp(ikr)$, where r is measured from the scattering nucleus. We first calculate the resultant amplitude of all the scattered waves at a point P at a distance d ($\gg t$) from the slab. This is a standard problem of Fresnel diffraction in optics.

Let O be the foot of the perpendicular from P to the slab. Consider two waves arriving at P, one scattered at O and one at X, where $OX = x$ ($\ll d$). The wave from X has to travel farther than the one from O by an amount

$$(d^2+x^2)^{1/2}-d \approx \frac{1}{2}\frac{x^2}{d}. \tag{6.2}$$

The phase difference between the waves at P is thus proportional to x^2. The number of nuclei in the slab in the disc of radius x is also proportional to x^2. So the phase–amplitude diagram is one for which

change of direction $\propto$ length along the curve.

The curve with this property is the circle. However, we have not taken account of the fact that as x increases the amplitudes of the individual waves at P decrease slightly. This causes the phase–amplitude curve

Fig. 6.1 Geometry for scattering in the forward direction.

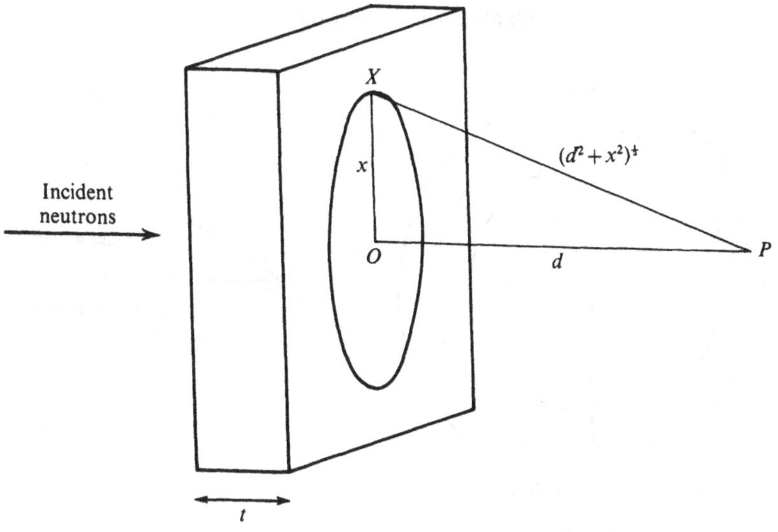

to curl up slightly faster than a circle, and the result is the spiral shown in Fig. 6.2.

Let $\boldsymbol{S}$ be the vector representing the amplitude of the total scattered wave at P. We assume the slab is large enough for the spiral to make a large number of turns and end near the centre. Then

$$S = \frac{L}{\pi}, \tag{6.3}$$

where L is the length along the curve of the first semicircle of the spiral in Fig. 6.2. Consider the 1st Fresnel zone, i.e. a disc centred on O with radius x_0, where

$$\frac{1}{2}\frac{x_0^2}{d} = \frac{1}{2}\lambda. \tag{6.4}$$

Then L represents what would be the total amplitude of the waves scattered by the nuclei in the 1st zone if these waves were all in phase. The number of nuclei in the 1st zone is

$$\pi x_0^2 t\rho = \pi d\lambda t\rho. \tag{6.5}$$

The amplitude of a single scattered wave at P is

$$-\frac{\bar{b}}{r} \approx -\frac{\bar{b}}{d}. \tag{6.6}$$

Therefore

$$S = \rho\lambda\bar{b}t. \tag{6.7}$$

Fig. 6.2 Phase–amplitude diagram for the scattered waves at P. L is the length along the curve from A to C.

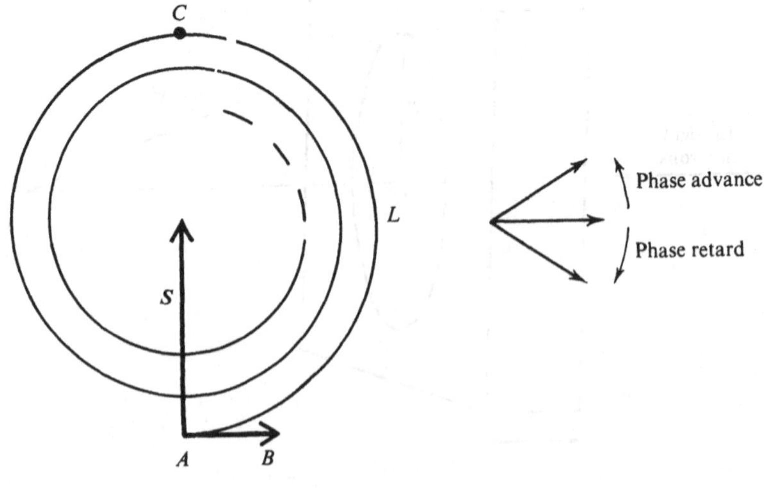

It can be seen from Fig. 6.2 that the phase of $\boldsymbol{S}$ is $\frac{1}{2}\pi$ in advance of the phase of the waves scattered at O, which are represented by AB. If $\bar{b}$ is positive, AB is π out of phase with $\boldsymbol{I}$, the incident wave at P. Thus $\boldsymbol{S}$ is $\frac{1}{2}\pi$ behind $\boldsymbol{I}$ (Fig. 6.3). If $\bar{b}$ is negative, $\boldsymbol{S}$ is $\frac{1}{2}\pi$ ahead of $\boldsymbol{I}$.

The amplitude of the incident wave is

$$I = 1, \tag{6.8}$$

since the incident wave function is $\exp(\mathrm{i}kz)$. The phase of $\boldsymbol{R}$, the resultant of $\boldsymbol{I}$ and $\boldsymbol{S}$, is behind that of $\boldsymbol{I}$ by the angle

$$\phi = \frac{S}{I} = \rho\lambda\bar{b}t. \tag{6.9}$$

We have assumed $S \ll I$, which is true for sufficiently small t. The refractive index n is related to ϕ by

$$\phi = \frac{2\pi}{\lambda}(1-n)t. \tag{6.10}$$

From (6.9) and (6.10) we have the required result

$$n = 1 - \frac{1}{2\pi}\rho\lambda^2\bar{b}. \tag{6.11}$$

We can readily see that for most substances

$$\frac{1}{2\pi}\rho\lambda^2\bar{b} \ll 1. \tag{6.12}$$

The quantity ρ is of the order of $1/a^3$, where a is the mean distance between the atoms. For thermal neutrons $\lambda \sim a \sim 10^{-10}$ m. For most nuclei $\bar{b} \sim 10^{-14}$ m. Therefore

$$\frac{1}{2\pi}\rho\lambda^2\bar{b} \sim \frac{1}{2\pi}\frac{\bar{b}}{a} \sim 10^{-5}. \tag{6.13}$$

Some values of $1-n$ for nickel are shown in Table 6.1.

Several points may be noted in the present discussion. First, the refractive index of the system does not depend on its structure, so (6.11) is true for any medium – crystalline, amorphous solid, or liquid. The reason the structure is irrelevant is that the refractive

Fig. 6.3 Relative phase of the incident wave $\boldsymbol{I}$ and the scattered wave $\boldsymbol{S}$ for positive scattering length.

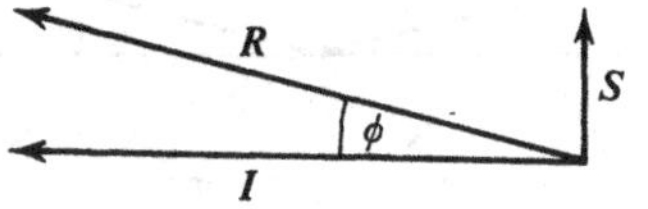

Table 6.1 Values of the refractive index n and the critical angle γ_c for nickel ($\rho = 9\times10^{28}\ \mathrm{m}^{-3}$, $\bar{b} = 10\times10^{-15}$ m)

$\lambda/10^{-10}$ m	$1-n$	γ_c
1	1.5×10^{-6}	6′
5	3.7×10^{-5}	30′

index depends on the scattering close to the forward direction, and the structure has no effect on the phase of this scattering. For the same reason the thermal motion of the scattering particles has no effect. Secondly, the wavelengths of the scattered and incident waves are the same, i.e. the relevant scattering is elastic. Thirdly, the scattered waves come from different nuclei, i.e. the scattering is coherent, and hence it is the coherent scattering length that comes into the expression for the refractive index.

6.2 Neutron reflection

We see from (6.11) that for positive $\bar{b}$ the refractive index is slightly less than 1, and for negative $\bar{b}$ it is slightly greater. Thus for positive $\bar{b}$ the neutrons may be totally reflected by the sample (Fig. 6.4). This will happen if the glancing angle γ is less than the critical value

$$\gamma_c = \cos^{-1} n. \tag{6.14}$$

Since n is close to 1, γ_c is small. Thus

$$\cos\gamma_c \approx 1 - \frac{1}{2}\gamma_c^2 \approx n = 1 - \frac{1}{2\pi}\rho\lambda^2\bar{b}, \tag{6.15}$$

i.e.

$$\gamma_c = \lambda(\rho\bar{b}/\pi)^{1/2}. \tag{6.16}$$

Values for nickel are given in Table 6.1.

Fig. 6.4 Total external reflection of neutrons. The value of $\bar{b}$ for the scattering material must be positive.

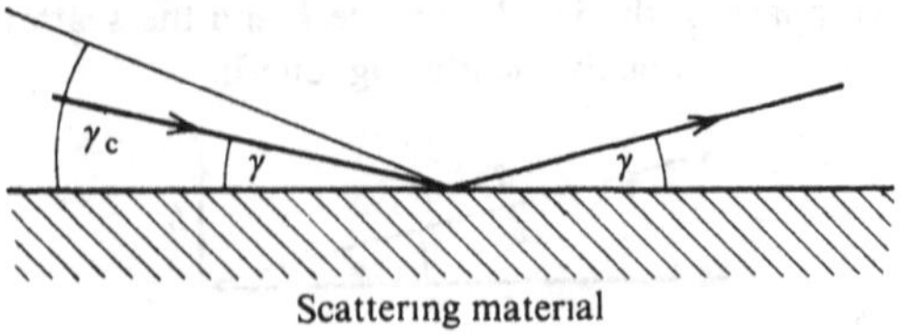

Neutron reflection is used to obtain accurate values of the coherent scattering length $\bar{b}$. A classic experiment to measure the scattering length of hydrogem was carried out by Burgy *et al.* (1951). The coherent scattering length $\bar{b}$ is negative for hydrogen, so a mirror of pure hydrogen cannot be used. However, carbon has a positive scattering length, and the experimenters used liquid hydrocarbon mirrors with positive mean values of the scattering length. The measurements gave the ratio of $\bar{b}$ for hydrogen and carbon, which was then combined with the known value of $\bar{b}$ for carbon. A refinement of the reflection method by Maier-Leibnitz (1962), known as the *gravity refractometer*, now gives scattering lengths to an accuracy of about 0.02%; see Koester (1977).

Another important application of neutron reflection is in neutron guide tubes. The reflecting material is usually nickel, which is evaporated on to a flat glass surface, and four such surfaces are mounted to form a tube of rectangular cross-section. Neutrons can be passed down the tube, being reflected from the internal surfaces with very little loss. They can thus be guided for long distances away from the reactor face to regions where the background is low, without the usual inverse-square law loss of intensity. Lengths up to 150 m have been achieved. Fig. 6.5 shows the transmission factor of a typical guide tube as a function of wavelength.

Fig. 6.5 Transmission as a function of wavelength for a neutron guide tube at Garching. The length of the tube is 48 m, and its cross-section is 140 × 25 mm. (After Maier-Leibnitz, 1972.)

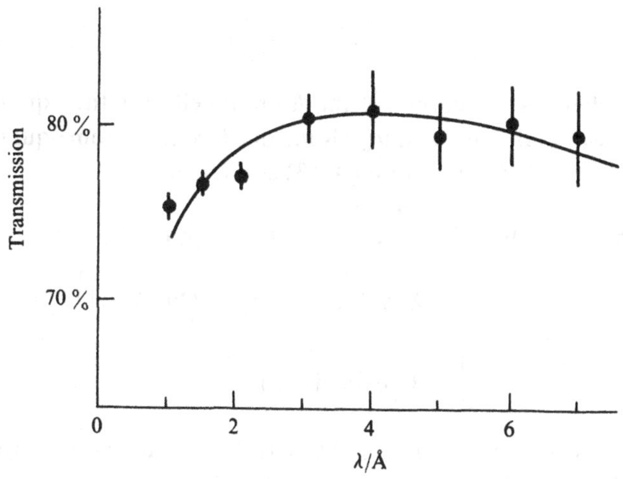

6.3 Dynamical theory of scattering

Basic theory

The theory in the previous chapters, known as *kinematic* theory, is based on the assumption that the incident neutron wave within the scattering system is the same as the incident wave outside. The theory in the present chapter, which takes account of the change in the incident wave within the scattering system, is known as *dynamical* theory. We now give a more systematic treatment of this theory. Dynamical theory was first developed by Darwin and Ewald for X-rays† and subsequently applied to the scattering of thermal neutrons by Goldberger and Seitz (1947). More comprehensive treatments of dynamical theory for neutrons have been given by Stassis and Oberteuffer (1974) and Rauch and Petrascheck (1978).

The theory proceeds by solving the Schrödinger equation for the wavefunction of the neutrons within the scattering system and matching the outside and inside wavefunctions at the surface. We start with the time-independent Schrödinger equation and assume that the potential seen by the neutron is due to a crystal whose nuclei are in positions smeared out by their thermal motion. From (2.33) the potential function is then

$$V(\boldsymbol{r}) = \frac{2\pi\hbar^2}{m}\langle\sum_j b_j \delta(\boldsymbol{r}-\boldsymbol{R}_j)\rangle, \tag{6.17}$$

where $\boldsymbol{R}_j$ is the position of the jth nucleus, and b_j its scattering length. $V(\boldsymbol{r})$ is periodic in the crystal lattice and can therefore be expressed as a Fourier series. As in Section 3.6 we put

$$\boldsymbol{R}_j = \boldsymbol{l} + \boldsymbol{d} + \boldsymbol{u}\binom{l}{d}, \tag{6.18}$$

where $\boldsymbol{l}$ denotes the corner of the lth unit cell, $\boldsymbol{d}$ is the equilibrium position within the cell, and $\boldsymbol{u}\binom{l}{d}$ is the displacement from equilibrium of the nucleus l, d. Then from (A.13) and (A.17)

$$\begin{aligned}\langle\sum_j b_j\delta(\boldsymbol{r}-\boldsymbol{R}_j)\rangle &= \frac{1}{(2\pi)^3}\int \exp(-\mathrm{i}\boldsymbol{\kappa}\,.\,\boldsymbol{r})\sum_l \exp(\mathrm{i}\boldsymbol{\kappa}\,.\,\boldsymbol{l}) \\ &\qquad \times\sum_d \bar{b}_d \exp(\mathrm{i}\boldsymbol{\kappa}\,.\,\boldsymbol{d})\left\langle\exp\left\{\mathrm{i}\boldsymbol{\kappa}\,.\,\boldsymbol{u}\binom{l}{d}\right\}\right\rangle \mathrm{d}\boldsymbol{\kappa} \\ &= \frac{1}{v_0}\sum_{\boldsymbol{\tau}} F_{\boldsymbol{\tau}} \exp(-\mathrm{i}\boldsymbol{\tau}\,.\,\boldsymbol{r}),\end{aligned} \tag{6.19}$$

† See James (1963) and Batterman and Cole (1964) for accounts of the dynamical theory of X-ray scattering.

where

$$F_{\boldsymbol{\tau}} = \sum_{d} \bar{b}_d \exp(\mathrm{i}\boldsymbol{\tau} \, . \, \boldsymbol{d}) \exp(-W_d), \tag{6.20}$$

$$W_d = \tfrac{1}{2}\left\langle \left\{ \boldsymbol{\tau} \, . \, \boldsymbol{u}\binom{l}{d} \right\}^2 \right\rangle, \tag{6.21}$$

$\boldsymbol{\tau}$ is a vector in the reciprocal lattice, and v_0 is the volume of the unit cell. $F_{\boldsymbol{\tau}}$ is the nuclear unit-cell structure factor previously defined in (3.76).

The Schrödinger equation is

$$\frac{\hbar^2}{2m}\nabla^2\psi + \{E - V(\boldsymbol{r})\}\psi = 0. \tag{6.22}$$

If $\boldsymbol{k}_0$ is the wavevector of the neutrons outside the crystal

$$E = \frac{\hbar^2}{2m} k_0^2. \tag{6.23}$$

From (6.17) and (6.19)

$$\frac{2m}{\hbar^2} V(\boldsymbol{r}) = \sum_{\boldsymbol{\tau}} G_{\boldsymbol{\tau}} \exp(-\mathrm{i}\boldsymbol{\tau} \, . \, \boldsymbol{r}), \tag{6.24}$$

where

$$G_{\boldsymbol{\tau}} = \frac{4\pi}{v_0} F_{\boldsymbol{\tau}}. \tag{6.25}$$

Therefore (6.22) becomes

$$\nabla^2\psi + k_0^2\psi = \{\sum_{\boldsymbol{\tau}} G_{\boldsymbol{\tau}} \exp(-\mathrm{i}\boldsymbol{\tau} \, . \, \boldsymbol{r})\}\psi. \tag{6.26}$$

The solution of a Schrödinger equation with a periodic potential is a Bloch function with the form

$$\psi = \sum_{\boldsymbol{\tau}} a_{\boldsymbol{\tau}} \exp\{\mathrm{i}(\boldsymbol{k} - \boldsymbol{\tau}) \, . \, \boldsymbol{r}\}, \tag{6.27}$$

where the coefficients $a_{\boldsymbol{\tau}}$ and the vector $\boldsymbol{k}$ are to be determined. We substitute (6.27) into (6.26) and use the result

$$\nabla^2 \exp\{\mathrm{i}(\boldsymbol{k} - \boldsymbol{\tau}) \, . \, \boldsymbol{r}\} = -(\boldsymbol{k} - \boldsymbol{\tau})^2 \exp\{\mathrm{i}(\boldsymbol{k} - \boldsymbol{\tau}) \, . \, \boldsymbol{r}\}. \tag{6.28}$$

This gives

$$\sum_{\boldsymbol{\tau}} a_{\boldsymbol{\tau}}\{k_0^2 - (\boldsymbol{k} - \boldsymbol{\tau})^2\} \exp\{\mathrm{i}(\boldsymbol{k} - \boldsymbol{\tau}) \, . \, \boldsymbol{r}\}$$

$$= \sum_{\boldsymbol{\tau}'} G_{\boldsymbol{\tau}'} \exp(-\mathrm{i}\boldsymbol{\tau}' \, . \, \boldsymbol{r}) \sum_{\boldsymbol{\tau}''} a_{\boldsymbol{\tau}''} \exp\{\mathrm{i}(\boldsymbol{k} - \boldsymbol{\tau}'') \, . \, \boldsymbol{r}\}. \tag{6.29}$$

Equating terms in $\exp\{i(\boldsymbol{k}-\boldsymbol{\tau}).\boldsymbol{r}\}$ gives

$$a_{\boldsymbol{\tau}}\{k_0^2-(\boldsymbol{k}-\boldsymbol{\tau})^2\}=\sum_{\boldsymbol{\tau}'} G_{\boldsymbol{\tau}'}a_{\boldsymbol{\tau}-\boldsymbol{\tau}'}. \tag{6.30}$$

We now distinguish two cases.

No Bragg reflection. In this case k_0^2 and $(\boldsymbol{k}-\boldsymbol{\tau})^2$ are approximately equal for $\boldsymbol{\tau}=0$, but not for any other $\boldsymbol{\tau}$. Then from (6.30) all the $a_{\boldsymbol{\tau}}$s are small except a_0. There is effectively only one term on the right-hand side of (6.30), and we have

$$a_0(k_0^2-k^2)=G_0a_0. \tag{6.31}$$

Since the difference between k and k_0 is small this gives

$$k_0-k=\frac{G_0}{2k_0}. \tag{6.32}$$

The neutrons have wavevector $\boldsymbol{k}_0$ outside the crystal and $\boldsymbol{k}$ inside. The refractive index n is given by

$$n=\frac{k}{k_0}=1-\frac{G_0}{2k_0^2}. \tag{6.33}$$

From (6.20) and (6.25)

$$G_0=\frac{4\pi}{v_0}r\bar{b}=4\pi\rho\bar{b}, \tag{6.34}$$

where r is the number of atoms in a unit cell, and ρ is the number of atoms in unit volume of the crystal. Since $k_0=2\pi/\lambda$, (6.33) and (6.34) give

$$n=1-\frac{1}{2\pi}\rho\lambda^2\bar{b}, \tag{6.35}$$

which is the same result as we obtained previously (6.11).

Near Bragg reflection. When $(\boldsymbol{k}-\boldsymbol{\tau})^2$ is close to k_0^2 for a particular non-zero $\boldsymbol{\tau}$, then the value of $a_{\boldsymbol{\tau}}$ becomes comparable with a_0, i.e. the neutrons are Bragg reflected. The relation (6.30) now gives rise to two equations, each with two terms on the right-hand side.

$$a_0(k_0^2-k^2)=G_0a_0+G_{-\boldsymbol{\tau}}a_{\boldsymbol{\tau}}, \tag{6.36}$$

$$a_{\boldsymbol{\tau}}\{k_0^2-(\boldsymbol{k}-\boldsymbol{\tau})^2\}=G_{\boldsymbol{\tau}}a_0+G_0a_{\boldsymbol{\tau}}. \tag{6.37}$$

These equations are consistent only if

$$\frac{a_{\boldsymbol{\tau}}}{a_0}=\frac{k_0^2-k^2-G_0}{G_{-\boldsymbol{\tau}}}=\frac{G_{\boldsymbol{\tau}}}{k_0^2-(\boldsymbol{k}-\boldsymbol{\tau})^2-G_0}. \tag{6.38}$$

Since the potential $V(\boldsymbol{r})$ is real, $G_{-\boldsymbol{\tau}} = G_{\boldsymbol{\tau}}^*$ and we may rewrite (6.38) as

$$\{k_0^2 - G_0 - k^2\}\{k_0^2 - G_0 - (\boldsymbol{k} - \boldsymbol{\tau})^2\} = |G_{\boldsymbol{\tau}}|^2. \tag{6.39}$$

The vector $\boldsymbol{k}$ given by this equation lies on a pair of surfaces known as *dispersion surfaces.* A section through them in the plane of $\boldsymbol{k}_0$ and $\boldsymbol{\tau}$ is shown in Fig. 6.6.

Bragg reflection occurs when $\boldsymbol{k}_0$ is close to the Brillouin zone boundary, that is, the plane that perpendicularly bisects the reciprocal lattice vector $\boldsymbol{\tau}$. The shape of the dispersion surfaces away from the zone boundary follows from the fact that the quantity $G_{\boldsymbol{\tau}}$ is small compared to k_0^2. ($G_{\boldsymbol{\tau}}/k_0^2 \sim G_0/k_0^2 = \rho\lambda^2\bar{b}/\pi \sim 10^{-5}$.) Eq. (6.39) then shows that away from the zone boundary either k^2 or $(\boldsymbol{k} - \boldsymbol{\tau})^2$ is very nearly equal to $k_0^2 - G_0$. Thus the surfaces tend to a pair of spheres of radius $(k_0^2 - G_0)^{1/2}$, one centred on the origin O, and the other on the reciprocal lattice point T, where $TO = \boldsymbol{\tau}$. As the surface tends to the sphere centred on O, $a_{\boldsymbol{\tau}}$ becomes small compared with a_0, and the wavefunction is effectively the single wave $\exp(i\boldsymbol{k}\,.\,\boldsymbol{r})$. It corresponds to the previous solution for no Bragg reflection. As the surface tends to the sphere centred on T, both a_0 and $a_{\boldsymbol{\tau}}$ become vanishingly small, and the surface has no physical significance.

Fig. 6.6 Section of dispersion surface in the plane of $\boldsymbol{k}_0$ and $\boldsymbol{\tau}$. The dashed curves are parts of circles of radius $(k_0^2 - G_0)^{1/2}$ centred on the origin O and the reciprocal lattice point T. The departure of the dispersion curves from the circles, and of the wavevectors from the zone boundary, is greatly exaggerated.

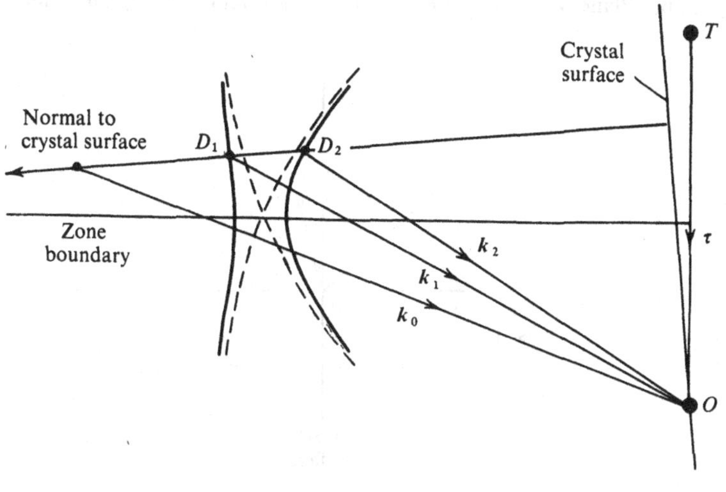

There are in general two solutions of the Schrödinger equation for a given $\boldsymbol{k}_0$, each with its own value of $\boldsymbol{k}$. To determine the two values $\boldsymbol{k}_1$ and $\boldsymbol{k}_2$ completely we require the boundary conditions for the wavefunction ψ. Consider the case when the orientation of the atomic planes with respect to the crystal surface is such that there is no Bragg reflected wave coming from the surface on which the neutrons are incident. The wavefunctions may then be expressed as follows:

outside the crystal

$$\psi = \psi_o = \exp(i\boldsymbol{k}_0 \,.\, \boldsymbol{r}), \tag{6.40}$$

inside the crystal

$$\begin{aligned}\psi = \psi_i = A_1[\exp(i\boldsymbol{k}_1 \,.\, \boldsymbol{r}) + \alpha_1 \exp\{i(\boldsymbol{k}_1 - \boldsymbol{\tau}) \,.\, \boldsymbol{r}\}] \\ + A_2[\exp(i\boldsymbol{k}_2 \,.\, \boldsymbol{r}) + \alpha_2 \exp\{i(\boldsymbol{k}_2 - \boldsymbol{\tau}) \,.\, \boldsymbol{r}\}],\end{aligned} \tag{6.41}$$

where α_1 and α_2, the values of a_τ/a_0 for the wavevectors $\boldsymbol{k}_1$ and $\boldsymbol{k}_2$, are given by (6.38). The coefficients A_1 and A_2 are determined by the boundary conditions. Suppose for simplicity that the normal to the crystal surface lies in the same plane S as $\boldsymbol{k}_0$ and $\boldsymbol{\tau}$. Take cartesian axes with x parallel to the surface in the plane S, y perpendicular to the surface, and the origin in the surface (Fig. 6.7). Now $\psi_i = \psi_o$ for all points in the plane $y = 0$. Therefore the variation with x is the same for ψ_i and ψ_o in this plane. This requires

$$k_{1x} = k_{2x} = k_{0x}, \tag{6.42}$$

Fig. 6.7 Plane S containing $\boldsymbol{k}_0$, $\boldsymbol{\tau}$, and the normal to the crystal surface.

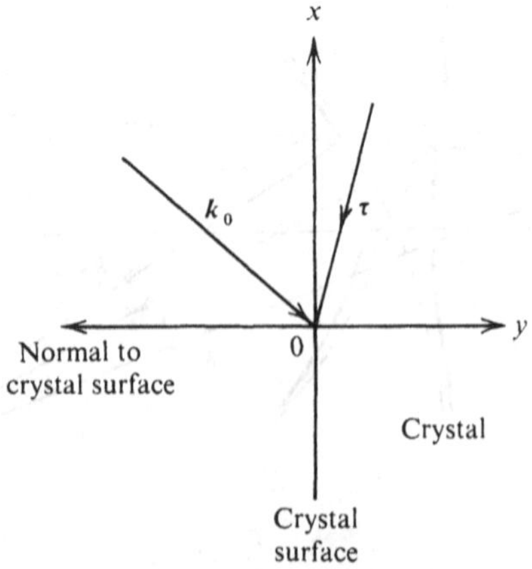

$$A_1 + A_2 = 1, \tag{6.43}$$

$$A_1\alpha_1 + A_2\alpha_2 = 0. \tag{6.44}$$

Thus

$$A_1 = -\frac{\alpha_2}{\alpha_1 - \alpha_2}, \quad \text{and} \quad A_2 = \frac{\alpha_1}{\alpha_1 - \alpha_2}. \tag{6.45}$$

It may be noted that although there is no wave outside the crystal due to Bragg reflection (more properly termed *diffraction*), there is a true reflected wave in that region. Thus there should be a second term on the right-hand side of (6.40) of the form $A_r \exp(\mathrm{i}\boldsymbol{k}_0' \cdot \boldsymbol{r})$, where $k_{0x}' = k_{0x}$ and $k_{0y}' = -k_{0y}$. The value of the coefficient A_r is determined by applying the second boundary condition, viz. that $\partial\psi/\partial y$ is continuous at the surface. However, provided the glancing angle is large compared to the critical angle γ_c, the coefficient A_r is very small compared to 1, and the reflected wave is negligible.

Eq. (6.42) shows that the components of $\boldsymbol{k}_1$, $\boldsymbol{k}_2$, and $\boldsymbol{k}_0$ parallel to the surface of the crystal are equal. This result, together with (6.39), determines the two vectors $\boldsymbol{k}_1$ and $\boldsymbol{k}_2$ for a given $\boldsymbol{k}_0$. The construction is shown in Fig. 6.6. A line is drawn through the end point of $\boldsymbol{k}_0$ normal to the surface of the crystal. The intersection of this line with the dispersion surfaces at the points D_1 and D_2 gives the vectors $\boldsymbol{k}_1$ and $\boldsymbol{k}_2$. Eq. (6.42) still holds when there is a Bragg reflected wave coming from the incident surface of the crystal; the construction in Fig. 6.6 is therefore generally valid.

If $\boldsymbol{\tau}$ is parallel to the crystal surface, then if $\boldsymbol{k}_0$ lies on the zone boundary, so do $\boldsymbol{k}_1$ and $\boldsymbol{k}_2$ (Fig. 6.8), i.e. the Bragg condition is satisfied for the two wavevectors simultaneously.† In this case we have from (6.39)

$$k^2 = k_0^2 - G_0 \pm |G_\tau|. \tag{6.46}$$

If the crystal has a centre of symmetry, which we take as the origin of $\boldsymbol{r}$, G_τ is real. Eqs. (6.38) and (6.45) then give

$$\frac{a_\tau}{a_0} = \mp 1, \quad A_1 = A_2 = \tfrac{1}{2}. \tag{6.47}$$

We may divide ψ_i in (6.41) into two wavefields $\psi_\nu (\nu = 1, 2)$, each consisting of a pair of waves travelling in the directions of $\boldsymbol{k}_\nu$ and

† For general geometry the Bragg condition cannot be satisfied for $\boldsymbol{k}_1$ and $\boldsymbol{k}_2$ simultaneously. This has the consequence that the maximum of a rocking curve does not occur at the exact Bragg condition, though the difference is only of the order of a second of arc. See Rauch and Petrascheck (1978) for details of the calculation.

$\boldsymbol{k}_\nu - \boldsymbol{\tau}$. It is shown in Example 6.1 that the neutron current, or flux vector, for each wavefield is

$$\boldsymbol{j}_\nu = \frac{\hbar}{m}|A_\nu|^2\{\boldsymbol{k}_\nu + |\alpha_\nu|^2(\boldsymbol{k}_\nu - \boldsymbol{\tau})\}. \tag{6.48}$$

An important property of the dispersion surfaces is that the normals to the surfaces at the points D_1 and D_2 in Fig. 6.6 are the directions of the neutron currents for the corresponding wavefields. (See Kato (1958) and James (1963) for proofs of this result.) Thus in the symmetric situation considered above with the vector $\boldsymbol{\tau}$ parallel to the crystal surface, when $\boldsymbol{k}_0$ lies on the zone boundary the neutron currents for both wavefields are perpendicular to $\boldsymbol{\tau}$, that is, parallel to the planes of the reflecting atoms. For any other orientation of $\boldsymbol{k}_0$, the two directions of $\boldsymbol{j}_\nu$ are at equal angles $\pm\Omega$ with respect to the atomic planes. As $\boldsymbol{k}_0$ departs from the zone boundary the angle Ω tends to the limiting value $\frac{1}{2}\theta$, the Bragg angle defined in (3.51). The range of currents between $-\frac{1}{2}\theta$ and $\frac{1}{2}\theta$ is known as the *Borrmann fan*. Only a small variation in the direction of $\boldsymbol{k}_0$ (of the order of a few seconds of arc) is required to reach the extremes of the fan, and this variation is usually present in what is regarded as a monochromatic incident beam. Thus the complete Borrmann fan is usually excited for a single

Fig. 6.8 Diagram showing wavevectors when $\boldsymbol{\tau}$ lies in the crystal surface, and the Bragg condition is satisfied.

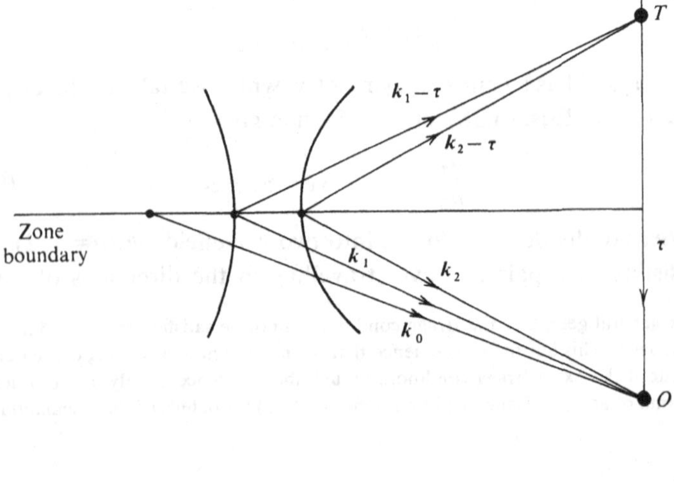

setting of $\boldsymbol{k}_0$ in the Bragg position. For the non-symmetric case the two directions of $\boldsymbol{j}_\nu$ are not at equal angles to the atomic planes, but they always lie within the Borrmann fan.

The neutron density for each wavefield is given by the function $|\psi_\nu|^2$. It is shown in Example 6.2 that one of these functions has a maximum at the sites of the atoms in the reflecting planes, and the other has a minimum at these sites. If the value of the mean scattering length $\bar{b}$ is positive, corresponding to a potential in (6.17) that is effectively repulsive, the wavefield with maxima at the atomic sites has the lower value of k_ν. Thus the neutrons of this wavefield have greater potential energy and less kinetic energy than those of the other wavefield. The total energy is the same for both wavefields, being equal to E, the kinetic energy of the neutrons outside the crystal. If the mean scattering length is negative, the wavefield with maxima at the atomic sites is the one with the larger value of k_ν.

Pendellösung fringes

We return to the expression for ψ_i in (6.41). Since the primary waves $\exp(\mathrm{i}\boldsymbol{k}_1 . \boldsymbol{r})$ and $\exp(\mathrm{i}\boldsymbol{k}_2 . \boldsymbol{r})$, propagating close to the incident direction, have wavevectors of slightly different magnitudes, beats occur between them, that is, the intensity of the resultant primary wave varies sinusoidally as it traverses the crystal. The same is true of the resultant of the reflected waves $\exp\{\mathrm{i}(\boldsymbol{k}_1-\boldsymbol{\tau}) . \boldsymbol{r}\}$ and $\exp\{\mathrm{i}(\boldsymbol{k}_2-\boldsymbol{\tau}) . \boldsymbol{r}\}$, the minima of the resultant reflected wave coinciding with the maxima of the resultant primary wave. This phenomenon of the neutron intensity oscillating between the primary and reflected waves is known as '*Pendellösung*' (pendulum solution).

The 'Pendellösung' phenomenon for neutrons has been demonstrated in some striking measurements by Shull (1968) and Shull and Oberteuffer (1972) for silicon. A beam of monoenergetic neutrons is incident through a narrow slit A on a thin slice of a single crystal of silicon with the reciprocal lattice vector for the 111 planes lying along the surface (Fig. 6.9). The angle of incidence $\frac{1}{2}\theta$ satisfies the Bragg condition. At the exit slit B, positioned opposite to A, the waves within the crystal give rise to two beams outside – one parallel to $\boldsymbol{k}_0$ and one to $\boldsymbol{k}_0-\boldsymbol{\tau}$. Measurements of the intensity of the reflected beam were made as a function of the wavelength λ of the incident neutrons. As λ varied, the angle $\frac{1}{2}\theta$ was varied to maintain the Bragg condition. The results are shown in Fig. 6.10.

Fig. 6.9 Arrangement for observing Pendellösung fringes in silicon. (After Shull, 1968.)

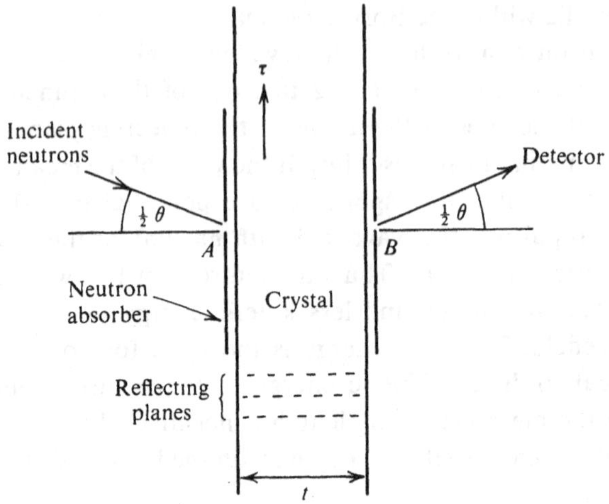

Fig. 6.10 Pendellösung fringes in silicon: reflected intensity at the centre of a Bragg peak as a function of neutron wavelength for different crystal thickness: (*a*) 10.000 mm, (*b*) 5.939 mm, (*c*) 3.315 mm. (After Shull, 1968.)

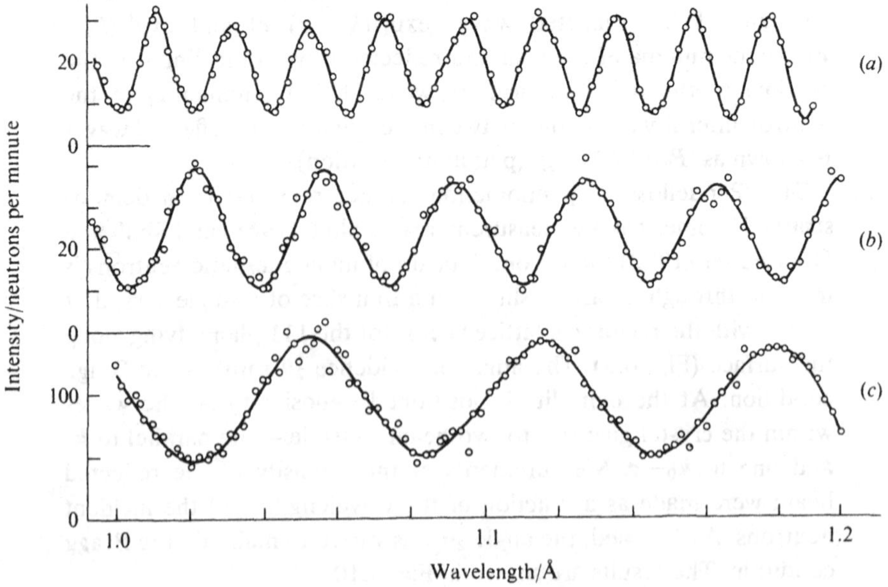

The variation of intensity with wavelength can be readily calculated. From (6.46)

$$k_1^2 - k_2^2 = 2G_\tau, \tag{6.49}$$

$$k_1 - k_2 \approx \frac{G_\tau}{k_0}. \tag{6.50}$$

The phase change between the two reflected (or incident) waves after traversing the crystal is therefore

$$\phi = (k_1 - k_2)\frac{t}{\cos\frac{1}{2}\theta} = \frac{G_\tau}{k_0}\frac{t}{\cos\frac{1}{2}\theta} = \frac{2\pi t}{\Delta_0}, \tag{6.51}$$

where t is the thickness of the crystal slice, and

$$\Delta_0 = \frac{\pi v_0 \cos\frac{1}{2}\theta}{F_\tau \lambda} \tag{6.52}$$

is a quantity known as the *Pendellösung length*. The intensity of the reflected beam is proportional to

$$|1 - \exp(i\phi)|^2 = 4\sin^2\tfrac{1}{2}\phi$$
$$= 4\sin^2(\pi t/\Delta_0). \tag{6.53}$$

The results in Fig. 6.10, together with the knowledge of the interatomic distances in silicon, provide an accurate value of F_τ. This, when corrected for the thermal motion in the Debye–Waller factor, yields a very accurate value of the scattering length of silicon. The measurements are so accurate that the contribution to $\bar{b}$ from the electric scattering (see Section 9.5) by the electrons and the nucleus is significantly greater than the experimental error. The value obtained for the atomic scattering length is $\bar{b}_{at} = 4.1534 \pm 0.0010$ fm. The contribution from the electric scattering is $b_{el} = 0.0043 \pm 0.0002$ fm. Therefore the scattering length for the pure nuclear-force interaction is

$$\bar{b}_{nuc} = 4.1491 \pm 0.0010 \text{ fm}. \tag{6.54}$$

It may be noted that, although this method yields a very accurate value of the scattering length, it is applicable only to substances which form large perfect single crystals, and moreover have low absorption and incoherent scattering cross-sections.

Primary extinction

For a very small crystal the number of neutrons scattered in a Bragg peak is a small fraction of the number striking the crystal. It is therefore a good assumption that the whole of the crystal is uniformly bathed in incident neutrons of the same intensity, which is the intensity of the incident beam outside the crystal. The intensity of the Bragg peak is then correctly given by one of the formulae in Chapter 3, for example (3.100) for the rotating crystal method. However, if the crystal is not small, only the part near the surface receives the full incident beam, and the intensity of the Bragg reflected beam is reduced below its theoretical level. This phenomenon is known as *primary extinction*.

We estimate the thickness of the crystal at which primary extinction begins to take effect. At the surface the two incident waves with wavevectors $\boldsymbol{k}_1$ and $\boldsymbol{k}_2$ are in phase. As the waves traverse the crystal they start to get out of phase. But for distances small compared to $1/(k_1-k_2)$ the phase difference is small, and the incident wave is almost unchanged. We define the extinction distance ξ by

$$(k_1-k_2)\frac{\xi}{\cos\frac{1}{2}\theta}=1. \tag{6.55}$$

From (6.51) and (6.55)

$$\xi=\frac{\Delta_0}{2\pi}=\frac{v_0\cos\frac{1}{2}\theta}{2F_\tau\lambda}. \tag{6.56}$$

Primary extinction is significant if the thickness of the crystal is $\gtrsim\xi$.

If a is the mean interatomic distance and b the scattering length, then $v_0\sim a^3$ and $F_\tau\sim b$. So for $\lambda\sim a$

$$\xi\sim\frac{a}{b}a\sim 10^4 a. \tag{6.57}$$

As a specific example, for the (111) reflection in nickel at room temperature, with

$$\lambda=2\text{ Å}\quad\text{and}\quad\cos\tfrac{1}{2}\theta=0.9 \tag{6.58}$$

$$\xi=2.2\times10^{-6}\text{ m}. \tag{6.59}$$

Extinction distances are of the same order for neutrons and X-rays (Bacon and Lowde, 1948).

From the same reasoning we may also obtain an estimate of the angular width of a Bragg peak for a single crystal. Consider the arrangement in Fig. 6.9. The neutrons in the Bragg peak are effectively diffracted by a crystal of thickness ξ rather than t. From basic diffraction theory, reducing the size of the scattering system causes a broadening of the diffraction peak. In this case the effect is to cause the reciprocal lattice point $\boldsymbol{\tau}$ to broaden into a line perpendicular to $\boldsymbol{\tau}$, with length of the order of $1/\xi$. This is equivalent to an angular spread in the crystal orientation of the order of $1/\tau\xi$, which leads to an angular width $\Delta\theta$ for the Bragg peak of a similar amount. Thus from (6.56) and the relation $\tau = 2k \sin\frac{1}{2}\theta$, we have

$$\Delta\theta \sim \frac{1}{\tau\xi} = \frac{\lambda^2 F_\tau}{\pi v_0 \sin\theta}. \tag{6.60}$$

For the (111) reflection in nickel, with $\xi = 2\times 10^{-6}$ m, this expression gives $\Delta\theta \sim 3$ seconds of arc. In general the values of $\Delta\theta$ for neutrons are comparable with, but somewhat smaller than, those for X-rays (see Bacon, 1975, Table 2).

Our derivation of the expressions for the extinction distance and the width of the Bragg peak is based on scattering obtained by transmission (Laue) geometry. The scattering may also be observed by reflection (Bragg) geometry (Fig. 6.11). The diffraction patterns – intensity as a function of θ – are somewhat different for the two geometries, but the extinction distances and the widths of the peaks are comparable in magnitude. For theoretical calculations of the patterns see Rauch and Petrascheck (1978). The diffraction patterns have been measured for both geometries by Shull (1973) and are in very good agreement with the theory.

The above discussion relates to a perfect single crystal, that is, one with a perfectly regular array of atomic positions throughout. In

Fig. 6.11 Geometries for Bragg scattering: (*a*) transmission (Laue) geometry, (*b*) reflection (Bragg) geometry. The dashed lines indicate the reflecting planes. The dotted lines in (*a*) show the directions of the neutron currents for the extremes of the Borrmann fan and for an incident beam that exactly satisfies the Bragg condition.

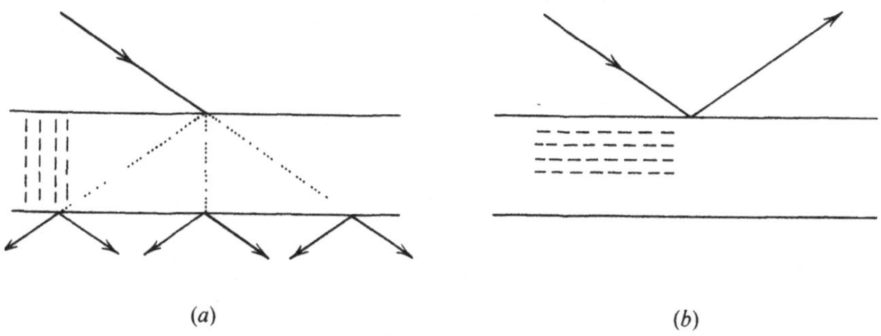

practice most single crystals are not perfect. Dislocations break the crystal into small regions or mosaic blocks, of the order of 10^{-6} m in size, in each of which the order is perfect, but which are slightly misaligned relative to each other by angles of the order of several minutes of arc, i.e. by angles much larger than the width $\Delta\theta$ of the Bragg peak for a single mosaic block. For this reason primary extinction may not occur even though the dimensions of the crystal are much larger than the extinction distance. In this case the neutron beam will eventually encounter mosaic blocks with orientations similar to those through which it has already passed. The resulting attenuation is known as *secondary extinction* (Bacon, 1975). The width of the Bragg peak is determined by the mosaic spread of the crystal.

Examples

6.1 Show that for the wavefield

$$\psi_\nu = A_\nu[\exp(\mathrm{i}\boldsymbol{k}_\nu \,.\, \boldsymbol{r}) + \alpha_\nu \exp\{\mathrm{i}(\boldsymbol{k}_\nu - \boldsymbol{\tau}) \,.\, \boldsymbol{r}\}],$$

the neutron flux, averaged over a unit cell, is

$$\boldsymbol{j}_\nu = \frac{\hbar}{m}|A_\nu|^2\{\boldsymbol{k}_\nu + |\alpha_\nu|^2(\boldsymbol{k}_\nu - \boldsymbol{\tau})\}.$$

6.2 A Bravais crystal with positive $\bar{b}$ has $\boldsymbol{\tau}$ parallel to its surface. Show that if the Bragg condition is satisfied and the origin of $\boldsymbol{r}$ is at an atomic site, then

$$|\psi_\nu|^2 = \tfrac{1}{2}(1 \mp \cos \boldsymbol{\tau} \,.\, \boldsymbol{r}),$$

where the negative sign corresponds to the larger and the positive sign to the smaller value of k_ν.

7

Magnetic scattering – basic theory

We now consider the scattering cross-sections due to the magnetic interaction between the neutron and unpaired electrons in the atom.

7.1 Preliminary results

We first derive an expression for the potential due to the magnetic interaction between a neutron and an electron, recalling some definitions and results from electromagnetic theory.

The operator corresponding to the magnetic dipole moment of the neutron is

$$\boldsymbol{\mu}_{\mathrm{n}} = -\gamma\mu_{\mathrm{N}}\boldsymbol{\sigma}, \tag{7.1}$$

where

$$\mu_{\mathrm{N}} = \frac{e\hbar}{2m_{\mathrm{p}}} \tag{7.2}$$

is the nuclear magneton. m_{p} is the mass of the proton and e its charge. γ is a positive constant whose value is

$$\gamma = 1.913. \tag{7.3}$$

$\boldsymbol{\sigma}$ is the Pauli spin operator for the neutron. The eigenvalues of its components are ± 1 (Appendix F.2).

The operator corresponding to the magnetic dipole moment of the electron is

$$\boldsymbol{\mu}_{\mathrm{e}} = -2\mu_{\mathrm{B}}\boldsymbol{s}, \tag{7.4}$$

where

$$\mu_{\mathrm{B}} = \frac{e\hbar}{2m_{\mathrm{e}}} \tag{7.5}$$

is the Bohr magneton. m_{e} is the mass of the electron. $\boldsymbol{s}$ is the spin angular momentum operator for the electron in units of $\hbar$. The eigenvalues of its components are $\pm\frac{1}{2}$. Note that though $\boldsymbol{\sigma}$ and $\boldsymbol{s}$

both relate to particles of spin $\frac{1}{2}$, their definitions differ by a factor of 2.

Consider an electron with momentum $\boldsymbol{p}$. The magnetic field at a point $\boldsymbol{R}$ from the electron due to its magnetic dipole moment is

$$\boldsymbol{B}_{\mathrm{S}} = \operatorname{curl} \boldsymbol{A}, \qquad \boldsymbol{A} = \frac{\mu_0}{4\pi} \frac{\boldsymbol{\mu}_{\mathrm{e}} \times \hat{\boldsymbol{R}}}{R^2}, \tag{7.6}$$

where $\hat{\boldsymbol{R}}$ is a unit vector in the direction of $\boldsymbol{R}$. The magnetic field due to the momentum of the electron is given by the Biot–Savart law, which states that the magnetic field at a point $\boldsymbol{R}$ from a current element $I\,\mathrm{d}\boldsymbol{l}$ is

$$\boldsymbol{B}_{\mathrm{L}} = \frac{\mu_0}{4\pi} I \frac{\mathrm{d}\boldsymbol{l} \times \hat{\boldsymbol{R}}}{R^2}. \tag{7.7}$$

The current element for the moving electron is

$$I\,\mathrm{d}\boldsymbol{l} = -\frac{e}{m_{\mathrm{e}}}\boldsymbol{p} = -\frac{2\mu_{\mathrm{B}}}{\hbar}\boldsymbol{p}. \tag{7.8}$$

The total magnetic field due to an electron of momentum $\boldsymbol{p}$ is therefore

$$\boldsymbol{B} = \boldsymbol{B}_{\mathrm{S}} + \boldsymbol{B}_{\mathrm{L}} = \frac{\mu_0}{4\pi}\left\{ \operatorname{curl}\left(\frac{\boldsymbol{\mu}_{\mathrm{e}} \times \hat{\boldsymbol{R}}}{R^2}\right) - \frac{2\mu_{\mathrm{B}}}{\hbar}\frac{\boldsymbol{p} \times \hat{\boldsymbol{R}}}{R^2}\right\}. \tag{7.9}$$

The potential of a neutron with dipole moment $\boldsymbol{\mu}_{\mathrm{n}}$ in this field is

$$-\boldsymbol{\mu}_{\mathrm{n}} \,.\, \boldsymbol{B} = -\frac{\mu_0}{4\pi}\gamma\mu_{\mathrm{N}} 2\mu_{\mathrm{B}} \boldsymbol{\sigma} \,.\, (\boldsymbol{W}_{\mathrm{S}} + \boldsymbol{W}_{\mathrm{L}}), \tag{7.10}$$

where

$$\boldsymbol{W}_{\mathrm{S}} = \operatorname{curl}\left(\frac{\boldsymbol{s} \times \hat{\boldsymbol{R}}}{R^2}\right), \tag{7.11}$$

and

$$\boldsymbol{W}_{\mathrm{L}} = \frac{1}{\hbar}\frac{\boldsymbol{p} \times \hat{\boldsymbol{R}}}{R^2}. \tag{7.12}$$

Thus the potential is the sum of two terms, the first arising from the spin of the electron and the second from its orbital motion.

We shall need the following mathematical results, which are proved in Appendix B.2.

$$\operatorname{curl}\left(\frac{\boldsymbol{s} \times \hat{\boldsymbol{R}}}{R^2}\right) = \frac{1}{2\pi^2}\int \hat{\boldsymbol{q}} \times (\boldsymbol{s} \times \hat{\boldsymbol{q}}) \exp(\mathrm{i}\boldsymbol{q} \,.\, \boldsymbol{R})\,\mathrm{d}\boldsymbol{q}, \tag{7.13}$$

$$\int \frac{\hat{\boldsymbol{R}}}{R^2} \exp(\mathrm{i}\boldsymbol{\kappa} \,.\, \boldsymbol{R})\,\mathrm{d}\boldsymbol{R} = 4\pi\mathrm{i}\frac{\hat{\boldsymbol{\kappa}}}{\kappa}. \tag{7.14}$$

$\hat{\boldsymbol{q}}$ and $\hat{\boldsymbol{\kappa}}$ are unit vectors in the directions of $\boldsymbol{q}$ and $\boldsymbol{\kappa}$.

7.2 Expression for $d^2\sigma/d\Omega\, dE'$

We start with (2.15) which gives the cross-section for a specific transition $\lambda \to \lambda'$ due to an interaction V between the neutron and the scattering system. This expression is correct for nuclear scattering of unpolarised neutrons. The spin state of the neutron does not appear; the dependence of the interaction on the spin state of the nucleus–neutron system is allowed for in the value of the scattering length.† However, the magnetic potential contains the spin operator $\boldsymbol{\sigma}$ explicitly, and it is therefore necessary to specify not only the wavevector $\boldsymbol{k}$ of the neutron but also its spin state σ. Eq. (2.15) is written in the form

$$\left(\frac{d^2\sigma}{d\Omega\, dE'}\right)_{\sigma\lambda\to\sigma'\lambda'} = \frac{k'}{k}\left(\frac{m}{2\pi\hbar^2}\right)^2 |\langle \boldsymbol{k}'\sigma'\lambda'|V_m|\boldsymbol{k}\sigma\lambda\rangle|^2 \delta(E_\lambda - E_{\lambda'} + \hbar\omega). \tag{7.15}$$

This is the cross-section for a process in which the system changes from the state λ to the state λ', and the neutron changes from the state $\boldsymbol{k}$, σ to the state $\boldsymbol{k}'$, σ'. V_m is the potential between the neutron and all the electrons in the scattering system.

Evaluation of $\langle \boldsymbol{k}'|V_m|\boldsymbol{k}\rangle$

We first evaluate $\langle \boldsymbol{k}'|V_m|\boldsymbol{k}\rangle$, i.e. we integrate over the space coordinates $\boldsymbol{r}$ of the neutron. It is convenient to treat the spin and orbital parts of V_m separately. Consider the spin contribution due to the ith electron with spin $\boldsymbol{s}_i$ and position vector $\boldsymbol{r}_i$ (Fig. 7.1). From (7.11) and (7.13)

$$\langle \boldsymbol{k}'|\boldsymbol{W}_{Si}|\boldsymbol{k}\rangle = \int \exp(-i\boldsymbol{k}'\,.\,\boldsymbol{r})\,\mathrm{curl}\left(\frac{\boldsymbol{s}_i \times \hat{\boldsymbol{R}}}{R^2}\right)\exp(i\boldsymbol{k}\,.\,\boldsymbol{r})\, d\boldsymbol{r}$$

$$= \frac{1}{2\pi^2}\int \exp(i\boldsymbol{\kappa}\,.\,\boldsymbol{r})\,\hat{\boldsymbol{q}} \times (\boldsymbol{s}_i \times \hat{\boldsymbol{q}})\exp(i\boldsymbol{q}\,.\,\boldsymbol{R})\, d\boldsymbol{q}\, d\boldsymbol{r}. \tag{7.16}$$

Now
$$\boldsymbol{r} = \boldsymbol{r}_i + \boldsymbol{R}. \tag{7.17}$$

The integration variable may be changed from $\boldsymbol{r}$ to $\boldsymbol{R}$, because $\boldsymbol{r}_i$ is constant, and the integration with respect to $\boldsymbol{r}$ over all space is equal to the corresponding integration with respect to $\boldsymbol{R}$.

$$\int \exp\{i(\boldsymbol{\kappa} + \boldsymbol{q})\,.\,\boldsymbol{R}\}\, d\boldsymbol{R} = (2\pi)^3\delta(\boldsymbol{\kappa} + \boldsymbol{q}). \tag{7.18}$$

† This procedure is justified explicitly in Section 9.2.

Thus $\langle \boldsymbol{k}'|\boldsymbol{W}_{Si}|\boldsymbol{k}\rangle = 4\pi \exp(i\boldsymbol{\kappa}\,.\,\boldsymbol{r}_i)\{\hat{\boldsymbol{\kappa}}\times(\boldsymbol{s}_i\times\hat{\boldsymbol{\kappa}})\}.$ (7.19)

For the orbital contribution we have, from (7.12) and (7.14),

$$\langle \boldsymbol{k}'|\boldsymbol{W}_{Li}|\boldsymbol{k}\rangle = \frac{1}{\hbar}\int \exp(i\boldsymbol{\kappa}\,.\,\boldsymbol{r})\frac{\boldsymbol{p}_i\times\hat{\boldsymbol{R}}}{R^2}\,d\boldsymbol{r}$$

$$= \frac{1}{\hbar}\exp(i\boldsymbol{\kappa}\,.\,\boldsymbol{r}_i)\int \exp(i\boldsymbol{\kappa}\,.\,\boldsymbol{R})\frac{\boldsymbol{p}_i\times\hat{\boldsymbol{R}}}{R^2}\,d\boldsymbol{R}$$

$$= \frac{4\pi i}{\hbar\kappa}\exp(i\boldsymbol{\kappa}\,.\,\boldsymbol{r}_i)(\boldsymbol{p}_i\times\hat{\boldsymbol{\kappa}}). \qquad (7.20)$$

Note that although the operators $\boldsymbol{p}_i$ and $\boldsymbol{r}_i$ do not commute, the operators $(\boldsymbol{p}_i\times\hat{\boldsymbol{\kappa}})$ and $(\boldsymbol{\kappa}\,.\,\boldsymbol{r}_i)$ do commute, so there is no problem about the order of the operators in (7.20).

From (7.19) and (7.20)

$$\sum_i \langle \boldsymbol{k}'|\boldsymbol{W}_{Si}+\boldsymbol{W}_{Li}|\boldsymbol{k}\rangle = 4\pi\boldsymbol{Q}_\perp \qquad (7.21)$$

where $$\boldsymbol{Q}_\perp = \sum_i \exp(i\boldsymbol{\kappa}\,.\,\boldsymbol{r}_i)\left\{\hat{\boldsymbol{\kappa}}\times(\boldsymbol{s}_i\times\hat{\boldsymbol{\kappa}})+\frac{i}{\hbar\kappa}(\boldsymbol{p}_i\times\hat{\boldsymbol{\kappa}})\right\}. \qquad (7.22)$$

Collecting up the multiplying factors in (7.10), (7.15), and (7.21) we have

$$-\frac{\mu_0}{4\pi}\gamma\mu_N 2\mu_B\frac{m}{2\pi\hbar^2}4\pi = -\frac{\mu_0}{4\pi}\gamma\frac{e\hbar}{2m_p}\frac{e\hbar}{2m_e}\frac{m}{\pi\hbar^2}4\pi \qquad (7.23)$$

$$= -\gamma r_0, \qquad (7.24)$$

where $$r_0 = \frac{\mu_0}{4\pi}\frac{e^2}{m_e}. \qquad (7.25)$$

Fig. 7.1 Neutron and electron position vectors.

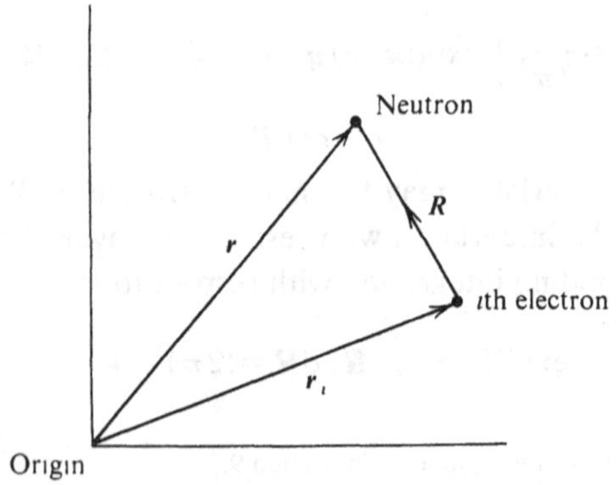

r_0 is known as the classical radius of the electron and has the value 2.818×10^{-15} m. In the step from (7.23) to (7.24) the mass of the neutron is taken to be equal to that of the proton. We thus arrive at the result

$$\left(\frac{d^2\sigma}{d\Omega\, dE'}\right)_{\sigma\lambda\to\sigma'\lambda'} = (\gamma r_0)^2\frac{k'}{k}|\langle\sigma'\lambda'|\boldsymbol{\sigma}\,.\,\boldsymbol{Q}_\perp|\sigma\lambda\rangle|^2\delta(E_\lambda - E_{\lambda'} + \hbar\omega). \tag{7.26}$$

It is instructive to compare the form of the matrix element $\langle\boldsymbol{k}'|V|\boldsymbol{k}\rangle$ for nuclear and magnetic scattering. For nuclear scattering from the jth nucleus, the matrix element is

$$b_j \exp(\mathrm{i}\boldsymbol{\kappa}\,.\,\boldsymbol{R}_j).$$

The factor b_j is a constant, independent of $\boldsymbol{\kappa}$, because the nuclear potential has a short range. For magnetic scattering from the ith electron, the matrix element is

$$-\gamma r_0\boldsymbol{\sigma}\,.\left\{\hat{\boldsymbol{\kappa}}\times(\boldsymbol{s}_i\times\hat{\boldsymbol{\kappa}}) + \frac{\mathrm{i}}{\hbar\kappa}(\boldsymbol{p}_i\times\hat{\boldsymbol{\kappa}})\right\}\exp(\mathrm{i}\boldsymbol{\kappa}\,.\,\boldsymbol{r}_i).$$

The expression is more complicated than its nuclear counterpart, partly because the magnetic interaction has a long range, and partly because both the dipole–dipole interaction for spin and the dipole–current interaction for the orbital motion are non-central forces.

Operators $\boldsymbol{Q}_\perp$ and $\boldsymbol{Q}$

The operator $\boldsymbol{Q}_\perp$ is related to the *magnetisation* of the scattering system as we now show. Consider first the spin part of $\boldsymbol{Q}_\perp$, which from (7.22) is

$$\boldsymbol{Q}_{\perp S} = \sum_i \exp(\mathrm{i}\boldsymbol{\kappa}\,.\,\boldsymbol{r}_i)\{\hat{\boldsymbol{\kappa}}\times(\boldsymbol{s}_i\times\hat{\boldsymbol{\kappa}})\}. \tag{7.27}$$

It is convenient to define an operator $\boldsymbol{Q}_S$ by

$$\boldsymbol{Q}_{\perp S} = \hat{\boldsymbol{\kappa}}\times(\boldsymbol{Q}_S\times\hat{\boldsymbol{\kappa}}). \tag{7.28}$$

Thus

$$\boldsymbol{Q}_S = \sum_i \exp(\mathrm{i}\boldsymbol{\kappa}\,.\,\boldsymbol{r}_i)\boldsymbol{s}_i. \tag{7.29}$$

The vector operator

$$\boldsymbol{\rho}_S(\boldsymbol{r}) = \sum_i \delta(\boldsymbol{r} - \boldsymbol{r}_i)\boldsymbol{s}_i \tag{7.30}$$

gives the electron spin density. Eqs. (7.29) and (7.30) show that $\boldsymbol{Q}_S$ is the Fourier transform of $\boldsymbol{\rho}_S(\boldsymbol{r})$, i.e.

$$\boldsymbol{Q}_S = \int \boldsymbol{\rho}_S(\boldsymbol{r})\exp(\mathrm{i}\boldsymbol{\kappa}\,.\,\boldsymbol{r})\,d\boldsymbol{r}. \tag{7.31}$$

The spin magnetisation operator is defined by

$$\boldsymbol{M}_{\mathrm{S}}(\boldsymbol{r}) = -2\mu_{\mathrm{B}}\boldsymbol{\rho}_{\mathrm{S}}(\boldsymbol{r}). \tag{7.32}$$

Thus

$$\boldsymbol{Q}_{\mathrm{S}} = -\frac{1}{2\mu_{\mathrm{B}}}\boldsymbol{M}_{\mathrm{S}}(\boldsymbol{\kappa}), \tag{7.33}$$

where

$$\boldsymbol{M}_{\mathrm{S}}(\boldsymbol{\kappa}) = \int \boldsymbol{M}_{\mathrm{S}}(\boldsymbol{r}) \exp(\mathrm{i}\boldsymbol{\kappa}\,.\,\boldsymbol{r})\,\mathrm{d}\boldsymbol{r}. \tag{7.34}$$

The corresponding calculation for the orbital term is somewhat more complicated and is given in Appendix H.1. It is shown that

$$\boldsymbol{Q}_{\perp\mathrm{L}} = \frac{\mathrm{i}}{\hbar\kappa}\sum_i \exp(\mathrm{i}\boldsymbol{\kappa}\,.\,\boldsymbol{r}_i)(\boldsymbol{p}_i \times \hat{\boldsymbol{\kappa}}) = -\frac{1}{2\mu_{\mathrm{B}}}\hat{\boldsymbol{\kappa}} \times \{\boldsymbol{M}_{\mathrm{L}}(\boldsymbol{\kappa}) \times \hat{\boldsymbol{\kappa}}\}, \tag{7.35}$$

where

$$\boldsymbol{M}_{\mathrm{L}}(\boldsymbol{\kappa}) = \int \boldsymbol{M}_{\mathrm{L}}(\boldsymbol{r}) \exp(\mathrm{i}\boldsymbol{\kappa}\,.\,\boldsymbol{r})\,\mathrm{d}\boldsymbol{r}. \tag{7.36}$$

The quantity $\boldsymbol{M}_{\mathrm{L}}(\boldsymbol{r})$ is the orbital magnetisation operator, i.e. it gives the magnetisation due to the orbital magnetic moments. The following is an outline of the reasoning leading to the result in (7.35). The quantity $\sum_i \exp(\mathrm{i}\boldsymbol{\kappa}\,.\,\boldsymbol{r}_i)\boldsymbol{p}_i$ is expressed as the Fourier transform of $\boldsymbol{j}(\boldsymbol{r})$, the operator for the current density due to the orbital motion. The latter can be written as

$$\boldsymbol{j}(\boldsymbol{r}) = \operatorname{curl} \boldsymbol{M}_{\mathrm{L}}(\boldsymbol{r}) + \operatorname{grad} \phi(\boldsymbol{r}). \tag{7.37}$$

The term grad ϕ is known as the *longitudinal* or *conduction* current density, and is due to the net motion of the electrons in some direction. The Fourier transform of grad ϕ is a vector in the direction of $\hat{\boldsymbol{\kappa}}$. Therefore, since $\hat{\boldsymbol{\kappa}} \times \hat{\boldsymbol{\kappa}} = 0$, this term gives no contribution to $\boldsymbol{Q}_{\perp\mathrm{L}}$.

Eq. (7.35) shows that $\boldsymbol{Q}_{\perp\mathrm{L}}$ has the same form as $\boldsymbol{Q}_{\perp\mathrm{S}}$, i.e. it may be written as

$$\boldsymbol{Q}_{\perp\mathrm{L}} = \hat{\boldsymbol{\kappa}} \times (\boldsymbol{Q}_{\mathrm{L}} \times \hat{\boldsymbol{\kappa}}), \tag{7.38}$$

where

$$\boldsymbol{Q}_{\mathrm{L}} = -\frac{1}{2\mu_{\mathrm{B}}}\boldsymbol{M}_{\mathrm{L}}(\boldsymbol{\kappa}). \tag{7.39}$$

We bring the spin and orbital terms together and put

$$\begin{aligned}\boldsymbol{M}(\boldsymbol{\kappa}) &= \boldsymbol{M}_{\mathrm{S}}(\boldsymbol{\kappa}) + \boldsymbol{M}_{\mathrm{L}}(\boldsymbol{\kappa}) \\ &= \int \boldsymbol{M}(\boldsymbol{r}) \exp(\mathrm{i}\boldsymbol{\kappa}\,.\,\boldsymbol{r})\,\mathrm{d}\boldsymbol{r},\end{aligned} \tag{7.40}$$

where

$$\boldsymbol{M}(\boldsymbol{r}) = \boldsymbol{M}_{\mathrm{S}}(\boldsymbol{r}) + \boldsymbol{M}_{\mathrm{L}}(\boldsymbol{r}) \tag{7.41}$$

is the operator for the total magnetisation – spin and orbital. Then from (7.28), (7.33), (7.38), (7.39)

$$\boldsymbol{Q}_{\perp} = \boldsymbol{Q}_{\perp\mathrm{S}} + \boldsymbol{Q}_{\perp\mathrm{L}} = \hat{\boldsymbol{\kappa}} \times (\boldsymbol{Q} \times \hat{\boldsymbol{\kappa}}), \tag{7.42}$$

where $$\boldsymbol{Q} = \boldsymbol{Q}_S + \boldsymbol{Q}_L = -\frac{1}{2\mu_B}\boldsymbol{M}(\boldsymbol{\kappa}). \tag{7.43}$$

We see that the operator $\boldsymbol{Q}$ is effectively the Fourier transform of $\boldsymbol{M}(\boldsymbol{r})$. The physical interpretation of (7.42) and (7.43) is that the magnetic scattering of neutrons is due to the interaction of the magnetic dipole moment of the neutron with the magnetic field produced by the unpaired electrons in the ion. This field is determined by the total magnetisation, i.e. the magnetic moments due to spin and orbital motion.

We now consider the geometrical relation between $\boldsymbol{Q}_\perp$ and $\boldsymbol{Q}$. Eq. (7.42) shows that $\boldsymbol{Q}_\perp$ is the vector projection of $\boldsymbol{Q}$ on to the plane perpendicular to $\hat{\boldsymbol{\kappa}}$ (Fig. 7.2). It follows from the diagram that

$$\boldsymbol{Q}_\perp = \boldsymbol{Q} - (\boldsymbol{Q}\,.\,\hat{\boldsymbol{\kappa}})\hat{\boldsymbol{\kappa}}. \tag{7.44}$$

Therefore

$$\begin{aligned}\boldsymbol{Q}_\perp^+\,.\,\boldsymbol{Q}_\perp &= \{\boldsymbol{Q}^+ - (\boldsymbol{Q}^+\,.\,\hat{\boldsymbol{\kappa}})\hat{\boldsymbol{\kappa}}\}\,.\,\{\boldsymbol{Q} - (\boldsymbol{Q}\,.\,\hat{\boldsymbol{\kappa}})\hat{\boldsymbol{\kappa}}\}\\ &= \boldsymbol{Q}^+\,.\,\boldsymbol{Q} - (\boldsymbol{Q}^+\,.\,\hat{\boldsymbol{\kappa}})(\boldsymbol{Q}\,.\,\hat{\boldsymbol{\kappa}})\\ &= \sum_{\alpha\beta}(\delta_{\alpha\beta} - \hat{\kappa}_\alpha\hat{\kappa}_\beta)Q_\alpha^+ Q_\beta,\end{aligned} \tag{7.45}$$

where α and β stand for x, y, z, and $\delta_{\alpha\beta}$ is the Kronecker delta.

Sum and average over the spin states of the neutron

The cross-section in (7.26) must be summed over the final states σ', λ', and averaged over the initial states σ, λ. We do this first for the

Fig. 7.2 Relation between $\boldsymbol{Q}$ and $\boldsymbol{Q}_\perp$.

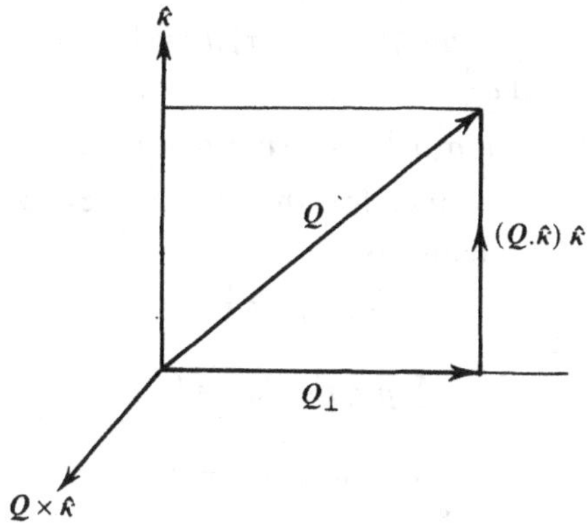

spin state of the neutron, i.e. we calculate

$$\sum_{\sigma\sigma'} p_\sigma |\langle\sigma'\lambda'|\boldsymbol{\sigma}\,.\,\boldsymbol{Q}_\perp|\sigma\lambda\rangle|^2,$$

where p_σ is the probability that the neutron is initially in the state σ.

Now
$$\boldsymbol{\sigma}\,.\,\boldsymbol{Q}_\perp = \sigma_x Q_{\perp x} + \sigma_y Q_{\perp y} + \sigma_z Q_{\perp z}. \tag{7.46}$$

The operator $\boldsymbol{\sigma}$ depends only on the spin coordinates of the neutron, while the operator $\boldsymbol{Q}_\perp$ depends only on the coordinates (space and spin) of the electron. The neutron and electron coordinates are independent. Therefore

$$\langle\sigma'\lambda'|\sigma_x Q_{\perp x}|\sigma\lambda\rangle = \langle\sigma'|\sigma_x|\sigma\rangle\langle\lambda'|Q_{\perp x}|\lambda\rangle. \tag{7.47}$$

When we multiply the matrix element by its complex conjugate we have two types of term, one involving only one cartesian component, e.g.

$$\langle\sigma|\sigma_x|\sigma'\rangle\langle\sigma'|\sigma_x|\sigma\rangle\langle\lambda|Q^+_{\perp x}|\lambda'\rangle\langle\lambda'|Q_{\perp x}|\lambda\rangle,$$

and the other involving two components, e.g.

$$\langle\sigma|\sigma_x|\sigma'\rangle\langle\sigma'|\sigma_y|\sigma\rangle\langle\lambda|Q^+_{\perp x}|\lambda'\rangle\langle\lambda'|Q_{\perp y}|\lambda\rangle.$$

Summing over σ' gives, for the two types,

$$\sum_{\sigma'} \langle\sigma|\sigma_x|\sigma'\rangle\langle\sigma'|\sigma_x|\sigma\rangle = \langle\sigma|\sigma_x^2|\sigma\rangle, \tag{7.48}$$

$$\sum_{\sigma'} \langle\sigma|\sigma_x|\sigma'\rangle\langle\sigma'|\sigma_y|\sigma\rangle = \langle\sigma|\sigma_x\sigma_y|\sigma\rangle. \tag{7.49}$$

The neutron has spin $\frac{1}{2}$, so it has two spin states which we denote by u and v. They may be regarded as corresponding to 'spin up' and 'spin down' relative to a specified axis which we take to be the z axis. The index σ stands for u or v. The results of operating with σ_x, σ_y, σ_z on u and v are given in Appendix F.2. From (F.12) and (F.16)

$$\langle u|\sigma_x^2|u\rangle = \langle v|\sigma_x^2|v\rangle = 1. \tag{7.50}$$

Similarly for σ_y^2 and σ_z^2.

$$\langle u|\sigma_x\sigma_y|u\rangle = -\langle v|\sigma_x\sigma_y|v\rangle = \mathrm{i}, \tag{7.51}$$

with corresponding results for the yz and zx components. For unpolarised incident neutrons

$$p_u = p_v = \tfrac{1}{2}. \tag{7.52}$$

Therefore
$$\sum_\sigma p_\sigma\langle\sigma|\sigma_x^2|\sigma\rangle = 1, \tag{7.53}$$

and
$$\sum_\sigma p_\sigma\langle\sigma|\sigma_x\sigma_y|\sigma\rangle = 0. \tag{7.54}$$

Thus, in evaluating the matrix elements of $\boldsymbol{Q}_\perp$, we have only terms like

$$\langle\lambda|Q^+_{\perp x}|\lambda'\rangle\langle\lambda'|Q_{\perp x}|\lambda\rangle.$$

Therefore

$$\begin{aligned}\frac{\mathrm{d}^2\sigma}{\mathrm{d}\Omega\,\mathrm{d}E'} &= (\gamma r_0)^2\frac{k'}{k}\sum_{\lambda\lambda'} p_\lambda \sum_\alpha \langle\lambda|Q^+_{\perp\alpha}|\lambda'\rangle\langle\lambda'|Q_{\perp\alpha}|\lambda\rangle\delta(E_\lambda - E_{\lambda'}+\hbar\omega)\\ &= (\gamma r_0)^2\frac{k'}{k}\sum_{\alpha\beta}(\delta_{\alpha\beta}-\hat{\kappa}_\alpha\hat{\kappa}_\beta)\\ &\quad\times\sum_{\lambda\lambda'} p_\lambda\langle\lambda|Q^+_\alpha|\lambda'\rangle\langle\lambda'|Q_\beta|\lambda\rangle\delta(E_\lambda - E_{\lambda'}+\hbar\omega). \qquad (7.55)\end{aligned}$$

The last step follows from (7.45).

7.3 Scattering due to spin only

The expressions we have derived so far relate to a general system. We now make the following restrictions. First we assume that the scattering system is a crystal, and that the unpaired electrons are localised close to the equilibrium positions of the ions in the lattice. This is known as the *Heitler–London* model. Secondly we assume LS coupling, i.e. in each ion the individual orbital angular momenta $\boldsymbol{l}_i$ of the unpaired electrons combine to form a resultant orbital angular momentum characterised by the quantum number L, and the individual spin angular momenta $\boldsymbol{s}_i$ combine to form a resultant spin angular momentum characterised by the quantum number S. In the present section we consider scattering due to spin only. This is the case when $L=0$, or when the resultant orbital angular momentum is quenched by the internal electric field of the crystal as in the elements in the iron group of the periodic table.

Consider a non-Bravais crystal with nucleus l, d at position $\boldsymbol{R}_{ld}$. l denotes the unit cell in which the ion is located, and d is the index specifying the ion within the unit cell (see Section 3.6). Let $\boldsymbol{r}_\nu$ be the position of the νth unpaired electron in the ion l, d relative to the nucleus (Fig. 7.3). Then

$$\boldsymbol{r}_i = \boldsymbol{R}_{ld} + \boldsymbol{r}_\nu. \qquad (7.56)$$

From (7.29) and (7.56)

$$\boldsymbol{Q} = \boldsymbol{Q}_\mathrm{S} = \sum_{ld}\exp(\mathrm{i}\boldsymbol{\kappa}\,.\,\boldsymbol{R}_{ld})\sum_{\nu(d)}\exp(\mathrm{i}\boldsymbol{\kappa}\,.\,\boldsymbol{r}_\nu)\boldsymbol{s}_\nu. \qquad (7.57)$$

Consider the contribution to $\langle\lambda'|\boldsymbol{Q}|\lambda\rangle$ from the ion l, d.

$$\langle\lambda'|\boldsymbol{Q}|\lambda\rangle_{ld} = \left\langle\lambda'\middle|\exp(\mathrm{i}\boldsymbol{\kappa}.\,\boldsymbol{R}_{ld})\sum_{\nu(d)}\exp(\mathrm{i}\boldsymbol{\kappa}\,.\,\boldsymbol{r}_\nu)\boldsymbol{s}_\nu\middle|\lambda\right\rangle. \qquad (7.58)$$

The state $|\lambda\rangle$ depends on the spin quantum numbers S_d of the ions, the orientations of the spin vectors $\boldsymbol{S}_{ld}$, the space states of the electrons, and the space states of the nuclei. The energy of the neutron is too small to change the values of S_d or the space states of the electrons; so the states $|\lambda\rangle$ and $|\lambda'\rangle$ differ only in the functions giving the orientations of the spins and the positions of the nuclei. In these circumstances

$$\left\langle\lambda'\left|\exp(\mathrm{i}\boldsymbol{\kappa}\,.\,\boldsymbol{R}_{ld})\sum_{\nu(d)}\exp(\mathrm{i}\boldsymbol{\kappa}\,.\,\boldsymbol{r}_\nu)\boldsymbol{s}_\nu\right|\lambda\right\rangle$$

$$=F_d(\boldsymbol{\kappa})\langle\lambda'|\exp(\mathrm{i}\boldsymbol{\kappa}\,.\,\boldsymbol{R}_{ld})\boldsymbol{S}_{ld}|\lambda\rangle, \tag{7.59}$$

where

$$F_d(\boldsymbol{\kappa})=\int s_d(\boldsymbol{r})\exp(\mathrm{i}\boldsymbol{\kappa}\,.\,\boldsymbol{r})\,\mathrm{d}\boldsymbol{r}. \tag{7.60}$$

The scalar function $s_d(\boldsymbol{r})$ is the normalised density of the unpaired electrons in the ion d, i.e. the density of the unpaired electrons divided by their number. $F_d(\boldsymbol{\kappa})$ is known as the *magnetic form factor.* The result in (7.59) is proved in Appendix H.2.

From (7.55), (7.57), and (7.59) the cross-section for spin-only scattering by ions with localised electrons is

$$\frac{\mathrm{d}^2\sigma}{\mathrm{d}\Omega\,\mathrm{d}E'}=(\gamma r_0)^2\frac{k'}{k}\sum_{\alpha\beta}(\delta_{\alpha\beta}-\hat{\kappa}_\alpha\hat{\kappa}_\beta)\sum_{l'd'}\sum_{ld}F^*_{d'}(\boldsymbol{\kappa})F_d(\boldsymbol{\kappa})$$

$$\times\sum_{\lambda\lambda'}p_\lambda\langle\lambda|\exp(-\mathrm{i}\boldsymbol{\kappa}\,.\,\boldsymbol{R}_{l'd'})S^\alpha_{l'd'}|\lambda'\rangle$$

$$\times\langle\lambda'|\exp(\mathrm{i}\boldsymbol{\kappa}\,.\,\boldsymbol{R}_{ld})S^\beta_{ld}|\lambda\rangle\delta(E_\lambda-E_{\lambda'}+\hbar\omega). \tag{7.61}$$

S^β_{ld} is the operator corresponding to the β component of spin for the ion l, d.

Fig. 7.3 Nucleus and electron position vectors.

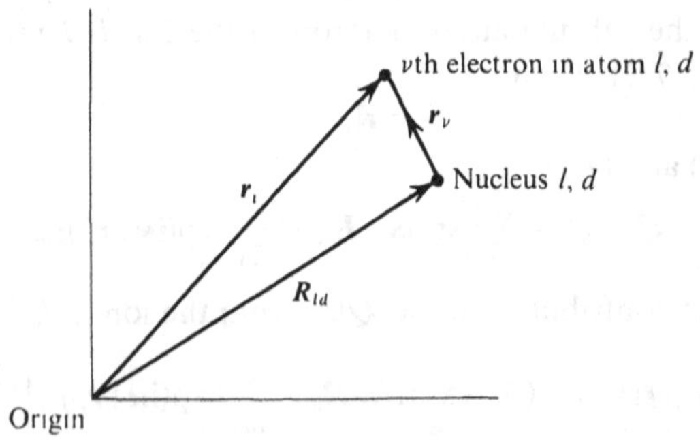

7.4 Scattering by ions with spin and orbital angular momentum

The ions of the rare earth elements have both spin and unquenched orbital angular momentum. A theoretical treatment of the scattering by such ions has been given by Johnston (1966). The calculation is complicated, and we simply quote the result for the case where $|\boldsymbol{\kappa}|^{-1}$ is large compared to the mean radius of the orbital wavefunctions of the unpaired electrons. The result is only approximate and is based on what is known as the *dipole approximation*, by analogy with a similar approximation in the theory of atomic radiation.

To simplify the notation we consider the case where all the ions in the crystal are identical. As before we assume LS coupling, so the angular momentum state of the ion is specified by the quantum numbers L, S, J. The calculation shows that when the ion has orbital angular momentum the previous expression for the cross-section can be used with two modifications. First, the term $F(\boldsymbol{\kappa})$ – (7.59) – is replaced by

$$\tfrac{1}{2}gF(\boldsymbol{\kappa})=\tfrac{1}{2}g_S\mathcal{J}_0+\tfrac{1}{2}g_L(\mathcal{J}_0+\mathcal{J}_2), \tag{7.62}$$

where

$$g=g_S+g_L, \tag{7.63}$$

$$g_S=1+\frac{S(S+1)-L(L+1)}{J(J+1)}, \tag{7.64}$$

$$g_L=\tfrac{1}{2}+\frac{L(L+1)-S(S+1)}{2J(J+1)}, \tag{7.65}$$

$$\mathcal{J}_n=4\pi\int_0^{\infty} j_n(\kappa r)\,s(r)\,r^2\,\mathrm{d}r. \tag{7.66}$$

g is the Landé splitting factor. $j_n(\kappa r)$ is a spherical Bessel function of order n, and $s(r)$ is the normalised density of the unpaired electrons averaged over all directions in space. The second change is that the operator $\boldsymbol{S}$ is to be regarded as the total angular momentum operator $\boldsymbol{J}$, or some effective spin operator if the orbital angular momentum is partially quenched.

7.5 Time-dependent operators

We may express the magnetic cross-section in terms of time-dependent angular momentum operators in a manner analogous to that for the nuclear cross-section. We first do the calculation for a model with localised electrons and consider LS coupling with either no orbital contribution or an orbital contribution that can be calculated on the dipole approximation.

Consider the expression for the cross-section in (7.61) and use the results

$$\delta(E_\lambda - E_{\lambda'} + \hbar\omega) = \frac{1}{2\pi\hbar}\int_{-\infty}^{\infty} \exp\{i(E_{\lambda'} - E_\lambda)t/\hbar\} \exp(-i\omega t)\, dt, \tag{7.67}$$

$$\exp(iHt/\hbar)|\lambda\rangle = \exp(iE_\lambda t/\hbar)|\lambda\rangle. \tag{7.68}$$

Then

$$\begin{aligned}
&\sum_{\lambda\lambda'} p_\lambda \langle\lambda| \exp(-i\boldsymbol{\kappa}\,.\,\boldsymbol{R}_{l'd'}) S^{\alpha}_{l'd'} |\lambda'\rangle \\
&\qquad \times \langle\lambda'| \exp(i\boldsymbol{\kappa}\,.\,\boldsymbol{R}_{ld}) S^{\beta}_{ld} |\lambda\rangle \delta(E_\lambda - E_{\lambda'} + \hbar\omega) \\
&= \frac{1}{2\pi\hbar}\int_{-\infty}^{\infty} \sum_{\lambda\lambda'} p_\lambda \langle\lambda| \exp(-i\boldsymbol{\kappa}\,.\,\boldsymbol{R}_{l'd'}) S^{\alpha}_{l'd'} |\lambda'\rangle \\
&\qquad \times \langle\lambda'| \exp(iHt/\hbar) \exp(i\boldsymbol{\kappa}\,.\,\boldsymbol{R}_{ld}) \exp(-iHt/\hbar) \\
&\qquad \times \exp(iHt/\hbar) S^{\beta}_{ld} \exp(-iHt/\hbar) |\lambda\rangle \exp(-i\omega t)\, dt \\
&= \frac{1}{2\pi\hbar}\int_{-\infty}^{\infty} \langle \exp\{-i\boldsymbol{\kappa}\,.\,\boldsymbol{R}_{l'd'}(0)\} S^{\alpha}_{l'd'}(0) \\
&\qquad \times \exp\{i\boldsymbol{\kappa}\,.\,\boldsymbol{R}_{ld}(t)\} S^{\beta}_{ld}(t)\rangle \exp(-i\omega t)\, dt,
\end{aligned} \tag{7.69}$$

where

$$S^{\beta}_{ld}(t) = \exp(iHt/\hbar) S^{\beta}_{ld} \exp(-iHt/\hbar). \tag{7.70}$$

The quantity $\langle\ \rangle$ in the last line of (7.69) is the thermal average of the operator enclosed at temperature T.

Now the orientations of the electron spins have only a small effect on the interatomic forces, and hence on the motion of the nuclei. If we assume the effect is zero the thermal average in (7.69) may be factorised giving

$$\begin{aligned}
&\langle \exp\{i\boldsymbol{\kappa}\,.\,\boldsymbol{R}_{l'd'}(0)\} S^{\alpha}_{l'd'}(0) \exp\{i\boldsymbol{\kappa}\,.\,\boldsymbol{R}_{ld}(t)\} S^{\beta}_{ld}(t)\rangle \\
&\quad = \langle \exp\{-i\boldsymbol{\kappa}\,.\,\boldsymbol{R}_{l'd'}(0)\} \exp\{i\boldsymbol{\kappa}\,.\,\boldsymbol{R}_{ld}(t)\}\rangle \langle S^{\alpha}_{l'd'}(0) S^{\beta}_{ld}(t)\rangle.
\end{aligned} \tag{7.71}$$

From (7.61), (7.69), and (7.71) we have for the magnetic cross-section

$$\begin{aligned}
\frac{d^2\sigma}{d\Omega\, dE'} &= \frac{(\gamma r_0)^2}{2\pi\hbar} \frac{k'}{k} \sum_{\alpha\beta} (\delta_{\alpha\beta} - \hat{\kappa}_\alpha \hat{\kappa}_\beta) \sum_{l'd'ld} \tfrac{1}{4} g_{d'} g_d F^*_{d'}(\boldsymbol{\kappa}) F_d(\boldsymbol{\kappa}) \\
&\qquad \times \int_{-\infty}^{\infty} \langle \exp\{-i\boldsymbol{\kappa}\,.\,\boldsymbol{R}_{l'd'}(0)\} \exp\{i\boldsymbol{\kappa}\,.\,\boldsymbol{R}_{ld}(t)\}\rangle \\
&\qquad \times \langle S^{\alpha}_{l'd'}(0) S^{\beta}_{ld}(t)\rangle \exp(-i\omega t)\, dt.
\end{aligned} \tag{7.72}$$

For a Bravais crystal this becomes

$$\frac{d^2\sigma}{d\Omega\, dE'} = \frac{(\gamma r_0)^2}{2\pi\hbar}\frac{k'}{k} N\{\tfrac{1}{2}gF(\boldsymbol{\kappa})\}^2 \sum_{\alpha\beta} (\delta_{\alpha\beta} - \hat{\kappa}_\alpha\hat{\kappa}_\beta) \sum_l \exp(i\boldsymbol{\kappa}\,.\,\boldsymbol{l})$$
$$\times \int_{-\infty}^{\infty} \langle \exp\{-i\boldsymbol{\kappa}\,.\,\boldsymbol{u}_0(0)\}\exp\{i\boldsymbol{\kappa}\,.\,\boldsymbol{u}_l(t)\}\rangle$$
$$\times \langle S_0^\alpha(0)S_l^\beta(t)\rangle \exp(-i\omega t)\, dt, \tag{7.73}$$

where $\boldsymbol{u}_l(t)$ is the displacement of nucleus l from its equilibrium position.

We may divide the cross-section in (7.72) into its elastic and inelastic components. The reasoning is similar to that in Section 4.7. Put

$$I_{jj'}(\boldsymbol{\kappa}, t) = \langle \exp\{-i\boldsymbol{\kappa}\,.\,\boldsymbol{R}_{l'd'}(0)\}\exp\{i\boldsymbol{\kappa}\,.\,\boldsymbol{R}_{ld}(t)\}\rangle, \tag{7.74}$$

$$J_{jj'}^{\alpha\beta}(t) = \langle S_{l'd'}^\alpha(0)S_{ld}^\beta(t)\rangle, \tag{7.75}$$

where j stands for the combination l,d. We express each of the functions $I_{jj'}(\boldsymbol{\kappa}, t)$ and $J_{jj'}^{\alpha\beta}(t)$ as the sum of its value at $t=\infty$ and a time-varying term, i.e. we put

$$I_{jj'}(\boldsymbol{\kappa}, t) = I_{jj'}(\boldsymbol{\kappa}, \infty) + I'_{jj'}(\boldsymbol{\kappa}, t), \tag{7.76}$$

$$J_{jj'}^{\alpha\beta}(t) = J_{jj'}^{\alpha\beta}(\infty) + J_{jj'}^{\prime\alpha\beta}(t). \tag{7.77}$$

Then (7.72) may be written in the form

$$\frac{d^2\sigma}{d\Omega\, dE'} = \frac{(\gamma r_0)^2}{2\pi\hbar}\frac{k'}{k} \sum_{\alpha\beta} (\delta_{\alpha\beta} - \hat{\kappa}_\alpha\hat{\kappa}_\beta) \sum_{jj'} \tfrac{1}{4} g_{d'}g_d F_{d'}^*(\boldsymbol{\kappa})F_d(\boldsymbol{\kappa})$$
$$\times \int_{-\infty}^{\infty} \{I_{jj'}(\boldsymbol{\kappa}, \infty) + I'_{jj'}(\boldsymbol{\kappa}, t)\}\{J_{jj'}^{\alpha\beta}(\infty) + J_{jj'}^{\prime\alpha\beta}(t)\}\exp(-i\omega t)\, dt. \tag{7.78}$$

This expression for the cross-section may be divided into four components.

The term $I_{jj'}(\boldsymbol{\kappa}, \infty)J_{jj'}^{\alpha\beta}(\infty)$ gives elastic magnetic scattering. The term $I'_{jj'}(\boldsymbol{\kappa}, t)J_{jj'}^{\alpha\beta}(\infty)$ gives what is known as *magnetovibrational* scattering. This is scattering which is inelastic in the phonon system but elastic in the spin system, i.e. the orientation of the electron spins remain unchanged, but the neutron excites or de-excites phonons in the crystal lattice via the magnetic interaction. We can readily evaluate the cross-section for this type of scattering. Consider a single domain of a Bravais ferromagnet with the spins aligned in the z direction. From (7.75)

$$\begin{aligned} J_{ll'}^{\alpha\beta}(\infty) &= \langle S^z\rangle^2 \qquad \alpha=\beta=z, \\ &= 0 \qquad\qquad \text{otherwise.} \end{aligned} \tag{7.79}$$

Therefore the magnetovibrational cross-section is the same as the cross-section for coherent inelastic nuclear scattering with the factor $\sigma_{\text{coh}}/4\pi$ replaced by

$$\tfrac{1}{4}(\gamma r_0)^2 g^2 F^2(\boldsymbol{\kappa})(1-\hat{\kappa}_z^2)\langle S^z\rangle^2. \tag{7.80}$$

If a magnetic field sufficiently large to align the domains is applied in the direction of $\boldsymbol{\kappa}$, then $\hat{\kappa}_z = -1$, and the magnetovibrational cross-section is zero. This serves to distinguish magnetovibrational scattering from its nuclear counterpart, as the latter is unaffected by the application of a magnetic field.

Of the other terms in (7.78), $I_{jj'}(\boldsymbol{\kappa},\infty)J_{jj'}^{\prime\alpha\beta}(t)$ gives inelastic magnetic scattering with no change in the phonon system, while $I'_{jj'}(\boldsymbol{\kappa},t)J_{jj'}^{\prime\alpha\beta}(t)$ gives scattering which is inelastic in both the spin and phonon systems.

The above results are for a localised model. To obtain an expression for the cross-section that is independent of a model we go back to (7.55). The relations (7.67) and (7.68) then give

$$\frac{\mathrm{d}^2\sigma}{\mathrm{d}\Omega\,\mathrm{d}E'} = \frac{(\gamma r_0)^2}{2\pi\hbar}\frac{k'}{k}\sum_{\alpha\beta}(\delta_{\alpha\beta}-\hat{\kappa}_\alpha\hat{\kappa}_\beta)\int\langle Q_\alpha(-\boldsymbol{\kappa},0)Q_\beta(\boldsymbol{\kappa},t)\rangle\exp(-\mathrm{i}\omega t)\,\mathrm{d}t, \tag{7.81}$$

where

$$Q_\beta(\boldsymbol{\kappa},t) = \exp(\mathrm{i}Ht/\hbar)Q_\beta(\boldsymbol{\kappa})\exp(-\mathrm{i}Ht/\hbar). \tag{7.82}$$

We may obtain the elastic cross-section from (7.81) by taking the thermal average at $t=\infty$, and integrating with respect to t and E'. This gives

$$\left(\frac{\mathrm{d}\sigma}{\mathrm{d}\Omega}\right)_{\text{el}} = (\gamma r_0)^2\sum_{\alpha\beta}(\delta_{\alpha\beta}-\hat{\kappa}_\alpha\hat{\kappa}_\beta)\langle Q_\alpha(-\boldsymbol{\kappa})\rangle\langle Q_\beta(\boldsymbol{\kappa})\rangle. \tag{7.83}$$

We showed in Section 7.2 that $\boldsymbol{Q}(\boldsymbol{\kappa})$ is related to the magnetisation operator $\boldsymbol{M}(\boldsymbol{r})$ by

$$\boldsymbol{Q}(\boldsymbol{\kappa}) = -\frac{1}{2\mu_{\text{B}}}\int\boldsymbol{M}(\boldsymbol{r})\exp(\mathrm{i}\boldsymbol{\kappa}\,.\,\boldsymbol{r})\,\mathrm{d}\boldsymbol{r} = -\frac{1}{2\mu_{\text{B}}}\boldsymbol{M}(\boldsymbol{\kappa}) \tag{7.84}$$

An alternative expression to (7.83), obtained via (7.42) and (7.45), is therefore

$$\left(\frac{\mathrm{d}\sigma}{\mathrm{d}\Omega}\right)_{\text{el}} = \left(\frac{\gamma r_0}{2\mu_{\text{B}}}\right)^2|\hat{\boldsymbol{\kappa}}\times\{\langle\boldsymbol{M}(\boldsymbol{\kappa})\rangle\times\hat{\boldsymbol{\kappa}}\}|^2. \tag{7.85}$$

The required product of the integral with its complex conjugate is the scalar product. $\boldsymbol{M}(\boldsymbol{r})$ here refers to the total magnetisation – spin plus orbital. If the scattering is due to spin alone (7.85) may be written as

$$\left(\frac{d\sigma}{d\Omega}\right)_{el} = (\gamma r_0)^2 |\hat{\boldsymbol{\kappa}} \times \{\langle \boldsymbol{\rho}_S(\boldsymbol{\kappa})\rangle \times \hat{\boldsymbol{\kappa}}\}|^2, \qquad (7.86)$$

where $\boldsymbol{\rho}_S(\boldsymbol{\kappa})$ is the Fourier transform of $\boldsymbol{\rho}_S(\boldsymbol{r})$, the vector electron spin density.

7.6 Cross-section for a paramagnet

Zero magnetic field. As an example we evaluate the elastic magnetic cross-section for a paramagnetic Bravais crystal with localised electrons in zero external magnetic field. The spins of the ions are randomly oriented, Hence there is no internal magnetic field. So if the orientation of the spin on a particular ion changes there is no change in the energy of the system. This means that the spin operators $\boldsymbol{S}_l$ commute with the Hamiltonian of the system and hence that the spin matrix element in (7.73) is time-independent, i.e.

$$\langle S_0^\alpha(0) S_l^\beta(t)\rangle = \langle S_0^\alpha S_l^\beta \rangle. \qquad (7.87)$$

Since we are considering elastic scattering we evaluate the matrix element for the nuclear displacements at $t = \infty$. As before this leads to the Debye–Waller factor. Inserting these results in (7.73) and integrating with respect to E' gives

$$\frac{d\sigma}{d\Omega} = (\gamma r_0)^2 \{\tfrac{1}{2} g F(\boldsymbol{\kappa})\}^2 \exp(-2W) \sum_{\alpha\beta} (\delta_{\alpha\beta} - \hat{\kappa}_\alpha \hat{\kappa}_\beta) N \sum_l \exp(i\boldsymbol{\kappa}\,.\,\boldsymbol{l}) \langle S_0^\alpha S_l^\beta \rangle. \qquad (7.88)$$

In a paramagnet there is no correlation between the spins of different ions. Therefore, for $l \neq 0$

$$\langle S_0^\alpha S_l^\beta \rangle = \langle S_0^\alpha \rangle \langle S_l^\beta \rangle = 0. \qquad (7.89)$$

For $l = 0$

$$\begin{aligned}\langle S_0^\alpha S_l^\beta \rangle &= \delta_{\alpha\beta} \langle (S_0^\alpha)^2 \rangle = \tfrac{1}{3} \delta_{\alpha\beta} \langle \boldsymbol{S}^2 \rangle \\ &= \tfrac{1}{3} \delta_{\alpha\beta} S(S+1). \end{aligned} \qquad (7.90)$$

Thus the cross-section is only non-zero for $l = 0$, $\alpha = \beta$; whence

$$\sum_{\alpha\beta} (\delta_{\alpha\beta} - \hat{\kappa}_\alpha \hat{\kappa}_\beta) = \sum_\alpha (1 - \hat{\kappa}_\alpha^2) = 2. \qquad (7.91)$$

These results give

$$\frac{d\sigma}{d\Omega}=\tfrac{2}{3}(\gamma r_0)^2 N\{\tfrac{1}{2}gF(\boldsymbol{\kappa})\}^2\exp(-2W)S(S+1). \tag{7.92}$$

Magnetic field. We now consider the case when a magnetic field $\boldsymbol{B}$ is applied to the paramagnet in the $-z$ direction. If the scattering process causes a change in the spin orientation of an ion, the change in the energy of the system is of the order of

$$\mu_B B=9\times10^{-24}\,\mathrm{J}\qquad \text{for } B=1\,\mathrm{T}. \tag{7.93}$$

The energy of a thermal neutron is of the order of

$$k_B T=4\times10^{-21}\,\mathrm{J}\qquad \text{for } T=300\,\mathrm{K}. \tag{7.94}$$

So for a thermal neutron and a magnetic field that is not very large, the change in the energy of the scattering system is small compared to the energy of the incident neutron. We therefore ignore the energy change and again take the spin matrix element to be time-independent. However, the values of $\langle S_0^\alpha S_l^\beta\rangle$ are not the same as before. The calculation is done in Example 7.2. The result is

$$\frac{d\sigma}{d\Omega}=(\gamma r_0)^2 N\{\tfrac{1}{2}gF(\boldsymbol{\kappa})\}^2\exp(-2W)\Big[(1-\hat{\kappa}_z^2)\frac{(2\pi)^3}{v_0}\langle S^z\rangle^2\sum_{\boldsymbol{\tau}}\delta(\boldsymbol{\kappa}-\boldsymbol{\tau})$$
$$+\hat{\kappa}_z^2\{\tfrac{1}{2}S(S+1)-\tfrac{3}{2}\langle(S^z)^2\rangle+\langle S^z\rangle^2\}+\tfrac{1}{2}S(S+1)+\tfrac{1}{2}\langle(S^z)^2\rangle-\langle S^z\rangle^2\Big], \tag{7.95}$$

$$\langle S^z\rangle=(S+\tfrac{1}{2})\coth\{(S+\tfrac{1}{2})u\}-\tfrac{1}{2}\coth(\tfrac{1}{2}u), \tag{7.96}$$

$$\langle(S^z)^2\rangle=S(S+1)-\coth(\tfrac{1}{2}u)\langle S^z\rangle, \tag{7.97}$$

$$u=g\mu_B B\beta. \tag{7.98}$$

For zero magnetic field the scattering is entirely diffuse, i.e. it is distributed continuously over scattering directions. When a magnetic field is applied, part of the scattering – that proportional to $\langle S^z\rangle^2$ – occurs in Bragg peaks; the rest is diffuse.

Examples

7.1 A symmetric or antisymmetric function of the space coordinates of two electrons has the form

$$\phi(\boldsymbol{r}_1,\boldsymbol{r}_2)=\{\psi_a(\boldsymbol{r}_1)\psi_b(\boldsymbol{r}_2)\pm\psi_b(\boldsymbol{r}_1)\psi_a(\boldsymbol{r}_2)\}/\sqrt{2},$$

where ψ_a and ψ_b are an orthonormal pair of single-electron state functions. Show that if $f(\boldsymbol{r})$ is an operator that depends on position

$$\langle\phi|f(\boldsymbol{r}_1)|\phi\rangle=\langle\phi|f(\boldsymbol{r}_2)|\phi\rangle$$

$$=\tfrac{1}{2}\int\{|\psi_a(\boldsymbol{r})|^2+|\psi_b(\boldsymbol{r})|^2\}f(\boldsymbol{r})\,\mathrm{d}\boldsymbol{r}.$$

7.2 (*a*) Derive the expression given in (7.95) to (7.98) for the elastic magnetic cross-section when a magnetic field $\boldsymbol{B}$ is applied in the $-z$ direction to a paramagnet.

(*b*) Show that

$$\text{as}\quad u\to\infty,\quad \langle S^z\rangle\to S,\qquad \langle(S^z)^2\rangle\to S^2,$$

$$\text{as}\quad u\to 0,\quad \langle S^z\rangle\to\tfrac{1}{3}S(S+1)u,\quad \langle(S^z)^2\rangle\to\tfrac{1}{3}S(S+1).$$

8

Scattering from magnetically ordered crystals

In the present chapter we consider the scattering from a magnetic crystal in which the spins are ordered, e.g. a ferromagnet, or an antiferromagnet, or a crystal with a helical arrangement of spins.

8.1 Elastic magnetic scattering

For a Bravais crystal with localised electrons the elastic cross-section is obtained from (7.73) by replacing the matrix elements by their limiting values as $t \to \infty$. As $t \to \infty$ $\langle S_0^\alpha(0) S_l^\beta(t)\rangle$ becomes independent of time. Thus

$$\lim_{t\to\infty} \langle S_0^\alpha(0) S_l^\beta(t)\rangle = \langle S_0^\alpha\rangle\langle S_l^\beta\rangle. \tag{8.1}$$

Substituting (8.1) in (7.73) and integrating with respect to E' gives the elastic cross-section

$$\left(\frac{d\sigma}{d\Omega}\right)_{\text{el}} = (\gamma r_0)^2 N\{\tfrac{1}{2}gF(\boldsymbol{\kappa})\}^2 \exp(-2W) \sum_{\alpha\beta} (\delta_{\alpha\beta} - \hat{\kappa}_\alpha\hat{\kappa}_\beta) \times \sum_l \exp(i\boldsymbol{\kappa}\,.\,\boldsymbol{l})\langle S_0^\alpha\rangle\langle S_l^\beta\rangle. \tag{8.2}$$

Ferromagnet

In the absence of an external magnetic field a ferromagnetic crystal is composed of small regions, or domains, in each of which the electron spins tend to align in the same direction. Consider a single domain and take the z axis, i.e. the axis of quantisation for specifying the spin states, to be along the mean direction of the spins. Then

$$\langle S_l^x\rangle = \langle S_l^y\rangle = 0. \tag{8.3}$$

For a Bravais ferromagnet $\langle S_l^z \rangle$ is independent of the site position l, i.e.

$$\langle S_l^z \rangle = \langle S^z \rangle. \tag{8.4}$$

This is the average value for a single ion of the component of spin along the mean spin direction. The magnetisation of the domain is proportional to $\langle S^z \rangle$.

From (8.2), (8.3), and (8.4) the elastic cross-section for a single domain is

$$\left(\frac{d\sigma}{d\Omega}\right)_{el} = (\gamma r_0)^2 N\{\tfrac{1}{2}gF(\boldsymbol{\kappa})\}^2 \exp(-2W)(1-\hat{\kappa}_z^2)\langle S^z \rangle^2 \sum_l \exp(i\boldsymbol{\kappa}\,.\,\boldsymbol{l}). \tag{8.5}$$

As usual

$$\sum_l \exp(i\boldsymbol{\kappa}\,.\,\boldsymbol{l}) = \frac{(2\pi)^3}{v_0}\sum_{\boldsymbol{\tau}} \delta(\boldsymbol{\kappa}-\boldsymbol{\tau}). \tag{8.6}$$

When

$$\boldsymbol{\kappa} = \boldsymbol{\tau}, \qquad \hat{\kappa}_z = \hat{\boldsymbol{\tau}}\,.\,\hat{\boldsymbol{\eta}}, \tag{8.7}$$

where $\hat{\boldsymbol{\tau}}$ is a unit vector in the direction of $\boldsymbol{\tau}$, and $\hat{\boldsymbol{\eta}}$ is a unit vector in the mean direction of the spins. Thus the cross-section for a sample with many domains is

$$\left(\frac{d\sigma}{d\Omega}\right)_{el} = (\gamma r_0)^2 N \frac{(2\pi)^3}{v_0} \langle S^{\eta} \rangle^2 \sum_{\boldsymbol{\tau}} \{\tfrac{1}{2}gF(\boldsymbol{\tau})\}^2 \exp(-2W) \times\{1-(\hat{\boldsymbol{\tau}}\,.\,\hat{\boldsymbol{\eta}})^2_{av}\}\delta(\boldsymbol{\kappa}-\boldsymbol{\tau}). \tag{8.8}$$

The quantity $\langle S^{\eta} \rangle$ is the mean value of the component of the spin in the direction of $\hat{\boldsymbol{\eta}}$ for each domain.

The average of $(\hat{\boldsymbol{\tau}}\,.\,\hat{\boldsymbol{\eta}})^2$ is taken over all the domains. If all directions in space are equally likely for $\hat{\boldsymbol{\eta}}$ then

$$\{1-(\hat{\boldsymbol{\tau}}\,.\,\hat{\boldsymbol{\eta}})^2\}_{av} = \tfrac{2}{3}. \tag{8.9}$$

The same result holds if $\hat{\boldsymbol{\eta}}$ is equally likely to be along axes related by cubic symmetry, e.g. 100, 010, 001 or 111, $\bar{1}11$, $1\bar{1}1$, $11\bar{1}$.

We see from (8.8) that for a ferromagnetic crystal the magnetic Bragg peaks occur at the same points in reciprocal space as the nuclear Bragg peaks. However, there are several important differences between magnetic and nuclear Bragg scattering. Firstly, the magnetic scattering, being proportional to $\langle S^{\eta} \rangle^2$, is very temperature dependent, and falls to zero at the Curie temperature T_C (see Fig. 8.1). The nuclear scattering varies little with temperature; the only term in the cross-section that is temperature dependent is the Debye–Waller factor.

Secondly, for magnetic scattering, the magnetic form factor $F(\boldsymbol{\tau})$ falls rapidly with increasing $|\boldsymbol{\tau}|$. This is because the form factor is the

Fourier transform of the magnetic potential, and the latter has a long range. The nuclear potential on the other hand has a very short range, and its Fourier transform is independent of $\boldsymbol{\kappa}$. For a Bravais crystal the only term that causes the intensity of the nuclear peaks to vary with $\boldsymbol{\tau}$ is again the Debye–Waller factor, and at moderate temperatures the variation is small.

If an external magnetic field is applied in the direction of $\boldsymbol{\tau}$ for a given Bragg peak, the spin directions of all the domains tend to align so that $\hat{\boldsymbol{\eta}}$ is along $-\boldsymbol{\tau}$. Then $\hat{\boldsymbol{\tau}}\,.\,\hat{\boldsymbol{\eta}}=-1$, and the magnetic scattering vanishes. The difference in the cross-section with and without an external magnetic field gives the magnetic scattering alone, thus providing a convenient method for separating the magnetic and nuclear scattering at a Bragg peak.

These remarks about magnetic Bragg scattering are valid for a localised or an itinerant electron model. However, the expression for the cross-section in (8.8) applies only to a localised model. The corresponding result for an itinerant model has been derived by Izuyama *et al.* (1963).

We may also use the expressions in (7.85) and (7.86), which do not depend on a model. For a single domain in a ferromagnetic crystal,

Fig. 8.1 Variation of $\langle S^{\eta}\rangle$ with temperature for a typical ferromagnet.

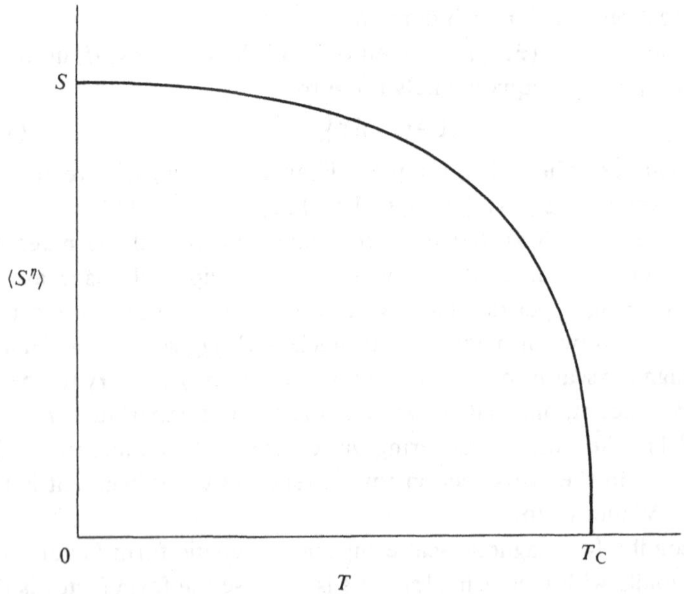

the magnetisation $\langle \boldsymbol{M}(\boldsymbol{r})\rangle$ is periodic in the unit cell of the crystal. Thus, provided the magnetisation does not affect the motion of the nuclei, we may put

$$\int \langle \boldsymbol{M}(\boldsymbol{r})\rangle \exp(\mathrm{i}\boldsymbol{\kappa}\,.\,\boldsymbol{r})\,\mathrm{d}\boldsymbol{r} = \mathscr{F}(\boldsymbol{\kappa})\sum_{l}\langle \exp(\mathrm{i}\boldsymbol{\kappa}\,.\,\boldsymbol{R}_l)\rangle, \tag{8.10}$$

where
$$\mathscr{F}(\boldsymbol{\kappa}) = \int_{\text{cell}} \langle \boldsymbol{M}(\boldsymbol{r})\rangle \exp(\mathrm{i}\boldsymbol{\kappa}\,.\,\boldsymbol{r})\,\mathrm{d}\boldsymbol{r}. \tag{8.11}$$

$\int_{\text{cell}}$ means integrate over the unit cell. Substituting (8.10) in (7.85) and using the result in Example 4.3 we obtain

$$\left(\frac{\mathrm{d}\sigma}{\mathrm{d}\Omega}\right)_{\text{el}} = \left(\frac{\gamma r_0}{2\mu_{\mathrm{B}}}\right)^2 N\frac{(2\pi)^3}{v_0}\sum_{\boldsymbol{\tau}} \exp(-2W)\delta(\boldsymbol{\kappa}-\boldsymbol{\tau})|\hat{\boldsymbol{\tau}}\times\{\mathscr{F}(\boldsymbol{\tau})\times\hat{\boldsymbol{\tau}}\}|^2. \tag{8.12}$$

From (B.15) $\langle \boldsymbol{M}(\boldsymbol{r})\rangle$ is given in terms of $\mathscr{F}(\boldsymbol{\tau})$ by

$$\langle \boldsymbol{M}(\boldsymbol{r})\rangle = \frac{1}{v_0}\sum_{\boldsymbol{\tau}} \mathscr{F}(\boldsymbol{\tau})\exp(-\mathrm{i}\boldsymbol{\tau}\,.\,\boldsymbol{r}). \tag{8.13}$$

Measurement of the intensity of the Bragg peak gives $\mathscr{F}(\boldsymbol{\tau})$, from which $\langle \boldsymbol{M}(\boldsymbol{r})\rangle$ may be calculated. These measurements are made most accurately with polarised neutrons and are discussed in the next chapter.

Antiferromagnet

In a ferromagnet all the spins tend to align in the same direction within a single domain. In an antiferromagnet each domain consists of two interpenetrating sublattices, A and B, the spins of the atoms in A being antiparallel to those in B. The term antiferromagnetism also covers more complicated spin patterns, but we shall consider only the simple case. An example of a simple antiferromagnet is $KMnF_3$. Its structure is shown in Fig. 8.2*a*.

Denote the mean spin direction in sublattice A by $\hat{\boldsymbol{\eta}}$. Obviously if the average $\langle S^{\eta}\rangle$ was taken over all the ions in the domain the result would be zero, as there are equal numbers of ions pointing up and down. We therefore define $\langle S^{\eta}\rangle$ to be the *staggered* mean spin, i.e. the value of $\langle S^{\eta}\rangle$ for the ions in sublattice A alone. The unit cell in sublattice A is called the *magnetic* unit cell. The unit cell obtained by ignoring the spin of the Mn^{2+} ion is called the *nuclear* unit cell.

To calculate the cross-section for a single domain we use the magnetic unit cell and treat the crystal as non-Bravais. Then

$$\sum_{ll'} \exp\{i\boldsymbol{\kappa} \cdot (\boldsymbol{l}-\boldsymbol{l}')\}\langle S_{l'}^{\eta}\rangle\langle S_{l}^{\eta}\rangle$$
$$= \langle S^{\eta}\rangle^2 N_{\mathrm{m}} \sum_{A} \exp(i\boldsymbol{\kappa} \cdot \boldsymbol{l}) \sum_{d} \sigma_d \exp(i\boldsymbol{\kappa} \cdot \boldsymbol{d}), \tag{8.14}$$

$$\sum_{A} \exp(i\boldsymbol{\kappa} \cdot \boldsymbol{l}) = \frac{(2\pi)^3}{v_{0\mathrm{m}}} \sum_{\boldsymbol{\tau}_{\mathrm{m}}} \delta(\boldsymbol{\kappa} - \boldsymbol{\tau}_{\mathrm{m}}). \tag{8.15}$$

In these equations $N_{\mathrm{m}}(=\frac{1}{2}N)$ is the number of magnetic unit cells in the crystal, $v_{0\mathrm{m}}$ $(=2v_0)$ is the volume of the magnetic unit cell, and $\boldsymbol{\tau}_{\mathrm{m}}$ is a vector in the magnetic reciprocal lattice. $\sum_A$ means sum over the ions in sublattice A. $\sum_d$ means sum over the ions in the magnetic unit cell. σ_d has the value $+1$ for an A ion and -1 for a B ion. The expression for the cross-section is thus

$$\left(\frac{d\sigma}{d\Omega}\right)_{\mathrm{el}} = (\gamma r_0)^2 N_{\mathrm{m}} \frac{(2\pi)^3}{v_{0\mathrm{m}}} \sum_{\boldsymbol{\tau}_{\mathrm{m}}} |F_{\mathrm{M}}(\boldsymbol{\tau}_{\mathrm{m}})|^2 \exp(-2W)$$
$$\times \{1-(\hat{\boldsymbol{\tau}}_{\mathrm{m}} \cdot \hat{\boldsymbol{\eta}})^2_{\mathrm{av}}\}\delta(\boldsymbol{\kappa}-\boldsymbol{\tau}_{\mathrm{m}}), \tag{8.16}$$

where
$$F_{\mathrm{M}}(\boldsymbol{\tau}_{\mathrm{m}}) = \tfrac{1}{2} g \langle S^{\eta}\rangle F(\boldsymbol{\tau}_{\mathrm{m}}) \sum_{d} \sigma_d \exp(i\boldsymbol{\tau}_{\mathrm{m}} \cdot \boldsymbol{d}). \tag{8.17}$$

Consider $KMnF_3$. The nuclear lattice is simple cubic (Fig. 8.2*a*); hence the nuclear reciprocal lattice is also simple cubic (Fig. 8.2*b*). Denote the unit-cell vectors of the nuclear reciprocal lattice by $\boldsymbol{\tau}_1$, $\boldsymbol{\tau}_2$, $\boldsymbol{\tau}_3$. A vector in the nuclear reciprocal lattice is given by

$$\boldsymbol{\tau} = t_1\boldsymbol{\tau}_1 + t_2\boldsymbol{\tau}_2 + t_3\boldsymbol{\tau}_3, \tag{8.18}$$

where t_1, t_2, t_3 are all integers. Nuclear Bragg scattering occurs when $\boldsymbol{\kappa} = \boldsymbol{\tau}$.

Fig. 8.2 (*a*) Structure of $KMnF_3$: ● Mn^{2+} (sublattice *A*), ○ Mn^{2+} (sublattice *B*), ⊘ K^+, • F^-. (*b*) Nuclear reciprocal lattice (obtained if all the Mn^{2+} ions are treated as identical). (*c*) Magnetic reciprocal lattice (obtained by taking account of the spin directions of the Mn^{2+} ions).

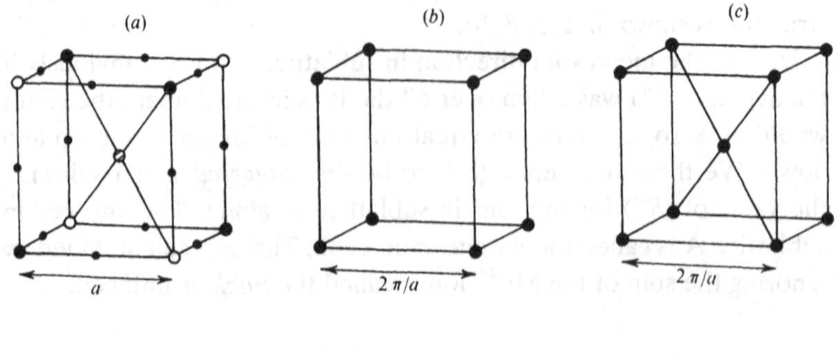

The magnetic lattice is face-centred cubic; hence the magnetic reciprocal lattice is body-centred cubic (Fig. 8.2*c*). A vector $\boldsymbol{\tau}_m$ in the magnetic reciprocal lattice has coordinates (in terms of $\boldsymbol{\tau}_1$, $\boldsymbol{\tau}_2$, $\boldsymbol{\tau}_3$)

$$t_1, t_2, t_3 \quad \text{or} \quad t_1+\tfrac{1}{2}, t_2+\tfrac{1}{2}, t_3+\tfrac{1}{2},$$

where t_1, t_2, t_3 are all integers. Summing over d in (8.17) for a pair of neighbouring A and B ions we find

$$\begin{aligned}\sum_d \sigma_d \exp(\mathrm{i}\boldsymbol{\tau}_m \,.\, \boldsymbol{d}) &= 0 \quad \text{for } \boldsymbol{\tau}_m = t_1, t_2, t_3,\\ &= 2 \quad \text{for } \boldsymbol{\tau}_m = t_1+\tfrac{1}{2}, t_2+\tfrac{1}{2}, t_3+\tfrac{1}{2}.\end{aligned} \tag{8.19}$$

Thus the nuclear and magnetic Bragg scattering occur at different points in reciprocal space. The neutron diffraction pattern (number of neutrons scattered as a function of scattering angle) for a powder

Fig. 8.3 Neutron diffraction pattern of $KMnF_3$ at 4.2 K. The numbers by the peaks are the coordinates in reciprocal space multiplied by 2. A trio of odd numbers represents a magnetic peak, and a trio of even numbers a nuclear peak. The (310) and (321) peaks are nuclear in origin and involve scattering from the K^+ and F^- ions, which at 4.2 K are slightly displaced from their positions in the ideal perovskite structure shown in Fig. 8.2*a*. The peaks marked Al arise from the aluminium sample holder. (After Scatturin *et al.*, 1961.)

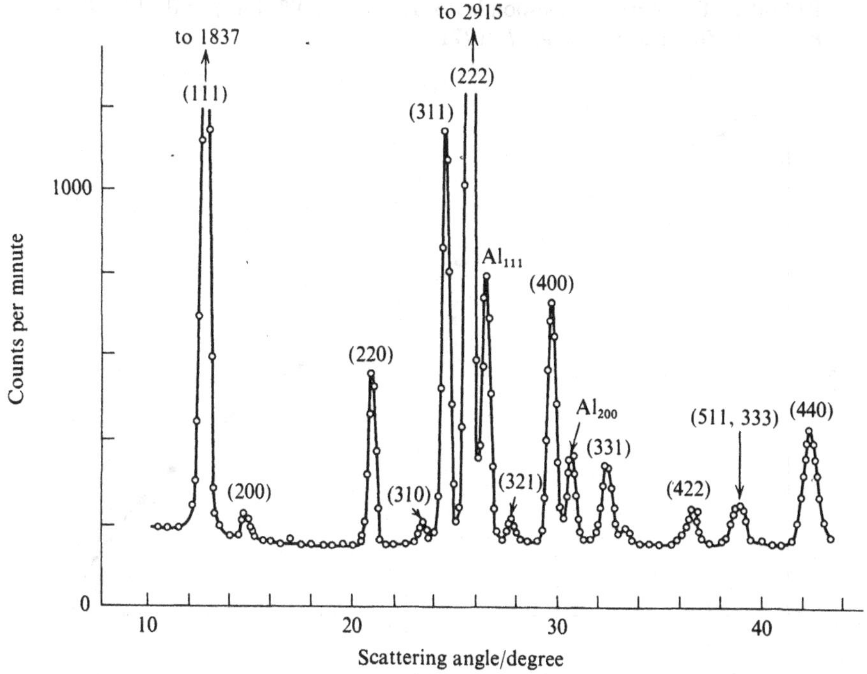

sample of $KMnF_3$ at 4.2 K is shown in Fig. 8.3. It is not always the case that nuclear and magnetic Bragg peaks occur at different points in reciprocal space for an antiferromagnet. In $MnTe_2$ for example, although the nuclear scattering from the manganese ions is zero at a magnetic Bragg peak (see Example 8.1) the nuclear scattering from the tellurium ions is not.

The staggered mean spin $\langle S^{\eta}\rangle$ varies with temperature, falling to zero at T_N, the Néel temperature. Since the intensity of a Bragg peak is proportional to $\langle S^{\eta}\rangle^2$, measurement of the intensity provides a method of measuring $\langle S^{\eta}\rangle$. Some results for $RbMnF_3$, which has the same structure as $KMnF_3$, are shown in Fig. 8.4, where the intensity of the $\frac{3}{2}, \frac{3}{2}, \frac{3}{2}$ Bragg peak is plotted as a function of temperature.

Helical arrangement of spins

Neutron diffraction has revealed the existence of other types of spin ordering besides ferromagnetic and antiferromagnetic. An example is provided by the alloy Au_2Mn where the spins have a helical

Fig. 8.4 Intensity I vs temperature T for the $(\frac{3}{2}, \frac{3}{2}, \frac{3}{2})$ Bragg reflection in $RbMnF_3$. The curve corresponds to $I \propto (T_N - T)^{2\beta}$ for $\beta = 0.318$, $T_N = 83.0$ K. (After Tucciarone *et al.*, 1971.)

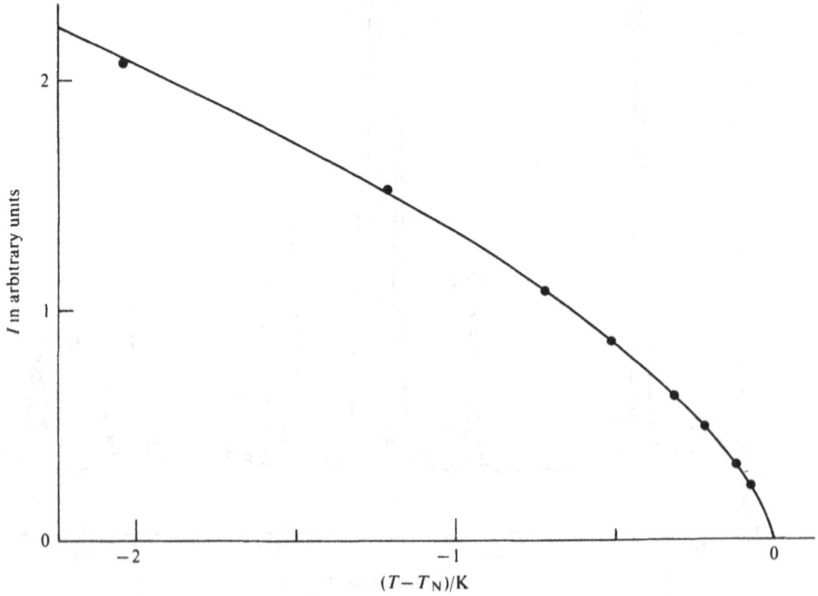

arrangement. The magnetic structure is shown in Fig. 8.5. The manganese ions form a body-centred tetragonal lattice. The spins of the atoms in a plane perpendicular to the z-axis (the tetrad) lie in the plane and point in the same direction, but this direction rotates about the z axis for successive planes of atoms. (In a general helical arrangement the axis of the helix, i.e. the axis perpendicular to the planes of constant spin direction, may not be perpendicular to the rotation plane of the spins. For a review article on helical spin ordering see Nagamiya, 1967.)

To calculate the scattering cross-section we write (8.2) in the form

$$\left(\frac{d\sigma}{d\Omega}\right)_{el} = (\gamma r_0)^2\{\tfrac{1}{2}gF(\boldsymbol{\kappa})\}^2 \exp(-2W)I, \tag{8.20}$$

where $$I = \sum_{\alpha\beta} (\delta_{\alpha\beta} - \hat{\kappa}_\alpha\hat{\kappa}_\beta) \sum_{ll'} \langle S^\alpha_{l'}\rangle\langle S^\beta_l\rangle \exp\{i\boldsymbol{\kappa}\,.\,(\boldsymbol{l}-\boldsymbol{l}')\}. \tag{8.21}$$

Fig. 8.5 Structure of Au_2Mn. Only the Mn atoms are shown. The tetrad axis is vertical.

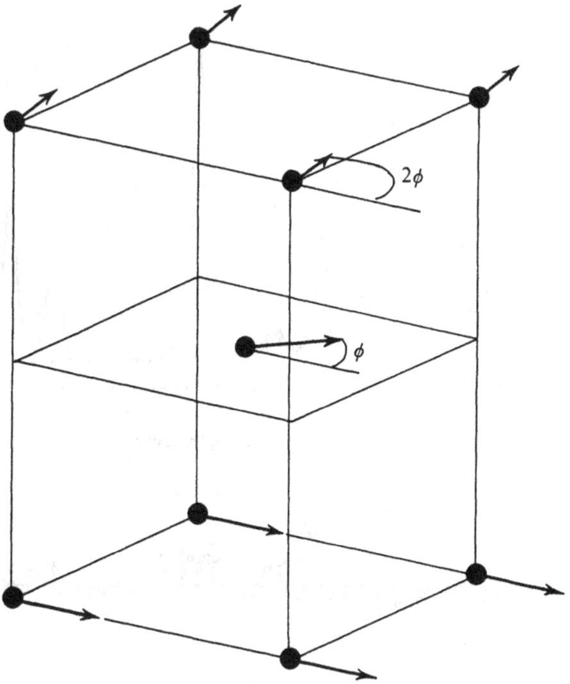

Let $\boldsymbol{Q}$ be a vector in the direction of the axis of the helix, of magnitude 2π divided by the pitch of the helix. Then the spin components have the form

$$\langle S_l^x\rangle = \langle S\rangle \cos(\boldsymbol{Q}\cdot\boldsymbol{l}), \qquad \langle S_l^y\rangle = \langle S\rangle \sin(\boldsymbol{Q}\cdot\boldsymbol{l}), \qquad \langle S_l^z\rangle = 0. \tag{8.22}$$

Inserting these values in (8.21) we obtain

$$I = \langle S\rangle^2 \sum_{ll'} \exp\{i\boldsymbol{\kappa}\cdot(\boldsymbol{l}-\boldsymbol{l}')\}[\{1-\tfrac{1}{2}(\hat{\kappa}_x^2+\hat{\kappa}_y^2)\}\cos\{\boldsymbol{Q}\cdot(\boldsymbol{l}-\boldsymbol{l}')\} + \tfrac{1}{2}(\hat{\kappa}_y^2-\hat{\kappa}_x^2)\cos\{\boldsymbol{Q}\cdot(\boldsymbol{l}+\boldsymbol{l}')\} - \hat{\kappa}_x\hat{\kappa}_y \sin\{\boldsymbol{Q}\cdot(\boldsymbol{l}+\boldsymbol{l}')\}]. \tag{8.23}$$

Unless $\boldsymbol{Q}$ is equal to a vector in the reciprocal lattice (which we can rule out as it would correspond to a collinear arrangement), the terms in $\cos\{\boldsymbol{Q}\cdot(\boldsymbol{l}+\boldsymbol{l}')\}$ and $\sin\{\boldsymbol{Q}\cdot(\boldsymbol{l}+\boldsymbol{l}')\}$ sum to zero. This follows because the sum over l and l' can be carried out by first summing over l with $\boldsymbol{l}-\boldsymbol{l}'$ constant and then summing over $\boldsymbol{l}-\boldsymbol{l}'$. The terms in

Fig. 8.6 Bragg scattering from Au_2Mn. (*a*) Spectrum at 293 K. (*b*) Difference between spectra at 293 K and 423 K. (After Herpin *et al.*, 1959. Reproduced by kind permission of Gauthier-Villars.)

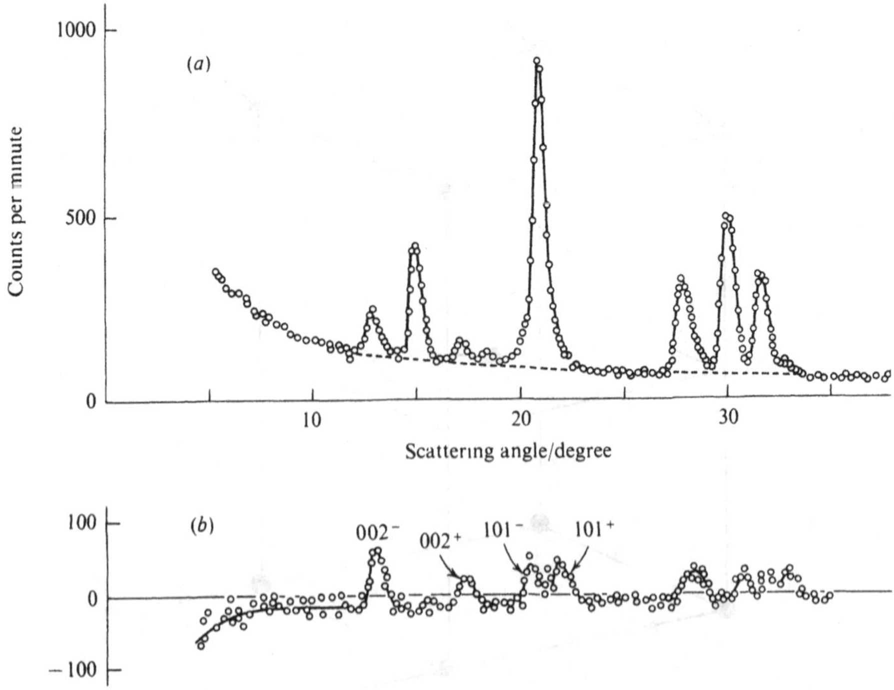

$Q.(l+l')$ give zero for the first summation. We therefore have

$$I=\langle S\rangle^2\{1-\tfrac{1}{2}(\hat{\kappa}_x^2+\hat{\kappa}_y^2)\}\sum_{ll'}\cos\{Q.(l-l')\}\exp\{i\kappa.(l-l')\}$$

$$=\tfrac{1}{4}\langle S\rangle^2(1+\hat{\kappa}_z^2)\sum_{ll'}\exp\{i(\kappa+Q).(l-l')\}+\exp\{i(\kappa-Q).(l-l')\}$$

$$=\frac{N}{4}\frac{(2\pi)^3}{v_0}\langle S\rangle^2(1+\hat{\kappa}_z^2)\sum_{\tau}\{\delta(\kappa+Q-\tau)+\delta(\kappa-Q-\tau)\}, \qquad (8.24)$$

which combined with (8.20) gives the cross-section.

Eq. (8.24) shows that magnetic Bragg scattering occurs when

$$\kappa=\tau\pm Q. \qquad (8.25)$$

Thus each nuclear Bragg peak $\kappa=\tau$ is flanked by a pair of magnetic satellites. To label the reflections we choose an orthogonal unit cell in the crystal, defined by the corner atoms in Fig. 8.5. If τ_1, τ_2, τ_3 are the unit vectors in the corresponding reciprocal lattice, the coordinates of κ for a nuclear peak are three integers t_1, t_2, t_3, where $t_1+t_2+t_3$ is even. (A body-centred tetragonal unit cell has a face-centred tetragonal reciprocal lattice.) The vector Q is in the z direction. If ϕ is the angle between the directions of the spins in successive planes (Fig. 8.5), it follows from (8.25) and the definition of Q that the coordinates of κ for a magnetic reflection are t_1, t_2, $t_3\pm\phi/\pi$.

The Néel temperature for Au_2Mn is 363 K. Herpin *et al.* (1959) measured the Bragg scattering at 293 K and 423 K. The former contains the nuclear and magnetic peaks, while the latter contains only the nuclear. The difference between the two sets of measurements thus gives the magnetic peaks. The results are shown in Fig. 8.6 and correspond to $\phi=51°$.

8.2 Scattering by spin waves

As an example of inelastic magnetic scattering of neutrons we consider scattering by spin waves in a ferromagnet.† We first review the main features of the results, taking a Bravais crystal and assuming a model of localised electrons. If each atom has spin S, the magnitude of its spin angular momentum is $\sqrt{S(S+1)}\,\hbar$. Any component of the

† For a comprehensive treatment of spin-wave theory see Keffer (1966). For further discussion and experimental results of neutron scattering by spin waves see Lowde (1965), Shirane *et al.* (1968), Marshall and Lovesey (1971), and Houmann and Møller (1976).

spin angular momentum has the value $M\hbar$, where $M = S, S-1, \ldots, -S$. As before we start by considering a single domain and take the mean spin direction as the z axis and axis of quantisation. At zero temperature all the spins are aligned, so $M = S$ for all the atoms. At finite temperatures other values of M occur, and the departure from $M = S$ is termed a *spin deviation.*

The spin deviations may be represented by the sum of the deviations due to a set of travelling sinusoidal waves. These are the spin waves. Although the spin deviations are discrete, the spin waves have continuous displacements. There is no inconsistency here, because the displacement of a spin wave, when squared, gives the *probability* of a spin deviation, and this is continuous.

The spin wave is the analogue of the normal mode of nuclear displacements. It is quantised, and its energy, relative to the ground state, is $n\hbar\omega$, where ω is the angular frequency of the wave, and n is an integer. The quantum of energy $\hbar\omega$ is known as a *magnon.* Just as in the nuclear case, the neutron can be scattered by a process in which $n \to n \pm 1$ for a particular spin wave, the change in energy being taken up by the kinetic energy of the neutron.

Linear spin-wave theory

We take the Hamiltonian of the magnetic system to be

$$H = -\sum_{ll'} J(\boldsymbol{l} - \boldsymbol{l}')\boldsymbol{S}_l \cdot \boldsymbol{S}_{l'}, \tag{8.26}$$

where $\boldsymbol{S}_l$ is the spin angular momentum operator in units of $\hbar$ for atom l. The quantity

$$J(\boldsymbol{l} - \boldsymbol{l}') = J(\boldsymbol{l}' - \boldsymbol{l}) \tag{8.27}$$

is known as the *exchange energy* or *exchange integral.* The form of the Hamiltonian in (8.26) was first put forward by Heisenberg and arises from the electrostatic interaction between the atoms, together with the requirement of antisymmetry for the electron wave functions. The sum in (8.26) is over all l, l' pairs, i.e. l, l' and l', l are counted as separate terms. (They are equal.) $J(0)$ is defined to be zero.

The exchange integral can be positive or negative. In a ferromagnet it is positive for neighbouring atoms. Therefore, to make the energy of the crystal as small as possible, neighbouring spins tend to align in the same direction. In an antiferromagnet the exchange integral is negative for neighbouring ions; so neighbouring spins tend to align in opposite directions.

Angular momentum operators. We summarise some results from quantum mechanics (see Appendix F). Let S_l^x, S_l^y, S_l^z be the operators for the atom l corresponding to the x, y, z components of spin angular momentum in units of $\hbar$. We restrict our attention for the moment to a specific atom and drop the subscript l. Denote the eigenstate of S^z with eigenvalue M by $|M\rangle$. We define operators

$$S^+ = S^x + \mathrm{i}S^y, \qquad S^- = S^x - \mathrm{i}S^y. \tag{8.28}$$

Then

$$\begin{aligned} S^+|M\rangle &= \{(S-M)(S+M+1)\}^{1/2}|M+1\rangle, \\ S^-|M\rangle &= \{(S+M)(S-M+1)\}^{1/2}|M-1\rangle. \end{aligned} \tag{8.29}$$

Thus S^+ converts the state $|M\rangle$ into the state $|M+1\rangle$, and S^- converts $|M\rangle$ into $|M-1\rangle$. Put

$$n = S - M. \tag{8.30}$$

This simply renumbers the states in terms of the spin deviations, and from now on the number inside the symbol $|\ \rangle$ refers to n. Eqs. (8.29) become

$$\begin{aligned} S^+|n\rangle &= (2Sn)^{1/2}\left(1-\frac{n-1}{2S}\right)^{1/2}|n-1\rangle, \\ S^-|n\rangle &= \{2S(n+1)\}^{1/2}\left(1-\frac{n}{2S}\right)^{1/2}|n+1\rangle. \end{aligned} \tag{8.31}$$

We define a pair of annihilation and creation operators for the spin deviations on the atom l by the relations

$$a|n\rangle = n^{1/2}|n-1\rangle, \qquad a^+|n\rangle = (n+1)^{1/2}|n+1\rangle. \tag{8.32}$$

Then

$$aa^+|n\rangle = (n+1)|n\rangle, \qquad a^+a|n\rangle = n|n\rangle, \tag{8.33}$$

and

$$[a, a^+] = 1. \tag{8.34}$$

These operators were first introduced into spin-wave theory by Holstein and Primakoff (1940). The operator a changes $|n\rangle$ into $|n-1\rangle$, so it produces the same effect as S^+ – only the constant multiplying the resulting function is different. Similarly a^+ produces the same effect as S^-.

Linear approximation. We now make an approximation, which is to neglect the terms $(n-1)/2S$ and $n/2S$ in (8.31). Thus

$$S_l^+ = (2S)^{1/2}a_l, \qquad S_l^- = (2S)^{1/2}a_l^+. \tag{8.35}$$

This is known as the *linear approximation*. We shall see its significance in the expression it gives for the Hamiltonian.

The linear approximation is a good one for $n \ll S$, i.e. for small spin deviations. In fact it is correct for S^+ operating on $|0\rangle$ and $|1\rangle$, and for S^- operating on $|0\rangle$. It gets worse as $n \to S$ and is completely wrong for $n > 2S$. The correct S^- operating on $|2S\rangle$ annihilates it, as it must do because $|2S\rangle$ is the state corresponding to the maximum possible spin deviation. But the approximate S^- converts $|2S\rangle$ into $|2S+1\rangle$, $|2S+1\rangle$ into $|2S+2\rangle$, and so on. It thus generates a non-existent set of eigenfunctions.

We shall require an expression for S_l^z. Since

$$S_l^z|n\rangle = (S-n)|n\rangle, \tag{8.36}$$

$$S_l^z = S - a_l^+ a_l. \tag{8.37}$$

From (8.35)

$$S_l^z = S - \frac{1}{2S} S_l^- S_l^+. \tag{8.38}$$

(8.37) is exact; (8.38) is approximate.

We introduce operators $b_{\boldsymbol{q}}$ and $b_{\boldsymbol{q}}^+$ by the Fourier expansions

$$\left.\begin{aligned} a_l &= N^{-1/2} \sum_{\boldsymbol{q}} \exp(\mathrm{i}\boldsymbol{q} \,.\, \boldsymbol{l}) b_{\boldsymbol{q}}, \\ a_l^+ &= N^{-1/2} \sum_{\boldsymbol{q}} \exp(-\mathrm{i}\boldsymbol{q} \,.\, \boldsymbol{l}) b_{\boldsymbol{q}}^+, \end{aligned}\right\} \tag{8.39}$$

where N is the number of atoms in the crystal. The sum over $\boldsymbol{q}$ runs over the N points in the 1st Brillouin zone, corresponding to waves periodic in the crystal.† The relations reciprocal to (8.39) are

$$\left.\begin{aligned} b_{\boldsymbol{q}} &= N^{-1/2} \sum_{l} \exp(-\mathrm{i}\boldsymbol{q} \,.\, \boldsymbol{l}) a_l, \\ b_{\boldsymbol{q}}^+ &= N^{-1/2} \sum_{l} \exp(\mathrm{i}\boldsymbol{q} \,.\, \boldsymbol{l}) a_l^+. \end{aligned}\right\} \tag{8.40}$$

They are obtained from (8.39) and (A.18) thus:

$$N^{-1/2} \sum_{l} \exp(-\mathrm{i}\boldsymbol{q}' \,.\, \boldsymbol{l}) a_l = N^{-1} \sum_{l\boldsymbol{q}} \exp\{\mathrm{i}(\boldsymbol{q}-\boldsymbol{q}') \,.\, \boldsymbol{l}\} b_{\boldsymbol{q}} = b_{\boldsymbol{q}'}. \tag{8.41}$$

Similarly for $b_{\boldsymbol{q}}^+$. The commutation relations for a_l and a_l^+ are

$$[a_l, a_{l'}^+] = \delta_{ll'}. \tag{8.42}$$

† A crystal with N atoms has $3N$ normal modes for the nuclear displacements. For a spin wave there is no counterpart to the polarisation of a normal mode. Thus there are only N spin waves.

The commutation relations for b_q and b_q^+ are obtained from (8.40), (8.42), and (A.18).

$$[b_q, b_{q'}^+] = \frac{1}{N}\left[\sum_l \exp(-\mathrm{i}\boldsymbol{q}\,.\,\boldsymbol{l})a_l, \sum_{l'} \exp(\mathrm{i}\boldsymbol{q}'\,.\,\boldsymbol{l}')a_{l'}^+\right]$$

$$= \frac{1}{N}\sum_l \exp\{\mathrm{i}(\boldsymbol{q}'-\boldsymbol{q})\,.\,\boldsymbol{l}\} = \delta_{qq'}. \tag{8.43}$$

We now use the linear approximation to derive an expression for the Hamiltonian in terms of the operators b_q, b_q^+. From (8.28)

$$\boldsymbol{S}_l\,.\,\boldsymbol{S}_{l'} = \tfrac{1}{2}(S_l^+S_{l'}^- + S_{l'}^+S_l^-) + S_l^zS_{l'}^z. \tag{8.44}$$

There are no terms with $l = l'$ in (8.26), and for $l \neq l'$ any S_l operator commutes with any $S_{l'}$ operator. We may therefore write

$$H = -\sum_{ll'} J(\boldsymbol{l}-\boldsymbol{l}')(S_l^+S_{l'}^- + S_l^zS_{l'}^z). \tag{8.45}$$

The sum of the ll' and $l'l$ terms, which have the same values for $J(\boldsymbol{l}-\boldsymbol{l}')$, correctly gives the sum of the corresponding terms in (8.26).

Substitute the expressions for S_l^+, S_l^-, S_l^z given in (8.35) and (8.37), and ignore terms in the 4th power of a. (This is correct for the linear approximation where such terms are implicitly omitted in S_l^+, $S_{l'}^-$.) This gives

$$H = -NS^2\mathcal{J}(0) + 2S\mathcal{J}(0)\sum_l a_l^+a_l - 2S\sum_{ll'} J(\boldsymbol{l}-\boldsymbol{l}')a_la_{l'}^+, \tag{8.46}$$

where
$$\mathcal{J}(\boldsymbol{q}) = \sum_{\boldsymbol{\rho}} J(\boldsymbol{\rho})\exp(\mathrm{i}\boldsymbol{q}\,.\,\boldsymbol{\rho}), \tag{8.47}$$

and
$$\boldsymbol{\rho} = \boldsymbol{l}-\boldsymbol{l}'. \tag{8.48}$$

$\sum_{\boldsymbol{\rho}}$ means sum over l' with l constant. From (8.39) and (A.18)

$$\sum_l a_l^+a_l = \frac{1}{N}\sum_{qq'}\sum_l \exp\{\mathrm{i}(\boldsymbol{q}'-\boldsymbol{q})\,.\,\boldsymbol{l}\}b_q^+b_{q'} = \sum_q b_q^+b_q. \tag{8.49}$$

Similarly

$$\sum_{ll'} J(\boldsymbol{l}-\boldsymbol{l}')a_la_{l'}^+$$

$$= \frac{1}{N}\sum_{qq'}\sum_{\boldsymbol{\rho}} J(\boldsymbol{\rho})\exp(\mathrm{i}\boldsymbol{q}'\,.\,\boldsymbol{\rho})\sum_l \exp\{\mathrm{i}(\boldsymbol{q}-\boldsymbol{q}')\,.\,\boldsymbol{l}\}b_qb_{q'}^+$$

$$= \sum_q \mathcal{J}(\boldsymbol{q})b_qb_q^+ = \sum_q \mathcal{J}(\boldsymbol{q})b_q^+b_q. \tag{8.50}$$

The last step follows from (8.43), together with the result

$$\sum_q \mathcal{J}(\boldsymbol{q}) = \sum_{\boldsymbol{\rho}} J(\boldsymbol{\rho})\sum_q \exp(\mathrm{i}\boldsymbol{q}\,.\,\boldsymbol{\rho}) = 0. \tag{8.51}$$

Inserting (8.49) and (8.50) in (8.46) gives

$$H = H^0 + \sum_{\boldsymbol{q}} \hbar\omega_{\boldsymbol{q}} b_{\boldsymbol{q}}^+ b_{\boldsymbol{q}}, \tag{8.52}$$

where

$$H^0 = -S^2 N \mathscr{J}(0) = -S^2 N \sum_{\boldsymbol{\rho}} J(\boldsymbol{\rho}), \tag{8.53}$$

and

$$\hbar\omega_{\boldsymbol{q}} = 2S\{\mathscr{J}(0) - \mathscr{J}(\boldsymbol{q})\}. \tag{8.54}$$

This is the required expression for the Hamiltonian. It is the sum of a set of terms, each one depending on a particular $\boldsymbol{q}$. There are no cross-terms like $b_{\boldsymbol{q}}^+ b_{\boldsymbol{q}'}$ where $\boldsymbol{q} \neq \boldsymbol{q}'$. The form of the Hamiltonian in (8.52) is identical to that in (G.31) for the nuclear displacements of a crystal with harmonic forces. Moreover, the commutation relations in (8.43) for the operators $b_{\boldsymbol{q}}$, $b_{\boldsymbol{q}}^+$ are identical to those in (G.30) for the operators a_s, a_s^+. All the results for a_s, a_s^+ that follow from the commutation relations and the form of the Hamiltonian (see Appendix E.1) therefore apply to the operators $b_{\boldsymbol{q}}$, $b_{\boldsymbol{q}}^+$. Thus

$$[b_{\boldsymbol{q}}, H] = \hbar\omega_{\boldsymbol{q}} b_{\boldsymbol{q}}, \qquad [b_{\boldsymbol{q}}^+, H] = -\hbar\omega_{\boldsymbol{q}} b_{\boldsymbol{q}}^+, \tag{8.55}$$

$$b_{\boldsymbol{q}}(t) = b_{\boldsymbol{q}} \exp(-\mathrm{i}\omega_{\boldsymbol{q}} t), \qquad b_{\boldsymbol{q}}^+(t) = b_{\boldsymbol{q}}^+ \exp(\mathrm{i}\omega_{\boldsymbol{q}} t). \tag{8.56}$$

If $|n\rangle$ is an eigenfunction of $H_{\boldsymbol{q}} = \hbar\omega_{\boldsymbol{q}} b_{\boldsymbol{q}}^+ b_{\boldsymbol{q}}$, then

$$b_{\boldsymbol{q}}|n\rangle = n^{1/2}|n-1\rangle, \qquad b_{\boldsymbol{q}}^+|n\rangle = (n+1)^{1/2}|n+1\rangle, \tag{8.57}$$

$$b_{\boldsymbol{q}} b_{\boldsymbol{q}}^+|n\rangle = (n+1)|n\rangle, \qquad b_{\boldsymbol{q}}^+ b_{\boldsymbol{q}}|n\rangle = n|n\rangle, \tag{8.58}$$

$$H_{\boldsymbol{q}}|n\rangle = n\hbar\omega_{\boldsymbol{q}}|n\rangle. \tag{8.59}$$

We see then that the Hamiltonian in (8.52) represents a set of independent sinusoidal waves – the spin waves. The operators $b_{\boldsymbol{q}}$ and $b_{\boldsymbol{q}}^+$ are the annihilation and creation operators for the wave $\boldsymbol{q}$. The fact that the spin waves are independent, i.e. non-interacting, is a consequence of the linear approximation, just as the independence of the normal modes for nuclear displacements is a consequence of the harmonic approximation.

Eqs. (8.39) and (8.56) give the time-varying form of the operators a_l and a_l^+

$$\begin{aligned} a_l(t) &= N^{-1/2} \sum_{\boldsymbol{q}} \exp\{\mathrm{i}(\boldsymbol{q}\,.\,\boldsymbol{l} - \omega_{\boldsymbol{q}} t)\} b_{\boldsymbol{q}}, \\ a_l^+(t) &= N^{-1/2} \sum_{\boldsymbol{q}} \exp\{-\mathrm{i}(\boldsymbol{q}\,.\,\boldsymbol{l} - \omega_{\boldsymbol{q}} t)\} b_{\boldsymbol{q}}^+. \end{aligned} \tag{8.60}$$

Magnon dispersion relation. The term H^0 in (8.52) represents the energy of the ground state of the spin system. It is the state when all the atoms have $M = S$. The energy of the spin wave with wavevector $\boldsymbol{q}$

is $n\hbar\omega_{\boldsymbol{q}}$, where

$$\hbar\omega_{\boldsymbol{q}} = 2S\{\mathscr{J}(0) - \mathscr{J}(\boldsymbol{q})\}. \tag{8.61}$$

This equation is the dispersion relation for spin waves.

Consider the simple case where the exchange integral has the constant value J for nearest neighbours and is zero for the rest. Let the number of nearest neighbours be r. Then

$$\mathscr{J}(0) = rJ, \tag{8.62}$$

and

$$\mathscr{J}(\boldsymbol{q}) = rJ\gamma_{\boldsymbol{q}}, \tag{8.63}$$

where

$$\gamma_{\boldsymbol{q}} = \frac{1}{r}\sum_{\boldsymbol{\rho}} \exp(\mathrm{i}\boldsymbol{q}\,.\,\boldsymbol{\rho}). \tag{8.64}$$

Suppose the crystal has a simple cubic structure with cube side a. For $qa \ll 1$ it can be shown (Example 8.3) that

$$\hbar\omega_{\boldsymbol{q}} = Dq^2, \tag{8.65}$$

where

$$D = 2JSa^2. \tag{8.66}$$

The form of the dispersion relation in (8.65) does not depend on a particular model, but is generally true at small q for a ferromagnet. The constant D is known as the *stiffness constant.* For a cubic crystal, D has the same value for all directions of $\boldsymbol{q}$; for a non-cubic crystal it varies with direction. These results may be contrasted with the results for phonons. For the latter, $\omega_{\boldsymbol{q}}$ is proportional to q at small q, and the constant of proportionality depends on the direction of $\boldsymbol{q}$ even for a cubic crystal.

One-magnon cross-sections

We now evaluate the matrix elements $\langle S_0^\alpha(0)S_l^\beta(t)\rangle$ in (7.73). From (8.35), (8.37), and (8.60)

$$\left.\begin{aligned} S_l^+(t) &= \left(\frac{2S}{N}\right)^{1/2} \sum_{\boldsymbol{q}} \exp\{\mathrm{i}(\boldsymbol{q}\,.\,\boldsymbol{l} - \omega_{\boldsymbol{q}}t)\}b_{\boldsymbol{q}}, \\ S_l^-(t) &= \left(\frac{2S}{N}\right)^{1/2} \sum_{\boldsymbol{q}} \exp\{-\mathrm{i}(\boldsymbol{q}\,.\,\boldsymbol{l} - \omega_{\boldsymbol{q}}t)\}b_{\boldsymbol{q}}^+, \end{aligned}\right\} \tag{8.67}$$

$$S_l^z(t) = S - \frac{1}{N}\sum_{\boldsymbol{q}\boldsymbol{q}'} \exp[\mathrm{i}\{(\boldsymbol{q}' - \boldsymbol{q})\,.\,\boldsymbol{l} - (\omega_{\boldsymbol{q}'} - \omega_{\boldsymbol{q}})t\}]b_{\boldsymbol{q}}^+ b_{\boldsymbol{q}'}. \tag{8.68}$$

Consider the matrix element $\langle\lambda|S_0^\alpha(0)S_l^\beta(t)|\lambda\rangle$. Because the Hamiltonian of the spin system has the form $H = H^0 + \sum_{\boldsymbol{q}} H_{\boldsymbol{q}}$, (8.52), the state $|\lambda\rangle$, which is an eigenfunction of H, is the product of eigenfunctions of the operators $H_{\boldsymbol{q}}$. By virtue of the relations in (8.57), the matrix element is zero unless, for each spin wave $\boldsymbol{q}$, there is an equal number of $b_{\boldsymbol{q}}$ and $b_{\boldsymbol{q}}^{+}$ in the operator. (The argument is identical to the one used for normal modes – see Appendix E.1.) Therefore

$$\langle + + \rangle = \langle - - \rangle = \langle + z \rangle = \langle z + \rangle = \langle - z \rangle = \langle z - \rangle = 0, \tag{8.69}$$

where $\langle \alpha\beta \rangle$ stands for $\langle \lambda | S_0^{\alpha}(0) S_l^{\beta}(t) | \lambda \rangle$. For example, the matrix element $\langle + + \rangle$ is – from (8.67) – the sum of terms like $\langle \lambda | b_{\boldsymbol{q}} b_{\boldsymbol{q}'} | \lambda \rangle$, and this is zero for all values of $\boldsymbol{q}, \boldsymbol{q}'$. The only non-zero matrix elements are

$$\langle z\, z \rangle, \qquad \langle + - \rangle, \qquad \langle - + \rangle.$$

We consider them in turn.

From (8.37)

$$\langle \lambda | S_0^z(0) S_l^z(t) | \lambda \rangle = \langle \lambda | S^2 - S a_0^{+}(0) a_0(0) - S a_l^{+}(t) a_l(t) | \lambda \rangle \tag{8.70}$$

in the linear approximation when the term in the 4th power of a is neglected. From (8.60)

$$\begin{aligned} \langle \lambda | a_l^{+}(t) a_l(t) | \lambda \rangle &= \frac{1}{N} \sum_{\boldsymbol{q}\boldsymbol{q}'} \exp[\mathrm{i}\{(\boldsymbol{q}' - \boldsymbol{q}) \,.\, \boldsymbol{l} - (\omega_{\boldsymbol{q}'} - \omega_{\boldsymbol{q}})t\}] \langle \lambda | b_{\boldsymbol{q}}^{+} b_{\boldsymbol{q}'} | \lambda \rangle \\ &= \frac{1}{N} \sum_{\boldsymbol{q}} \langle \lambda | b_{\boldsymbol{q}}^{+} b_{\boldsymbol{q}} | \lambda \rangle, \end{aligned} \tag{8.71}$$

since the only non-zero terms are those with $\boldsymbol{q} = \boldsymbol{q}'$. Thus $\langle \lambda | a_l^{+}(t) a_l(t) | \lambda \rangle$ is independent of l and t. From the last two equations and (8.58) we have

$$\langle S_0^z(0) S_l^z(t) \rangle = S^2 - \frac{2S}{N} \sum_{\boldsymbol{q}} \langle n_{\boldsymbol{q}} \rangle. \tag{8.72}$$

This expression is independent of t and therefore gives rise to elastic scattering. It is in fact the scattering we considered in Section 8.1.

From (8.67)

$$\langle \lambda | S_0^{+}(0) S_l^{-}(t) | \lambda \rangle = \frac{2S}{N} \sum_{\boldsymbol{q}} \exp\{-\mathrm{i}(\boldsymbol{q} \,.\, \boldsymbol{l} - \omega_{\boldsymbol{q}} t)\} \langle \lambda | b_{\boldsymbol{q}} b_{\boldsymbol{q}}^{+} | \lambda \rangle. \tag{8.73}$$

Thus

$$\langle S_0^{+}(0) S_l^{-}(t) \rangle = \frac{2S}{N} \sum_{\boldsymbol{q}} \exp\{-\mathrm{i}(\boldsymbol{q} \,.\, \boldsymbol{l} - \omega_{\boldsymbol{q}} t)\} \langle n_{\boldsymbol{q}} + 1 \rangle. \tag{8.74}$$

Similarly

$$\langle S_0^{-}(0) S_l^{+}(t) \rangle = \frac{2S}{N} \sum_{\boldsymbol{q}} \exp\{\mathrm{i}(\boldsymbol{q} \,.\, \boldsymbol{l} - \omega_{\boldsymbol{q}} t)\} \langle n_{\boldsymbol{q}} \rangle. \tag{8.75}$$

Expressing S_l^x and S_l^y in terms of S_l^+ and S_l^- we have

$$\langle S_0^x(0)S_l^x(t)\rangle = \langle S_0^y(0)S_l^y(t)\rangle$$
$$= \frac{S}{2N}\sum_{\boldsymbol{q}} \exp\{-\mathrm{i}(\boldsymbol{q}\,.\,\boldsymbol{l}-\omega_{\boldsymbol{q}}t)\}\langle n_{\boldsymbol{q}}+1\rangle + \exp\{\mathrm{i}(\boldsymbol{q}\,.\,\boldsymbol{l}-\omega_{\boldsymbol{q}}t)\}\langle n_{\boldsymbol{q}}\rangle, \tag{8.76}$$

$$\langle S_0^x(0)S_l^y(t)\rangle = -\langle S_0^y(0)S_l^x(t)\rangle$$
$$= \frac{\mathrm{i}S}{2N}\sum_{\boldsymbol{q}} \exp\{-\mathrm{i}(\boldsymbol{q}\,.\,\boldsymbol{l}-\omega_{\boldsymbol{q}}t)\}\langle n_{\boldsymbol{q}}+1\rangle - \exp\{\mathrm{i}(\boldsymbol{q}\,.\,\boldsymbol{l}-\omega_{\boldsymbol{q}}t)\}\langle n_{\boldsymbol{q}}\rangle. \tag{8.77}$$

The preceding reasoning shows that it is the terms with $\alpha, \beta = x, y$ in (7.73) which gives the inelastic scattering. They are sometimes called the *transverse* terms. The term with $\alpha=\beta=z$ is called the *longitudinal* term. The terms $\alpha=x$, $\beta=y$, and $\alpha=y$, $\beta=x$ are multiplied by the same factor $\hat{\kappa}_x\hat{\kappa}_y$. From (8.77) they sum to zero. We are left with the terms $\alpha=\beta=x$, and $\alpha=\beta=y$. Now

$$1-\hat{\kappa}_x^2+1-\hat{\kappa}_y^2 = 1+\hat{\kappa}_z^2. \tag{8.78}$$

From (7.73), (8.76), and (8.78), the inelastic magnetic cross-section for a single domain is

$$\frac{\mathrm{d}^2\sigma}{\mathrm{d}\Omega\,\mathrm{d}E'} = (\gamma r_0)^2\frac{k'}{k}\frac{1}{2\pi\hbar}\frac{1}{2}S(1+\hat{\kappa}_z^2)\{\tfrac{1}{2}gF(\boldsymbol{\kappa})\}^2\exp(-2W)\sum_{\boldsymbol{l}}\exp(\mathrm{i}\boldsymbol{\kappa}\,.\,\boldsymbol{l})$$
$$\times\int_{-\infty}^{\infty}\sum_{\boldsymbol{q}}[\exp\{-\mathrm{i}(\boldsymbol{q}\,.\,\boldsymbol{l}-\omega_{\boldsymbol{q}}t)\}\langle n_{\boldsymbol{q}}+1\rangle$$
$$+\exp\{\mathrm{i}(\boldsymbol{q}\,.\,\boldsymbol{l}-\omega_{\boldsymbol{q}}t)\}\langle n_{\boldsymbol{q}}\rangle]\exp(-\mathrm{i}\omega t)\,\mathrm{d}t$$
$$= (\gamma r_0)^2\frac{k'}{k}\frac{(2\pi)^3}{v_0}\frac{1}{2}S(1+\hat{\kappa}_z^2)\{\tfrac{1}{2}gF(\boldsymbol{\kappa})\}^2\exp(-2W)$$
$$\times\sum_{\boldsymbol{\tau},\boldsymbol{q}}\{\delta(\boldsymbol{\kappa}-\boldsymbol{q}-\boldsymbol{\tau})\delta(\hbar\omega_{\boldsymbol{q}}-\hbar\omega)\langle n_{\boldsymbol{q}}+1\rangle$$
$$+\delta(\boldsymbol{\kappa}+\boldsymbol{q}-\boldsymbol{\tau})\delta(\hbar\omega_{\boldsymbol{q}}+\hbar\omega)\langle n_{\boldsymbol{q}}\rangle\}. \tag{8.79}$$

The thermal average of $n_{\boldsymbol{q}}$ is

$$\langle n_{\boldsymbol{q}}\rangle = \{\exp(\hbar\omega_{\boldsymbol{q}}\beta)-1\}^{-1}. \tag{8.80}$$

The cross-section in (8.79) is the sum of two terms, the first corresponding to the creation and the second to the annihilation of one magnon. Consider the first term. The quantum number of the spin wave $\boldsymbol{q}$ increases by unity. The two δ-function terms in the cross-section show that the conditions for the process are

$$\frac{\hbar^2}{2m}(k^2 - k'^2) = \hbar\omega_{\boldsymbol{q}}, \tag{8.81}$$

$$\boldsymbol{k} - \boldsymbol{k}' = \boldsymbol{\tau} + \boldsymbol{q}. \tag{8.82}$$

The spin wave gains energy $\hbar\omega_{\boldsymbol{q}}$ and the kinetic energy of the neutron decreases by this amount. In addition the initial and final wavevectors of the neutron have to satisfy (8.82). In magnon annihilation the quantum number of the spin wave decreases by unity and the resulting decrease in energy is taken up by the neutron.

Eqs. (8.81) and (8.82) are identical with the equations for scattering with the creation of one phonon, and thus give rise to the same type of velocity spectrum for the scattered neutrons. With monoenergetic neutrons incident on a single crystal the velocity spectrum of the neutrons scattered in a given direction contains peaks. Measurement of $\boldsymbol{k}$ and $\boldsymbol{k}'$ for the peak enables the $\omega_{\boldsymbol{q}}$, $\boldsymbol{q}$ values of the spin wave to be calculated. In this way the magnon dispersion relation may be determined.

Some results for the rare-earth metal gadolinium are shown in Fig. 8.7. A model based on localised magnetic moments provides a good description of the magnetic properties of the rare-earth metals. In general the ions of these metals have unquenched orbital angular momentum, which produces anisotropy terms in the Hamiltonian. However, for gadolinium the orbital angular momentum is zero, and the Hamiltonian is well represented by the simple Heisenberg expression in (8.26).

Gadolinium, like the other heavy rare-earth metals, has the hcp (hexagonal close-packed) structure, which has two atoms per unit cell. Hence there is an optic as well as an acoustic branch for the magnon modes. The hcp structure may be regarded as two interpenetrating Bravais lattices. The dispersion relation in (8.61), derived for a single Bravais lattice, becomes, for the hcp structure†

$$\hbar\omega_{\boldsymbol{q}} = 2S\{\mathscr{J}(0) - \mathscr{J}(\boldsymbol{q}) + \mathscr{J}'(0) \pm |\mathscr{J}'(\boldsymbol{q})|\}, \tag{8.83}$$

where $\mathscr{J}(\boldsymbol{q})$ is the quantity defined in (8.47) with l and l' in the same Bravais lattice, and $\mathscr{J}'(\boldsymbol{q})$ is the same quantity with l and l' in different lattices. The positive sign in (8.83) gives the optic branch and the negative sign the acoustic branch.

† See Cooper (1968), p. 421.

For the transition metals a model based on localised magnetic moments would not seem appropriate. However, the Heisenberg Hamiltonian does appear to give a reasonable phenomenological description of spin waves in a transition metal, at any rate for small q. The question of the extent to which localised and itinerant electron models can account for the magnetic properties of the transition elements has been the source of much discussion – see Izuyama *et al.* (1963), Herring (1966), and Cooke (1976).

We return to (8.79), the expression for the inelastic cross-section for a single domain. In the absence of a magnetic field, the mean spin direction $\hat{\boldsymbol{\eta}}$ varies from one domain to another. For a cubic crystal the value of $(1+\hat{\kappa}_\eta^2)$ averaged over many domains is

$$(1+\hat{\kappa}_\eta^2)_{\text{av}} = \tfrac{4}{3}. \tag{8.84}$$

A sufficiently strong magnetic field $\boldsymbol{B}$ causes the mean spin directions to become aligned. Then

$$\begin{aligned} 1+\hat{\kappa}_\eta^2 &= 1 \quad \text{for } \boldsymbol{B} \perp \boldsymbol{\kappa}, \\ &= 2 \quad \text{for } \boldsymbol{B} \| \boldsymbol{\kappa}. \end{aligned} \tag{8.85}$$

Fig. 8.7 Magnon frequencies in gadolinium at 78 K along the direction ΓM and $MK\Gamma$ (see Fig. 3.14) and ΓA, the hexad axis. (After Koehler *et al.*, 1970.)

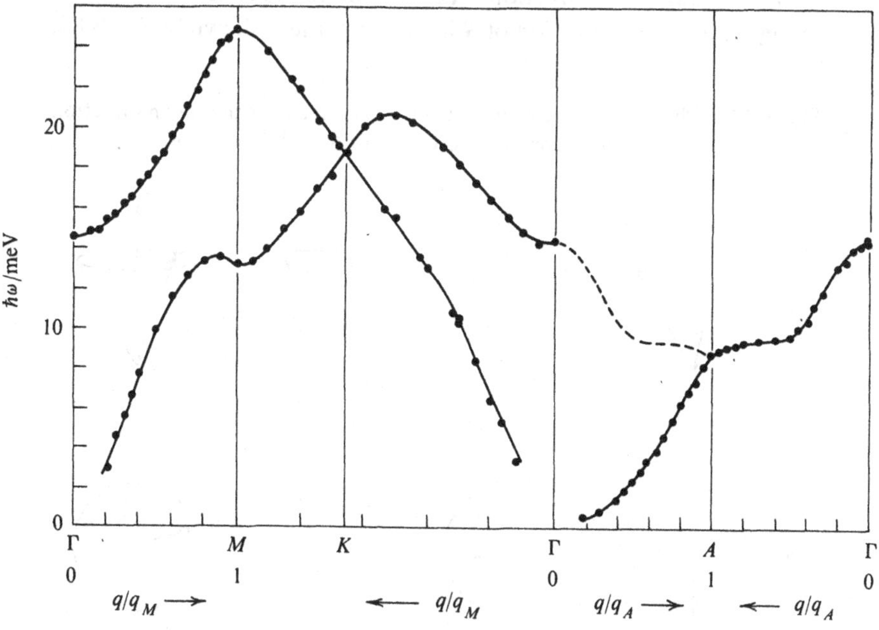

Thus the application of a magnetic field causes the cross-section for one-magnon scattering to vary by a factor of 2. This provides a method of distinguishing one-magnon, magnetovibrational, and nuclear phonon scattering. We recall that the magnetovibrational cross-section contains the factor $1-\hat{\kappa}_\eta^2$, and is therefore zero when $\boldsymbol{B}\|\boldsymbol{\kappa}$. A magnetic field has no effect on nuclear phonon scattering.

In the linear approximation only zero-magnon (elastic) and one-magnon processes occur. We have previously noted that the nuclear equivalent to the linear approximation is the harmonic approximation. However, in contrast to the magnetic case, this approximation leads to 0, 1, 2, . . . phonon processes.

Physical picture of a spin wave

The expressions for the matrix elements in (8.76) and (8.77) suggest a physical picture of a spin wave. Consider a picture in which the spin vectors $\boldsymbol{S}$ of all the atoms in a domain lie on cones of equal semi-angle. The axes of the cones are parallel to the mean direction of the spin vectors, and as before we take this to be the z axis. For a spin wave of wavevector $\boldsymbol{q}$, all the spin vectors of the atoms in a plane perpendicular to $\boldsymbol{q}$ are pointing in the same direction, but as we advance along $\boldsymbol{q}$ this direction precesses round the cone axis as shown in Fig. 8.8*a*. The projection of $\boldsymbol{S}$ in the xy plane is shown in Fig. 8.8*b*.

Fig. 8.8 (*a*) Physical picture of a spin wave. The z axis is the mean direction of the spin vector. (*b*) Projection of $\boldsymbol{S}$ in the xy plane.

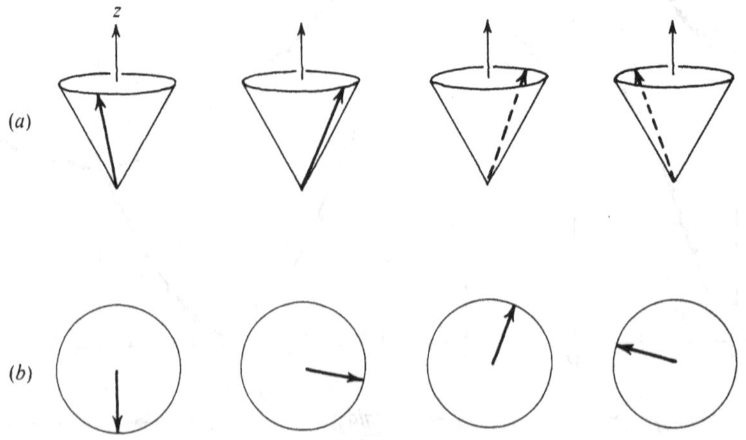

On this picture the z component of $\boldsymbol{S}$ is the same for all the atoms. The x and y components are

$$\begin{aligned} S_l^x &= A\cos(\boldsymbol{q}\,.\,\boldsymbol{l}-\omega t+\phi), \\ S_l^y &= A\sin(\boldsymbol{q}\,.\,\boldsymbol{l}-\omega t+\phi), \end{aligned} \tag{8.86}$$

where A is the magnitude of the component of $\boldsymbol{S}$ in the xy plane, and ϕ is a phase angle. The value of $S_0^x(0)S_l^x(t)$ averaged over all values of ϕ is

$$\begin{aligned} \langle S_0^x(0)S_l^x(t)\rangle_{\text{cl}} &= A^2\langle\cos\phi\,\cos(\boldsymbol{q}\,.\,\boldsymbol{l}-\omega t+\phi)\rangle \\ &= \tfrac{1}{2}A^2\cos(\boldsymbol{q}\,.\,\boldsymbol{l}-\omega t). \end{aligned} \tag{8.87}$$

The subscript cl denotes that the calculation is classical. $\langle S_0^y(0)S_l^y(t)\rangle_{\text{cl}}$ has the same value. Similarly

$$\langle S_0^x(0)S_l^y(t)\rangle_{\text{cl}} = \tfrac{1}{2}A^2\sin(\boldsymbol{q}\,.\,\boldsymbol{l}-\omega t). \tag{8.88}$$

The quantum values of these quantities are given by (8.76) and (8.77) which, for a single spin wave, may be written as

$$\begin{aligned} &\langle S_0^x(0)S_l^x(t)\rangle \\ &\quad=\frac{S}{N}[\langle n_{\boldsymbol{q}}\rangle\cos(\boldsymbol{q}\,.\,\boldsymbol{l}-\omega_{\boldsymbol{q}}t)+\tfrac{1}{2}\exp\{-\mathrm{i}(\boldsymbol{q}\,.\,\boldsymbol{l}-\omega_{\boldsymbol{q}}t)\}], \end{aligned} \tag{8.89}$$

$$\begin{aligned} &\langle S_0^x(0)S_l^y(t)\rangle \\ &\quad=\frac{S}{N}[\langle n_{\boldsymbol{q}}\rangle\sin(\boldsymbol{q}\,.\,\boldsymbol{l}-\omega_{\boldsymbol{q}}t)+\frac{\mathrm{i}}{2}\exp\{-\mathrm{i}(\boldsymbol{q}\,.\,\boldsymbol{l}-\omega_{\boldsymbol{q}}t)\}]. \end{aligned} \tag{8.90}$$

Ignoring the second term on the right-hand side of (8.89) and (8.90) we see that the classical and quantum results correspond, provided ω the angular frequency of precession of the spin vector in the classical picture is equal to $\omega_{\boldsymbol{q}}$ for the quantum spin wave, and

$$A^2 = \frac{2S}{N}\langle n_{\boldsymbol{q}}\rangle. \tag{8.91}$$

Thus the angle between $\boldsymbol{S}$ and the cone axis depends on $\langle n_{\boldsymbol{q}}\rangle$ and increases with the spin-wave excitation.

In the classical picture S_l^z is independent of l and constant in time. The value of $(S_l^z)^2$ is given by

$$(S_l^z)^2 = S^2-\{(S_l^x)^2+(S_l^y)^2\} = S^2-A^2 = S^2-\frac{2S}{N}\langle n_{\boldsymbol{q}}\rangle, \tag{8.92}$$

which is equal to the quantum value given in (8.72).

The classical picture therefore gives a good approximation to the quantum results for all the components of $\boldsymbol{S}$. The main discrepancy is that there is no classical counterpart to the second term on the right-hand side of (8.89) and (8.90). In the ground state, when $\langle n_q \rangle = 0$, the classical picture gives $S_l^x = S_l^y = 0$, i.e. $\boldsymbol{S}$ is along the z axis. This is contrary to a fundamental result in quantum mechanics that in general the three components of angular momentum cannot be known simultaneously. However, provided it is not relied upon for detailed numerical prediction, the physical picture of $\boldsymbol{S}$ precessing about the magnetisation axis is a useful one for a spin wave.

Interactions between spin waves

In the linear approximation there is no interaction between the spin waves. The effect of such interactions in a more refined theory was considered first by Dyson (1956). He divided the interactions into two kinds, called *kinematic* and *dynamic*. The kinematic interaction arises because the maximum value of the spin deviation n at any atom in the crystal is $2S$. The linear approximation allows further spin deviation. The fact that it does not occur may be regarded as due to a

Fig. 8.9 Stiffness constant D as a function of temperature: ○ nickel, ● iron. (After Stringfellow, 1968. © 1968, Institute of Physics.)

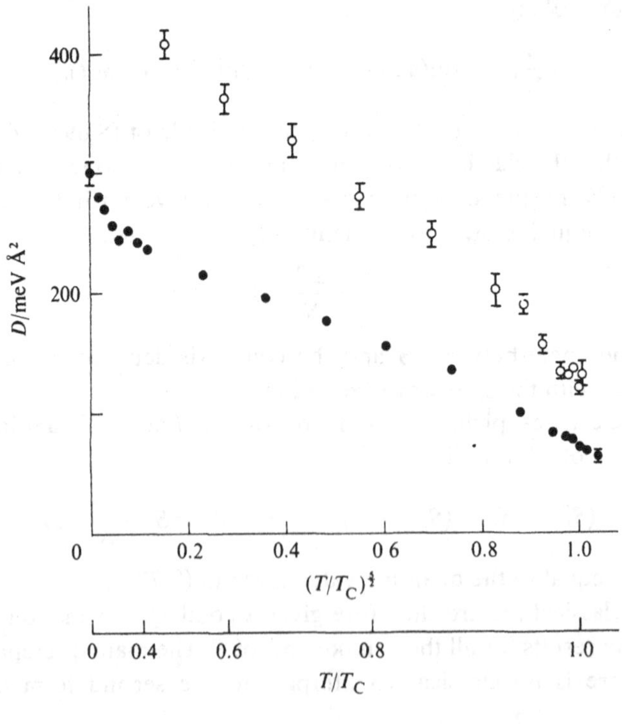

repulsive interaction between the spin waves, i.e. one spin wave acts to diminish the effect of another. The dynamic interaction comes from the fact that if an atom next to the atom l already has a spin deviation, less energy is required to produce a deviation on l. In the linear approximation the energy required is independent of the spin deviation of a neighbouring atom. This effect is equivalent to an attractive interaction between spin waves.

Dyson showed that at low temperatures the kinematic interaction is small and that only the dynamic interaction need be considered. The main results of the calculations are firstly that the magnon energies are reduced. The stiffness constant D in (8.65) is reduced by a factor proportional to $T^{5/2}$. Secondly, for a given q, the energy of the magnon is spread by an amount that increases with q and T. Some experimental results showing how D varies with temperature for iron and nickel are shown in Fig. 8.9.

Examples

8.1 In $MnTe_2$ the Mn ions have a face-centred cubic structure. The reciprocal lattice, with unit cell vectors $\boldsymbol{\tau}_1$, $\boldsymbol{\tau}_2$, $\boldsymbol{\tau}_3$, is defined for the simple cubic lattice. In the antiferromagnetic phase the spins in a plane perpendicular to the third cube axis are all aligned, and are opposite to those in the neighbouring parallel planes. If $\boldsymbol{\kappa} = t_1\boldsymbol{\tau}_1 + t_2\boldsymbol{\tau}_2 + t_3\boldsymbol{\tau}_3$, what are the conditions on t_1, t_2, t_3 for a magnetic Bragg reflection?

8.2 A one-dimensional ferromagnet with Heisenberg nearest-neighbour interaction consists of N atoms with spin $\frac{1}{2}$, distance a apart. If ϕ_n represents a state with all the spins except that of the nth atom pointing in the same direction, show that $\sum_n \exp(iqna)\phi_n$ is an eigenfunction of the Hamiltonian, and that its energy is

$$E = 2J(1 - \cos qa) + E_0,$$

where J is the exchange parameter, and $E_0 = -\frac{1}{2}J(N-1)$ is the energy of the ground state.

8.3 A ferromagnet with a Heisenberg nearest-neighbour interaction has a simple cubic structure with cube side a. Show that for $qa \ll 1$ the magnon dispersion relation is

$$\hbar\omega = Dq^2,$$

where $D = 2JSa^2$, J is the exchange parameter, and S the spin of an atom. Show that D is given by the same expression for a face-centred cubic and a body-centred cubic structure, where a is the cube side in each case.

8.4 Show that, for a ferromagnetic crystal, one-magnon scattering in the forward direction occurs at scattering angles less than

$$\theta_0 = \sin^{-1}\left(\frac{\hbar^2}{2mD}\right),$$

where D is the stiffness constant.

8.5 Show that, for a Heisenberg ferromagnet at a temperature $T \ll T_C$ (Curie temperature), the expectation value of the z component of spin over a single domain varies with temperature according to the relation

$$\langle S^z \rangle = S - cT^{3/2},$$

where S is the spin of an atom, and c is a constant.

9

Polarisation analysis

9.1 Introduction

So far we have considered scattering of neutrons only from one momentum state to another. However, experiments which determine the spin state as well as the momentum of the neutron give further information about the scattering system, and in the present chapter we calculate the cross-sections for these more detailed processes. We include a discussion of scattering due to electric forces between the neutron and the atoms in the scattering system. These forces are much smaller than the nuclear and magnetic forces, and polarisation experiments are necessary to detect them.

Spin-state cross-sections

The spin state of a neutron is defined relative to a direction known as the *polarisation direction.* The spin states of the scattered neutrons are normally analysed in the polarisation direction of the incident neutrons, and we take this common direction as the z axis. As before we denote the spin states of the neutron by u and v. u is the 'spin up' state with eigenvalue $+1$ for the operator σ_z, and v is the 'spin down' state with eigenvalue -1. Any previous cross-section $(\mathrm{d}^2\sigma/\mathrm{d}\Omega\,\mathrm{d}E')_{\boldsymbol{k}\to\boldsymbol{k}'}$ now gives rise to four cross-sections, which we call *spin-state* cross-sections. They correspond to

$$u \to u, \qquad v \to v, \qquad u \to v, \qquad v \to u.$$

The processes $u \to u$ and $v \to v$ involve no change of spin. The processes $u \to v$ and $v \to u$ involve a change of spin and are known as *spin-flip* processes. Any cross-section $\boldsymbol{k} \to \boldsymbol{k}'$ for unpolarised neutrons is related to the corresponding spin-state cross-sections by

$$\left(\frac{\mathrm{d}^2\sigma}{\mathrm{d}\Omega\,\mathrm{d}E'}\right)_{\boldsymbol{k}\to\boldsymbol{k}'} = \tfrac{1}{2} \ldots \text{(sum of the 4 spin-state cross-sections)}. \qquad (9.1)$$

The factor $\frac{1}{2}$ expresses the fact that the incident neutrons are in the states u and v with equal probability.

If a beam of neutrons has a fraction f of neutrons in the state u we define the *polarisation* of the beam to be a vector $\boldsymbol{P}$ in the z direction with magnitude

$$P = 2f - 1. \tag{9.2}$$

For an unpolarised beam $P = 0$. For a completely polarised beam, $P = 1$ if all the spins are up, and $P = -1$ if all the spins are down.

Polarisation spectrometer

An apparatus for polarisation measurements is shown schematically in Fig. 9.1. The polariser is a magnetic crystal that monochromates and polarises the incident beam by Bragg reflection; the analyser is a similar crystal that measures the energy and spin state of the scattered neutrons (see Section 9.4). A guide field is maintained along the path of the incident and scattered neutrons. This is a uniform magnetic field of a few millitesla which serves to preserve the spin state of the neutron.

Between the polariser and the scattering sample is a device, known as a *flipper*, which produces a radio-frequency field perpendicular to the guide field. When activated the flipper changes the spin state of the neutron. A second flipper is placed between the sample and the analyser. Suppose the polariser produces neutrons in the state u and the analyser reflects only neutrons in the state u. With both flippers off, the apparatus measures the cross-section for the spin-state transition $u \rightarrow u$. With the first flipper off and the second on, it measures the $u \rightarrow v$ cross-section, and so on for the other two combinations.

The sample is mounted between the poles of an electromagnet, whose magnetic field determines the polarisation axis. Provision is made for rotating the magnet (together with guide fields 2 and 3 in the figure) so that the polarisation axis can be made either parallel or perpendicular to the scattering vector.

9.2 Nuclear scattering

We first calculate the spin-state cross-sections for nuclear scattering. Consider a set of identical nuclei of non-zero spin I. The neutron plus nucleus form a system with total spin t, where

$$t = I + \tfrac{1}{2}, \quad \text{or} \quad t = I - \tfrac{1}{2}. \tag{9.3}$$

Each value of t has its own value of the scattering length, denoted by b^+ for $I+\frac{1}{2}$ and b^- for $I-\frac{1}{2}$.

Scattering length operator

We start by constructing an operator for the scattering length, whose eigenvalues are b^+ and b^-. If $|+\rangle$ and $|-\rangle$ are states corresponding to $t=I+\frac{1}{2}$ and $I-\frac{1}{2}$, the required operator $\hat{b}$ satisfies the relations

$$\hat{b}|+\rangle = b^+|+\rangle, \qquad \hat{b}|-\rangle = b^-|-\rangle. \tag{9.4}$$

Fig. 9.1 Schematic arrangement of the polarisation spectrometer used by Moon *et al.*, (1969) at the Oak Ridge National Laboratory.

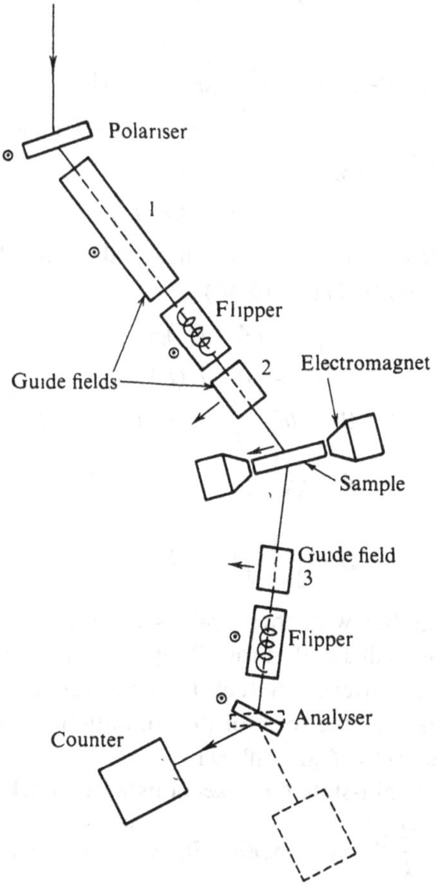

Denote the spin angular momentum of the nucleus (in units of $\hbar$) by the operator $\boldsymbol{I}$. The corresponding spin angular momentum of the neutron is $\frac{1}{2}\boldsymbol{\sigma}$, where $\boldsymbol{\sigma}$ is the Pauli spin operator. The operator for the spin of the nucleus–neutron system is

$$\boldsymbol{t}=\boldsymbol{I}+\tfrac{1}{2}\boldsymbol{\sigma}. \tag{9.5}$$

Therefore

$$\boldsymbol{t}^2=\boldsymbol{t}\,.\,\boldsymbol{t}=\boldsymbol{I}^2+\tfrac{1}{4}\boldsymbol{\sigma}^2+\boldsymbol{\sigma}\,.\,\boldsymbol{I}. \tag{9.6}$$

The states $|+\rangle$ and $|-\rangle$ are eigenfunctions of $\boldsymbol{I}^2$ with the same eigenvalues, namely $I(I+1)$. They are also eigenfunctions of $\frac{1}{4}\boldsymbol{\sigma}^2$ with the same eigenvalues, namely $\frac{1}{2}(\frac{1}{2}+1)=\frac{3}{4}$. The state $|+\rangle$ is an eigenfunction of $\boldsymbol{t}^2$ with eigenvalue

$$t(t+1)=(I+\tfrac{1}{2})(I+\tfrac{3}{2})=I^2+2I+\tfrac{3}{4}. \tag{9.7}$$

The state $|-\rangle$ is an eigenfunction of t^2 with eigenvalue

$$t(t+1)=(I-\tfrac{1}{2})(I+\tfrac{1}{2})=I^2-\tfrac{1}{4}. \tag{9.8}$$

Therefore

$$\begin{aligned}\boldsymbol{\sigma}\,.\,\boldsymbol{I}|+\rangle&=(\boldsymbol{t}^2-\boldsymbol{I}^2-\tfrac{1}{4}\boldsymbol{\sigma}^2)|+\rangle=I|+\rangle,\\ \boldsymbol{\sigma}\,.\,\boldsymbol{I}|-\rangle&=(\boldsymbol{t}^2-\boldsymbol{I}^2-\tfrac{1}{4}\boldsymbol{\sigma}^2)|-\rangle=-(I+1)|-\rangle.\end{aligned} \tag{9.9}$$

We write $\hat{b}$ in the form

$$\hat{b}=A+B\boldsymbol{\sigma}\,.\,\boldsymbol{I}, \tag{9.10}$$

where A and B are two constants whose values are chosen to make $\hat{b}$ satisfy (9.4). From (9.9) and (9.10)

$$\begin{aligned}\hat{b}|+\rangle&=(A+BI)|+\rangle,\\ \hat{b}|-\rangle&=\{A-B(I+1)\}|-\rangle.\end{aligned} \tag{9.11}$$

Therefore

$$A+BI=b^+,\qquad A-B(I+1)=b^-; \tag{9.12}$$

whence

$$A=\frac{1}{2I+1}\{(I+1)b^++Ib^-\}, \tag{9.13}$$

$$B=\frac{1}{2I+1}(b^+-b^-). \tag{9.14}$$

Eq. (9.10), together with these values of A and B, is the required operator $\hat{b}$. We shall see that for all types of interaction between the neutron and the scattering system, the operator corresponding to the scattering length has the form of (9.10); only the scalar quantities A and B, and the vector $\boldsymbol{I}$ are different.

To obtain the spin-state cross-sections we go back to (2.40).

$$\left(\frac{d^2\sigma}{d\Omega\, dE'}\right)_{\lambda\to\lambda'}=\frac{k}{k'}\left|\sum_j b_j\langle\lambda'|\exp(i\boldsymbol{\kappa}\,.\,\boldsymbol{R}_j)|\lambda\rangle\right|^2\delta(E_\lambda-E_{\lambda'}+\hbar\omega). \tag{9.15}$$

This must now be written

$$\left(\frac{d^2\sigma}{d\Omega\, dE'}\right)_{\sigma\lambda\to\sigma'\lambda'} = \frac{k'}{k}\left|\sum_j \langle\sigma'\lambda'|\hat{b}_j \exp(i\boldsymbol{\kappa}\,.\,\boldsymbol{R}_j)|\sigma\lambda\rangle\right|^2 \delta(E_\lambda - E_{\lambda'} + \hbar\omega), \tag{9.16}$$

where σ and σ', the initial and final spin states of the neutron, are either u or v. The matrix element for the jth nucleus is

$$\langle\sigma'\lambda'|\hat{b}_j \exp(i\boldsymbol{\kappa}\,.\,\boldsymbol{R}_j)|\sigma\lambda\rangle = \langle\lambda'|\exp(i\boldsymbol{\kappa}\,.\,\boldsymbol{R}_j)\langle\sigma'|\hat{b}_j|\sigma\rangle|\lambda\rangle. \tag{9.17}$$

We calculate $\langle\sigma'|\hat{b}_j|\sigma\rangle$ for a specific j and drop the subscript j for the moment. Then

$$\hat{b} = A + B\boldsymbol{\sigma}\,.\,\boldsymbol{I}. \tag{9.18}$$

From the relations in (F.16)

$$\begin{aligned}\hat{b}|u\rangle &= \{A + B(\sigma_x I_x + \sigma_y I_y + \sigma_z I_z)\}|u\rangle \\ &= A|u\rangle + B(I_x + iI_y)|v\rangle + BI_z|u\rangle.\end{aligned} \tag{9.19}$$

So

$$\langle u|\hat{b}|u\rangle = A + BI_z.$$

Similarly

$$\begin{aligned}\langle v|\hat{b}|v\rangle &= A - BI_z, \\ \langle v|\hat{b}|u\rangle &= B(I_x + iI_y), \\ \langle u|\hat{b}|v\rangle &= B(I_x - iI_y).\end{aligned} \tag{9.20}$$

The expressions on the right-hand side of these equations are the scattering lengths for the four spin-state transitions.

Coherent scattering

It is necessary at this stage to distinguish between an average over the nuclear spin states (i.e. the $2I+1$ eigenstates of the operator I_z), and an average over the isotopes in the scattering system. The former will be denoted by $(\ )_{sp}$ and the latter by $\langle\ \rangle_{iso}$. An average over both nuclear spin states and isotopes will be denoted by a bar over the quantity.

Coherent scattering is proportional to $\bar{b}$. Consider the spin-state transition $u \to u$. From (9.20)

$$\bar{b} = \langle (A + BI_z)_{sp}\rangle_{iso}. \tag{9.21}$$

The quantities A and B, given in (9.13) and (9.14), depend on the isotope. I_z depends on the nuclear spin state. We assume the nuclear spins are randomly oriented. Then

$$(I_x)_{sp} = (I_y)_{sp} = (I_z)_{sp} = 0. \tag{9.22}$$

Thus for

$$\begin{aligned} u \to u, &\quad \bar{b} = \langle A \rangle_{\text{iso}}; \\ v \to v, &\quad \bar{b} = \langle A \rangle_{\text{iso}}; \\ u \to v, &\quad \bar{b} = 0; \\ v \to u, &\quad \bar{b} = 0. \end{aligned} \tag{9.23}$$

We see that coherent nuclear scattering occurs with no change of neutron spin. The results of (9.23) are readily shown to be consistent with the previous results for coherent nuclear scattering. From (9.1) the value of $(\bar{b})^2$ for unpolarised neutrons is

$$(\bar{b})^2 = \tfrac{1}{2}(\langle A \rangle_{\text{iso}}^2 + \langle A \rangle_{\text{iso}}^2) = \langle A \rangle_{\text{iso}}^2. \tag{9.24}$$

Eq. (9.13) shows that A is the mean value of b for a single isotope – see (2.75). Thus $\langle A \rangle_{\text{iso}}$ is identical to the quantity $\bar{b}$ defined in (2.76).

Incoherent scattering

Incoherent scattering is proportional to $\overline{b^2} - (\bar{b})^2$. Consider the transition $u \to u$

$$\begin{aligned} \overline{b^2} &= \langle (\{A + BI_z\}^2)_{\text{sp}} \rangle_{\text{iso}} \\ &= \langle A^2 \rangle_{\text{iso}} + \langle B^2 (I_z^2)_{\text{sp}} \rangle_{\text{iso}} + 2\langle AB(I_z)_{\text{sp}} \rangle_{\text{iso}}. \end{aligned} \tag{9.25}$$

For random nuclear spins

$$(I_x^2)_{\text{sp}} = (I_y^2)_{\text{sp}} = (I_z^2)_{\text{sp}} = \tfrac{1}{3} I(I+1). \tag{9.26}$$

As before

$$(I_z)_{\text{sp}} = 0. \tag{9.27}$$

Therefore

$$\overline{b^2} - (\bar{b})^2 = \langle A^2 \rangle_{\text{iso}} - \langle A \rangle_{\text{iso}}^2 + \tfrac{1}{3}\langle B^2 I(I+1) \rangle_{\text{iso}}. \tag{9.28}$$

The corresponding values for the other three spin state transitions are calculated in the same way. For the four transitions the values are

$$\left.\begin{aligned} u \to u \\ v \to v \end{aligned}\right\} \quad \overline{b^2} - (\bar{b})^2 = \langle A^2 \rangle_{\text{iso}} - \langle A \rangle_{\text{iso}}^2 + \tfrac{1}{3}\langle B^2 I(I+1) \rangle_{\text{iso}}, \tag{9.29}$$

$$\left.\begin{aligned} u \to v \\ v \to u \end{aligned}\right\} \quad \overline{b^2} - (\bar{b})^2 = \tfrac{2}{3}\langle B^2 I(I+1) \rangle_{\text{iso}}. \tag{9.30}$$

Fig. 9.2 shows some results for the incoherent scattering from polycrystalline nickel. All its isotopes have $I = 0$. Thus there is no spin-flip scattering. Fig. 9.3 shows some results for vanadium, which has only one isotope. Thus

$$\langle A^2 \rangle_{\text{iso}} = \langle A \rangle_{\text{iso}}^2, \tag{9.31}$$

and according to (9.29) and (9.30) the cross-section for non spin-flip is one half the cross-section for spin-flip processes. The experiments confirm these conclusions.

9.3 Magnetic scattering

To calculate the spin-state cross-sections for magnetic scattering we go back to (7.26). The cross-section for the transition $\sigma\lambda \to \sigma'\lambda'$ is

$$\left(\frac{d^2\sigma}{d\Omega\, dE'}\right)_{\sigma\lambda\to\sigma'\lambda'} = (\gamma r_0)^2 \frac{k'}{k} |\langle\sigma'\lambda'|\boldsymbol{\sigma}\,.\,\boldsymbol{Q}_\perp|\sigma\lambda\rangle|^2 \delta(E_\lambda - E_{\lambda'} + \hbar\omega). \tag{9.32}$$

The operator $\boldsymbol{\sigma}\,.\,\boldsymbol{Q}_\perp$ has the same form as the term $B\boldsymbol{\sigma}\,.\,\boldsymbol{I}$ in (9.18). The spin-state matrix elements therefore follow immediately from (9.20) with BI_x replaced by $Q_{\perp x}$ etc.

$$\begin{aligned} \langle u|\boldsymbol{\sigma}\,.\,\boldsymbol{Q}_\perp|u\rangle &= Q_{\perp z}, \\ \langle v|\boldsymbol{\sigma}\,.\,\boldsymbol{Q}_\perp|v\rangle &= -Q_{\perp z}, \\ \langle v|\boldsymbol{\sigma}\,.\,\boldsymbol{Q}_\perp|u\rangle &= Q_{\perp x} + \mathrm{i}Q_{\perp y}, \\ \langle u|\boldsymbol{\sigma}\,.\,\boldsymbol{Q}_\perp|v\rangle &= Q_{\perp x} - \mathrm{i}Q_{\perp y}. \end{aligned} \tag{9.33}$$

Fig. 9.2 Isotopic incoherent scattering from nickel. In this figure and in Figs. 9.3 and 9.4 the measurements were made by rocking the analyser crystal through the elastic position with the scattering angle fixed. In the three figures the counting rates for 'flipper off' (•) and 'flipper on' (○) are proportional to the cross-sections $u \to u$ and $v \to u$ respectively. (After Moon *et al.*, 1969.)

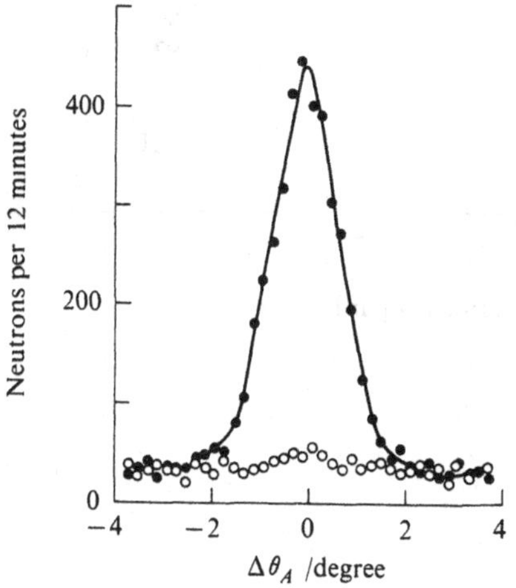

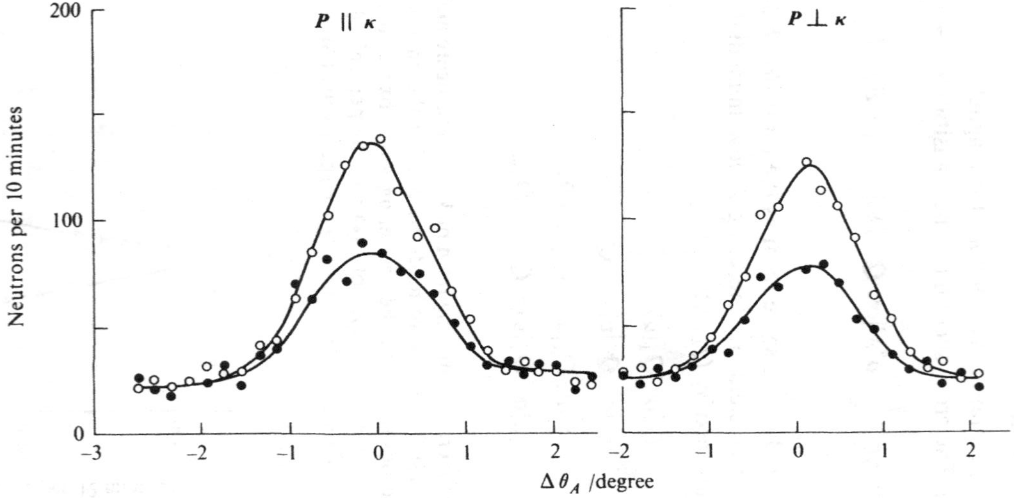

Fig. 9.3 Nuclear-spin incoherent scattering from vanadium. • flipper off, ∘ flipper on. The 'flipper on' values are effectively the same for $\boldsymbol{P}\|\boldsymbol{\kappa}$ and $\boldsymbol{P}\perp\boldsymbol{\kappa}$, showing that the spin-flip scattering is due to nuclear spin. (After Moon *et al.*, 1969.)

Paramagnet

As an example we consider the elastic magnetic scattering of polarised neutrons from a paramagnetic Bravais crystal with localised electrons. We have already discussed the scattering of unpolarised neutrons from the system (Section 7.6).

From (9.32) and (9.33) the cross-section for non spin-flip processes, i.e. $u \to u$ and $v \to v$, is

$$\left(\frac{d^2\sigma}{d\Omega\, dE'}\right)_{\substack{\lambda\to\lambda' \\ \text{nsf}}} = (\gamma r_0)^2 \frac{k'}{k} |\langle\lambda'|Q_{\perp z}|\lambda\rangle|^2 \delta(E_\lambda - E_{\lambda'} + \hbar\omega). \quad (9.34)$$

We sum over λ', average over λ, and integrate with respect to E'. This gives

$$\left(\frac{d\sigma}{d\Omega}\right)_{\text{nsf}} = (\gamma r_0)^2 \langle Q^+_{\perp z} Q_{\perp z}\rangle. \quad (9.35)$$

From (7.44)

$$Q_{\perp z} = -Q_x \hat{\kappa}_x \hat{\kappa}_z - Q_y \hat{\kappa}_y \hat{\kappa}_z + Q_z(1 - \hat{\kappa}_z^2). \quad (9.36)$$

From (7.58) and (7.59) we may write $\boldsymbol{Q}$ in the form

$$\boldsymbol{Q} = \tfrac{1}{2} g F(\boldsymbol{\kappa}) \sum_l \exp(i\boldsymbol{\kappa} \,.\, \boldsymbol{R}_l) \boldsymbol{S}_l, \quad (9.37)$$

where $\boldsymbol{R}_l$ is the position of nucleus l, and $\boldsymbol{S}_l$ is the spin or total angular momentum operator (see Section 7.4). Thus

$$\langle Q^+_\alpha Q_\beta\rangle = \{\tfrac{1}{2} g F(\boldsymbol{\kappa})\}^2 \exp(-2W) N \sum_l \exp(i\boldsymbol{\kappa} \,.\, \boldsymbol{l}) \langle S_0^\alpha S_l^\beta\rangle. \quad (9.38)$$

For a paramagnet (see Section 7.6)

$$\langle S_0^\alpha S_l^\beta\rangle = \tfrac{1}{3}\delta_{0l}\delta_{\alpha\beta} S(S+1). \quad (9.39)$$

Therefore, when $Q_{\perp z}$ is multiplied by its Hermitian conjugate in (9.35) only the terms in $Q_x^+ Q_x$, $Q_y^+ Q_y$, and $Q_z^+ Q_z$ are non-zero. Inserting these results in (9.35) and noting that

$$\hat{\kappa}_x^2 \hat{\kappa}_z^2 + \hat{\kappa}_y^2 \hat{\kappa}_z^2 + (1 - \hat{\kappa}_z^2)^2 = 1 - \hat{\kappa}_z^2, \quad (9.40)$$

we obtain

$$\left(\frac{d\sigma}{d\Omega}\right)_{\text{nsf}} = \tfrac{1}{3}(\gamma r_0)^2 N \{\tfrac{1}{2} g F(\boldsymbol{\kappa})\}^2 \exp(-2W) S(S+1)(1 - \hat{\kappa}_z^2). \quad (9.41)$$

The cross-section for spin-flip processes ($u \to v$, and $v \to u$) is

$$\left(\frac{d\sigma}{d\Omega}\right)_{\text{sf}} = (\gamma r_0)^2 \langle (Q_{\perp x} + iQ_{\perp y})^+ (Q_{\perp x} + iQ_{\perp y})\rangle$$
$$= (\gamma r_0)^2 \langle Q^+_{\perp x} Q_{\perp x} + Q^+_{\perp y} Q_{\perp y}\rangle, \quad (9.42)$$

the cross-terms being zero for a paramagnet. Comparison of (9.42) with (9.35) gives the result

$$\left(\frac{d\sigma}{d\Omega}\right)_{sf}=\tfrac{1}{3}(\gamma r_0)^2N\{\tfrac{1}{2}gF(\boldsymbol{\kappa})\}^2\exp(-2W)S(S+1)(1+\hat{\kappa}_z^2). \quad (9.43)$$

We may note as a check that the sum of the cross-sections for non spin-flip and spin-flip processes is equal to the cross-section for unpolarised neutrons (7.92).

The z axis in (9.41) and (9.43) is the direction of $\boldsymbol{P}$, the polarisation of the incident neutrons. If

$$\boldsymbol{P}\|\boldsymbol{\kappa}, \qquad \hat{\kappa}_z=1, \quad (9.44)$$

and the scattering is entirely spin-flip. If

$$\boldsymbol{P}\perp\boldsymbol{\kappa}, \qquad \hat{\kappa}_z=0, \quad (9.45)$$

and the scattering cross-sections for spin-flip and non spin-flip are equal. Some results for MnF_2 are shown in Fig. 9.4.

Polarisation measurements provide a method of separating paramagnetic scattering from other sources of diffuse scattering. The latter include multiple Bragg scattering, phonon scattering, and incoherent scattering due to isotopic disorder and nuclear spin. Only

Fig. 9.4 Paramagnetic scattering from MnF_2. • flipper off, ○ flipper on. The small peak in the 'flipper off' data for $\boldsymbol{P}\|\boldsymbol{\kappa}$, and the small difference between the two sets of data for $\boldsymbol{P}\perp\boldsymbol{\kappa}$, are due to multiple Bragg scattering. (After Moon *et al.*, 1969.)

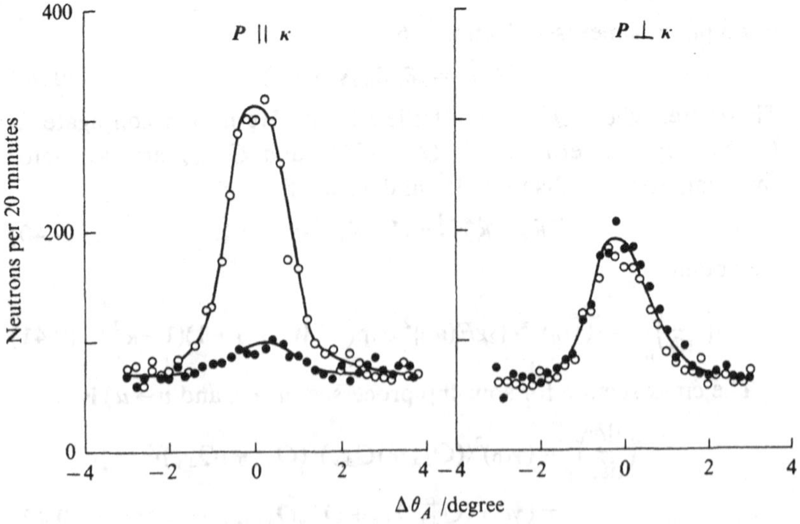

paramagnetic and nuclear spin scattering give rise to spin-flip processes. They may be distinguished by measuring the spin-flip cross-section with $\boldsymbol{P}$ first parallel and then perpendicular to $\boldsymbol{\kappa}$. The nuclear spin scattering is unchanged, whereas the paramagnetic scattering changes according to the above equations.

9.4 Bragg scattering from magnetically ordered crystals

We now consider coherent elastic scattering from crystals with magnetic order. We shall consider all the scattering – nuclear, magnetic, and interference between the two. For simplicity we restrict the discussion to Bravais crystals.

Effective scattering length

We first define a scattering length operator T_l that includes the nuclear and magnetic interaction between a neutron and the atom l. The cross-section for nuclear scattering is

$$\left(\frac{\mathrm{d}^2\sigma}{\mathrm{d}\Omega\,\mathrm{d}E'}\right)_{\sigma\lambda\to\sigma'\lambda'}=\frac{k'}{k}\left|\left\langle\sigma'\lambda'\left|\sum_l \hat{b}_l\exp(\mathrm{i}\boldsymbol{\kappa}\,.\,\boldsymbol{R}_l)\right|\sigma\lambda\right\rangle\right|^2\delta(E_\lambda-E_{\lambda'}+\hbar\omega),\tag{9.46}$$

where
$$\hat{b}_l=A_l+B_l\boldsymbol{\sigma}\,.\,\boldsymbol{I}_l.\tag{9.47}$$

The magnetic counterpart of the operator $\sum_l \hat{b}_l\exp(\mathrm{i}\boldsymbol{\kappa}\,.\,\boldsymbol{R}_l)$ is $-\gamma r_0\boldsymbol{\sigma}\,.\,\boldsymbol{Q}_\perp$. This result (apart from the minus sign) comes from a comparison of (9.32) and (9.46). The minus sign follows firstly from (2.33), which shows that a positive value of the scattering length b corresponds to a positive nuclear pseudopotential, and secondly from (7.10) and (7.21), which show that a positive value of $\boldsymbol{\sigma}\,.\,\boldsymbol{Q}_\perp$ corresponds to a negative magnetic potential. For a model with localised electrons, (7.42) and (9.37) give

$$\gamma r_0\boldsymbol{Q}_\perp=\tfrac{1}{2}\gamma r_0 gF(\boldsymbol{\kappa})\sum_l\hat{\boldsymbol{\kappa}}\times(\boldsymbol{S}_l\times\hat{\boldsymbol{\kappa}})\exp(\mathrm{i}\boldsymbol{\kappa}\,.\,\boldsymbol{R}_l).\tag{9.48}$$

We therefore arrive at the result that the cross-section for nuclear and magnetic scattering is the same as (9.46) with $\hat{b}_l$ replaced by

$$T_l=A_l+\boldsymbol{\sigma}\,.\,\{B_l\boldsymbol{I}_l-\tfrac{1}{2}\gamma r_0 gF(\boldsymbol{\kappa})\hat{\boldsymbol{\kappa}}\times(\boldsymbol{S}_l\times\hat{\boldsymbol{\kappa}})\}.\tag{9.49}$$

The matrix element $\langle\sigma'|T_l|\sigma\rangle$ – with $\boldsymbol{S}_l$ replaced by its thermal average – is the effective scattering length for atom l for the spin-state transition $\sigma\to\sigma'$. Since T_l has the same form as $\hat{b}$ in (9.10), the matrix elements have the same form as (9.20). They are averaged over nuclear spin and isotopes as in Section 9.2, again on the assumption that the nuclear spins are randomly oriented. The results of (9.20) and (9.23) then give the following expressions for the coherent scattering length for the four spin-state transitions

$$\begin{aligned}\langle u|T_l|u\rangle &= \bar{b} - C_l^z,\\ \langle v|T_l|v\rangle &= \bar{b} + C_l^z,\\ \langle v|T_l|u\rangle &= -(C_l^x + \mathrm{i}C_l^y),\\ \langle u|T_l|v\rangle &= -(C_l^x - \mathrm{i}C_l^y),\end{aligned} \tag{9.50}$$

where

$$\boldsymbol{C}_l = \tfrac{1}{2}\gamma r_0 g F(\boldsymbol{\kappa})\hat{\boldsymbol{\kappa}} \times \{\langle \boldsymbol{S}_l\rangle \times \hat{\boldsymbol{\kappa}}\}. \tag{9.51}$$

The cross-section for Bragg scattering for a particular spin-state transition $\sigma \to \sigma'$ is given by (3.48) with $\sigma_{\mathrm{coh}}/4\pi$ replaced by the square of the modulus of the corresponding scattering length in (9.50).

Although the expressions in (9.50) and (9.51) relate to a Bravais crystal they are readily generalised to the non-Bravais case. The nuclear scattering length $\bar{b}$ becomes the nuclear unit-cell structure factor

$$F_{\mathrm{N}}(\boldsymbol{\kappa}) = \sum_d \overline{b_d} \exp(\mathrm{i}\boldsymbol{\kappa} \,.\, \boldsymbol{d}) \exp(-W_d). \tag{9.52}$$

(See Section 3.6.) The expression for $\boldsymbol{C}_l$ becomes

$$\boldsymbol{C}_l = \tfrac{1}{2}\gamma r_0 \hat{\boldsymbol{\kappa}} \times \left\{\sum_d g_d F_d(\boldsymbol{\kappa}) \exp(\mathrm{i}\boldsymbol{\kappa} \,.\, \boldsymbol{d}) \exp(-W_d)\langle \boldsymbol{S}_{ld}\rangle \times \hat{\boldsymbol{\kappa}}\right\}. \tag{9.53}$$

We now consider the cross-sections for various scattering geometries and types of magnetic order.

Polarisation perpendicular to scattering vector

Consider Bragg scattering from a ferromagnetic Bravais crystal in which the mean spin directions $\hat{\boldsymbol{\eta}}$ of all the domains are aligned by a magnetic field. This field defines the direction of the polarisation vector $\boldsymbol{P}$, i.e. $\boldsymbol{P}$ must be parallel (or antiparallel) to $\hat{\boldsymbol{\eta}}$. We take it as parallel. Suppose the scattering vector $\boldsymbol{\kappa}$ is perpendicular to $\boldsymbol{P}$ (Fig. 9.5).

For a ferromagnet $\langle \boldsymbol{S}_l\rangle$ is independent of l and we drop the subscript l. Then, since $\langle \boldsymbol{S}\rangle$ is perpendicular to $\boldsymbol{\kappa}$,

$$\boldsymbol{C} = \tfrac{1}{2}\gamma r_0 g F(\boldsymbol{\kappa})\langle \boldsymbol{S}\rangle. \tag{9.54}$$

The vector $\langle \boldsymbol{S} \rangle$ is in the direction of $\hat{\boldsymbol{\eta}}$, i.e. of z. Therefore $\boldsymbol{C}$ has no component in the xy plane, and, from (9.50), the spin-flip scattering is zero. The cross-sections for scattering without spin-flip are

$$\left(\frac{\mathrm{d}\sigma}{\mathrm{d}\Omega}\right)_{u\to u} \propto (\bar{b}-C^{\eta})^2, \qquad \left(\frac{\mathrm{d}\sigma}{\mathrm{d}\Omega}\right)_{v\to v} \propto (\bar{b}+C^{\eta})^2, \tag{9.55}$$

where

$$C^{\eta} = \tfrac{1}{2}\gamma r_0 g F(\boldsymbol{\kappa})\langle S^{\eta}\rangle, \tag{9.56}$$

and $\langle S^{\eta}\rangle$ is the mean component of the spin† of an atom in the direction $\hat{\boldsymbol{\eta}}$. These results have two important applications.

Production and analysis of polarised neutrons. Suppose it happens that

$$\bar{b} = \tfrac{1}{2}\gamma r_0 g F(\boldsymbol{\kappa})\langle S^{\eta}\rangle. \tag{9.57}$$

Then the cross-section $u \to u$ is zero. If the incident neutrons are unpolarised, they may be regarded as a mixture with 50% in state u and 50% in state v. Since the cross-section $u \to u$ is zero, only neutrons in state v appear in the scattered beam. The spin vector of the scattered neutrons is in the direction of the applied magnetic field $\boldsymbol{B}$ (Fig. 9.5). (If the condition (9.57) holds but with a minus sign, the final spin vector is opposite to $\boldsymbol{B}$.) Even if the equality in (9.57) is not

Fig. 9.5 Geometry for non spin-flip scattering in magnetic Bragg reflection from a ferromagnet. The spin state for the scattered neutrons corresponds to $\bar{b}/F(\boldsymbol{\kappa})$ positive.

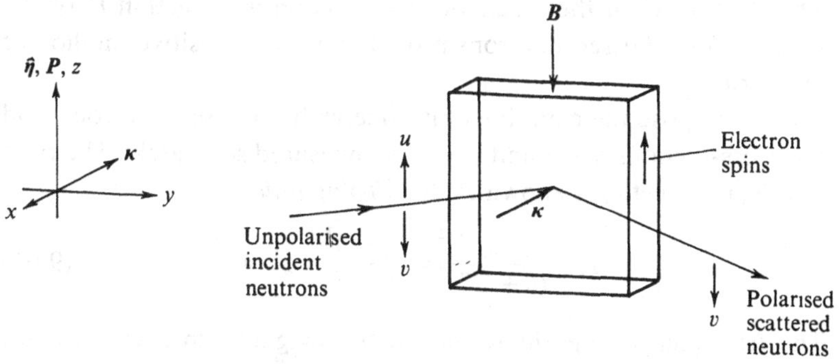

† For the negative electron the spin vector and the magnetic dipole vector are in opposite directions. Thus the magnetic field $\boldsymbol{B}$ and the magnetisation directions are along $-\hat{\boldsymbol{\eta}}$. The quantities γ, r_0, g, and $\langle S^{\eta}\rangle$ are all positive. $\bar{b}$ and $F(\boldsymbol{\kappa})$ may be positive or negative.

exactly satisfied the scattered beam is partly polarised. The equality is found to hold quite well for the (111) Bragg reflection in a cobalt–iron alloy (atomic composition 92% Co, 8% Fe). This is a commonly used method for producing polarised neutrons. Notice that the crystal monochromates as well as polarises the neutrons. The same device may be used to measure the polarisation of a beam of neutrons.

Another method of polarising neutrons is to reflect them from a mirror. Since the effective scattering lengths for the two spin-states are different, so are the critical angles of reflection (see Section 6.2). If the glancing angle of incidence lies between the two values of the critical angle, the reflected beam is polarised.†

Determination of spin densities. The magnetic form factor $F(\boldsymbol{\kappa})$ is related to the normalised density $s(\boldsymbol{r})$ of the unpaired electrons by

$$F(\boldsymbol{\kappa})=\int s(\boldsymbol{r})\exp(\mathrm{i}\boldsymbol{\kappa}\,.\,\boldsymbol{r})\,\mathrm{d}\boldsymbol{r}. \tag{9.58}$$

Thus $s(\boldsymbol{r})$ can be deduced from the values of $F(\boldsymbol{\kappa})$, i.e. from the values of C^{η}. If the measurements are made with unpolarised neutrons, the cross-section is proportional to

$$\tfrac{1}{2}\{(\bar{b}-C^{\eta})^2+(\bar{b}+C^{\eta})^2\}=(\bar{b})^2+(C^{\eta})^2. \tag{9.59}$$

C^{η} is usually much smaller than $\bar{b}$. Put

$$C^{\eta}=r\bar{b}, \tag{9.60}$$

where $r\ll 1$. Then the magnetic cross-section is a fraction r^2 of the total. So unpolarised neutrons provide a very insensitive method of measuring C^{η}.

Now suppose the experiment is done with polarised neutrons, and the cross-sections $u\to u$ and $v\to v$ are measured separately. The ratio of the cross-sections, known as the *flipping ratio*, is

$$R=\frac{(\bar{b}-C^{\eta})^2}{(\bar{b}+C^{\eta})^2}=\frac{(1-r)^2}{(1+r)^2}\approx 1-4r. \tag{9.61}$$

The fractional change in R due to the magnetic interaction is $4r$, which is much larger than r^2. Not only is the polarisation method more sensitive, but it provides the additional advantage of giving the

† See Williams (1975) and Hayter (1976) for accounts of polarising methods.

sign of C^{η} relative to $\bar{b}$. Since the scattering process involves no change in the spin state of the neutron it is not necessary to measure the polarisation of the scattered neutrons. Some results for the electron spin density in nickel, obtained by this method, are shown in Fig. 9.6.

The term $\pm 2\bar{b}C^{\eta}$ in the $u \to u$ and $v \to v$ cross-sections is known as the *nuclear–magnetic interference* term. It is only non-zero if both the nuclear and magnetic scattering are non-zero. For a ferromagnet these conditions are always satisfied at a Bragg peak, but for an antiferromagnet they may not be. For example, in $RbMnF_3$ the nuclear and magnetic Bragg peaks are at different points in reciprocal space, so nuclear–magnetic interference does not occur. In a rutile antiferromagnet, it occurs at some reciprocal lattice points and not at others (Example 9.2).

Fig. 9.6 Magnetic moment distribution in the (001) plane for nickel. ● nickel nucleus. The values of the contours are in units of μ_B Å^{-3}. (After Mook, 1966.)

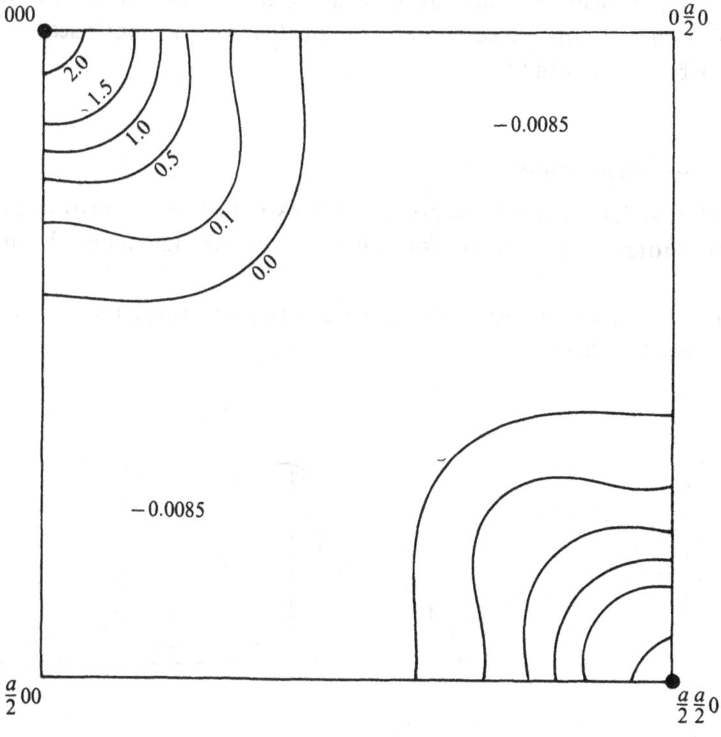

Polarisation parallel to scattering vector

We now consider Bragg scattering with $\boldsymbol{P}$ parallel to $\boldsymbol{\kappa}$. For a ferromagnet this arrangement is not in general useful, because, as we have seen, the mean spin direction $\hat{\boldsymbol{\eta}}$ must be parallel to $\boldsymbol{P}$. So the arrangement would correspond to $\hat{\boldsymbol{\eta}}$ parallel to $\boldsymbol{\kappa}$, which would give zero value for $\boldsymbol{C}$. (But see 'Non-collinear spins' below.) However for an antiferromagnet the spin directions $\pm\hat{\boldsymbol{\eta}}$ need not be parallel to $\boldsymbol{P}$, and this scattering geometry gives useful information.

From (9.51) $\boldsymbol{C}_l$ is perpendicular to $\boldsymbol{\kappa}$. Therefore, if $\boldsymbol{\kappa}$ is parallel to $\boldsymbol{P}$, the z component of $\boldsymbol{C}_l$ is zero, and the magnetic scattering is entirely spin-flip. This is a general result for $\boldsymbol{P}$ parallel to $\boldsymbol{\kappa}$, and is independent of the type of scattering (e.g. elastic or inelastic), the nature of the scattering system, and of the model used to describe it.

Suppose the scattering system is an antiferromagnet, and that the spin directions are perpendicular to $\boldsymbol{\kappa}$, thereby maximising the magnetic scattering (Fig. 9.7). As mentioned above, in some antiferromagnets nuclear and magnetic peaks occur at the same value of $\boldsymbol{\kappa}$. However, the former correspond to non spin-flip, and the latter to spin-flip processes. Thus by measuring the cross-sections for non spin-flip and spin-flip separately, the nuclear and magnetic scattering may be distinguished.

Non-collinear spins

So far we have considered polarisation experiments for ferromagnets and antiferromagnets, i.e. for collinear spin arrangements. We now

Fig. 9.7 Geometry for spin-flip scattering in magnetic Bragg reflection from an antiferromagnet.

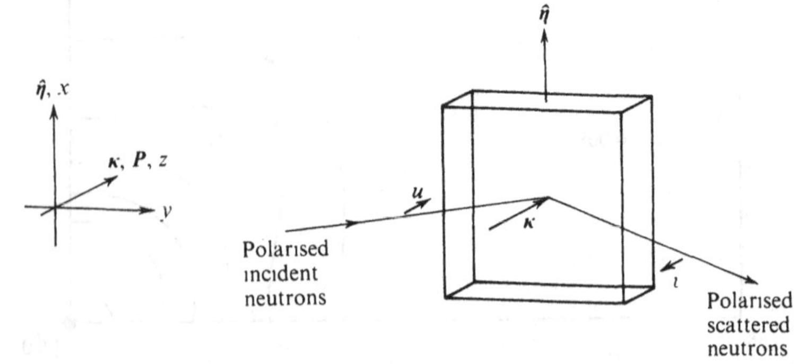

consider the spin-flip scattering for a helical arrangement of spins. Suppose we scatter polarised neutrons from the alloy Au_2Mn (see Section 8.1) with the polarisation $\boldsymbol{P}$ and the scattering vector $\boldsymbol{\kappa}$ both parallel to the tetrad axis. As usual we take the z axis in the direction of $\boldsymbol{P}$. Since $\boldsymbol{\kappa}$ is perpendicular to the xy plane, we have, from (8.22) and (9.54),

$$C_l^x = C\cos(\boldsymbol{Q}\,.\,\boldsymbol{l}), \qquad C_l^y = C\sin(\boldsymbol{Q}\,.\,\boldsymbol{l}), \tag{9.62}$$

where

$$C = \tfrac{1}{2}\gamma r_0 g F(\boldsymbol{\kappa})\langle S\rangle. \tag{9.63}$$

From (9.50)

$$\begin{aligned}\left(\frac{\mathrm{d}\sigma}{\mathrm{d}\Omega}\right)_{u\to v} &\propto \sum_{ll'} (C_{l'}^x - \mathrm{i}C_{l'}^y)(C_l^x + \mathrm{i}C_l^y)\exp\{\mathrm{i}\boldsymbol{\kappa}\,.\,(\boldsymbol{l}-\boldsymbol{l}')\}\\ &= C^2\sum_{ll'}\exp\{\mathrm{i}(\boldsymbol{\kappa}+\boldsymbol{Q})\,.\,(\boldsymbol{l}-\boldsymbol{l}')\}\\ &= NC^2\frac{(2\pi)^3}{v_0}\sum_{\boldsymbol{\tau}}\delta(\boldsymbol{\kappa}+\boldsymbol{Q}-\boldsymbol{\tau}).\end{aligned} \tag{9.64}$$

Similarly

$$\left(\frac{\mathrm{d}\sigma}{\mathrm{d}\Omega}\right)_{v\to u} \propto NC^2\frac{(2\pi)^3}{v_0}\sum_{\boldsymbol{\tau}}\delta(\boldsymbol{\kappa}-\boldsymbol{Q}-\boldsymbol{\tau}). \tag{9.65}$$

Thus scattering occurs when

$$\begin{aligned}\boldsymbol{\kappa} &= \boldsymbol{\tau}-\boldsymbol{Q}, \qquad u\to v,\\ \boldsymbol{\kappa} &= \boldsymbol{\tau}+\boldsymbol{Q}, \qquad v\to u.\end{aligned} \tag{9.66}$$

We therefore conclude that with polarised incident neutrons and the above geometry only one of the two magnetic satellite reflections occurs. If the incident neutrons are unpolarised, both reflections occur, and the scattered neutrons are polarised in opposite directions in the two reflections. The results in (9.66) are for a right-hand helix. For a left-hand helix the conditions are reversed. So in order to observe the polarisation effects it would be necessary to have a crystal consisting of a single domain with only one type of helix.

We return now to ferromagnetic and antiferromagnetic crystals. The discussion on pp. 181–6 was based on the expression (9.48) for a localised model. Consider the more general expression

$$\gamma r_0 \boldsymbol{Q}_\perp = -\frac{\gamma r_0}{2\mu_{\mathrm{B}}}\hat{\boldsymbol{\kappa}}\times\{\boldsymbol{M}(\boldsymbol{\kappa})\times\hat{\boldsymbol{\kappa}}\} \tag{9.67}$$

for the case of a ferromagnet. $\boldsymbol{M}(\boldsymbol{\kappa})$ is the Fourier transform of the magnetisation operator $\boldsymbol{M}(\boldsymbol{r})$. As before we put

$$\langle \boldsymbol{M}(\boldsymbol{\kappa})\rangle = \int \langle \boldsymbol{M}(\boldsymbol{r})\rangle \exp(\mathrm{i}\boldsymbol{\kappa}\,.\,\boldsymbol{r})\,\mathrm{d}\boldsymbol{r} = \mathscr{F}(\boldsymbol{\kappa})\sum_l \langle \exp(\mathrm{i}\boldsymbol{\kappa}\,.\,\boldsymbol{R}_l)\rangle, \quad (9.68)$$

where

$$\mathscr{F}(\boldsymbol{\kappa}) = \int_{\text{cell}} \langle \boldsymbol{M}(\boldsymbol{r})\rangle \exp(\mathrm{i}\boldsymbol{\kappa}\,.\,\boldsymbol{r})\,\mathrm{d}\boldsymbol{r}. \quad (9.69)$$

If $\langle \boldsymbol{M}(\boldsymbol{r})\rangle$ is along the magnetisation direction $-\hat{\boldsymbol{\eta}}$ for all values of $\boldsymbol{r}$, then $\mathscr{F}(\boldsymbol{\kappa})$ is along $-\hat{\boldsymbol{\eta}}$ for all $\boldsymbol{\kappa}$, and $\langle \gamma r_0 \boldsymbol{Q}_\perp \rangle$ is correctly represented by the thermal average of an expression such as (9.48) containing the *scalar* form factor $F(\boldsymbol{\kappa})$. However, although by definition the *mean* value of $\langle \boldsymbol{M}(\boldsymbol{r})\rangle$ (i.e. $\int_{\text{cell}} \langle \boldsymbol{M}(\boldsymbol{r})\rangle\,\mathrm{d}\boldsymbol{r}$) is along $-\hat{\boldsymbol{\eta}}$, it is possible for the direction of $\langle \boldsymbol{M}(\boldsymbol{r})\rangle$ to vary over the unit cell. In which case the direction of $\mathscr{F}(\boldsymbol{\kappa})$ varies with $\boldsymbol{\kappa}$. This possibility was first pointed out by Blume (1963).

Evidence for non-collinear magnetisation in ferromagnets and antiferromagnets may be obtained by polarisation experiments. On pp. 182–3 we deduced that for $\boldsymbol{P} \perp \boldsymbol{\kappa}$ the scattering is entirely non spin-flip. This followed because, for a localised model, the vector $\boldsymbol{C}$ has no component perpendicular to $\hat{\boldsymbol{\eta}}$, and hence to $\boldsymbol{P}$. Departure from collinearity would give components in the plane perpendicular to $\boldsymbol{P}$, which would give rise to spin-flip scattering.

Another method for investigating non-collinear effects is to arrange the scattering geometry so that $\boldsymbol{P}$, $\boldsymbol{\kappa}$, and $\hat{\boldsymbol{\eta}}$ are parallel to each other. Since $\boldsymbol{P}$ is parallel to $\boldsymbol{\kappa}$, the magnetic scattering is entirely spin-flip. The magnetic scattering depends on the component of $\mathscr{F}(\boldsymbol{\kappa})$ perpendicular to $\boldsymbol{\kappa}$. Therefore, since $\boldsymbol{\kappa}$ is parallel to $\hat{\boldsymbol{\eta}}$, any spin-flip scattering at a Bragg peak is due to non-collinear components of the spin. Moon and Koehler (1969) used this method to search for non-collinear spin density in hexagonal cobalt. They found none outside the limits of experimental error, and concluded that any contribution of non-collinear effects to form factors calculated from a collinear model would be considerably less than the experimental errors quoted for these quantities.

The polarisation effects described so far in the present chapter are summarised in Table 9.1.

9.5 Scattering by the atomic electric field

The expression for the electromagnetic potential between the neutron and the scattering system, based on the expression for $-\boldsymbol{\mu}_{\mathrm{n}}\,.\,\boldsymbol{B}$ in

Table 9.1 Summary of polarisation effects

		Fraction of scattering: Non spin-flip	Fraction of scattering: Spin-flip	Polarisation dependent
Nuclear				
coherent		1	0	no
incoherent	$I=0$	1	0	no
	single isotope	$\frac{1}{3}$	$\frac{2}{3}$	no
Magnetic				
paramagnet		$\frac{1}{2}(1-\hat{\kappa}_z^2)$	$\frac{1}{2}(1+\hat{\kappa}_z^2)$	no
Bragg scattering (magnetic + nuclear) from crystals with magnetic order				
$\boldsymbol{P} \perp \boldsymbol{\kappa}$				
c-ferromagnet	$\hat{\boldsymbol{\eta}} \parallel \boldsymbol{P}$	1	0	yes
n-ferromagnet	$\hat{\boldsymbol{\eta}} \parallel \boldsymbol{P}$	<1	>0	yes
$\boldsymbol{P} \parallel \boldsymbol{\kappa}$				
c-ferromagnet	$\hat{\boldsymbol{\eta}} \parallel \boldsymbol{P}$	The nuclear scattering is non slip-flip. The magnetic scattering is spin-flip. For a c-ferromagnet the spin-flip scattering is zero.		no
n-ferromagnet	$\hat{\boldsymbol{\eta}} \parallel \boldsymbol{P}$			yes
c-antiferromagnet	$\hat{\boldsymbol{\eta}} \perp \boldsymbol{P}$			no
helical	spins $\perp \boldsymbol{P}$			yes

The nuclear spins are assumed to be randomly oriented.
c = collinear, n = non-collinear.

(7.10), is not exact, but is the leading term in an expansion in powers of m_e/m, the ratio of the masses of the electron and neutron. The next term in the expansion, which arises from the electric field $\boldsymbol{E}$ due to the electrons and nuclei, is

$$V_{\mathrm{E}}(\boldsymbol{r}) = V_{\mathrm{SO}}(\boldsymbol{r}) + V_{\mathrm{F}}(\boldsymbol{r}), \tag{9.70}$$

where
$$V_{\mathrm{SO}}(\boldsymbol{r}) = \frac{\gamma\mu_{\mathrm{N}}}{mc^2}\boldsymbol{\sigma}\,.\,(\boldsymbol{E}\times\boldsymbol{p}), \tag{9.71}$$

and
$$V_{\mathrm{F}}(\boldsymbol{r}) = \frac{\hbar\gamma\mu_{\mathrm{N}}}{2mc^2}\operatorname{div}\boldsymbol{E} \tag{9.72}$$

(see Foldy, 1958). $\boldsymbol{p}$ is the momentum operator for the neutron, and the other symbols are defined in Section 7.1.

The first term in (9.70) is known as the *spin-orbit* interaction. When the neutron moves through an electric field $\boldsymbol{E}$ with velocity $\boldsymbol{v}$ then, in the rest frame of the neutron, there is a magnetic field

$$\boldsymbol{B} = \frac{1}{c^2}\boldsymbol{E}\times\boldsymbol{v}. \tag{9.73}$$

Since the neutron has magnetic moment

$$\boldsymbol{\mu}_{\mathrm{n}} = -\gamma\mu_{\mathrm{N}}\boldsymbol{\sigma}, \tag{9.74}$$

this gives the potential (9.71), which we note depends on the spin and velocity of the neutron.

The existence of the second term in (9.70) was first pointed out by Foldy and is known by his name. The neutron may be regarded as dissociated for part of the time into a proton and a π^- meson. Thus there is a separation of charge, which has an electrostatic interaction with the electrons and nuclei of the scattering system. The potential due to this effect is proportional to the magnetic moment μ_{n} (which arises from the dissociation), but is independent of the spin and velocity of the neutron.

To evaluate the cross-sections for these interactions we require – see (2.13) – the matrix element

$$\frac{m}{2\pi\hbar^2}\int \exp(-\mathrm{i}\boldsymbol{k}'\,.\,\boldsymbol{r})V_{\mathrm{E}}(\boldsymbol{r})\exp(\mathrm{i}\boldsymbol{k}\,.\,\boldsymbol{r})\,\mathrm{d}\boldsymbol{r},$$

where $\boldsymbol{r}$ is the neutron coordinate. We consider first the matrix element for the spin-orbit interaction. Denote the charge density at $\boldsymbol{r}$ by $n(\boldsymbol{r})$. This charge arises from the nuclei and all the electrons – paired and unpaired – in the scattering system. The Fourier transform of $n(\boldsymbol{r})$ is

$$n(\boldsymbol{\kappa}) = \int n(\boldsymbol{r})\exp(-\mathrm{i}\boldsymbol{\kappa}\,.\,\boldsymbol{r})\,\mathrm{d}\boldsymbol{r}, \tag{9.75}$$

with the inverse relation

$$n(\boldsymbol{r})=\frac{1}{(2\pi)^3}\int n(\boldsymbol{\kappa})\exp(\mathrm{i}\boldsymbol{\kappa}\,.\,\boldsymbol{r})\,\mathrm{d}\boldsymbol{\kappa}. \tag{9.76}$$

Note that since the scattering system is electrically neutral

$$n(\boldsymbol{\kappa}=0)=\int n(\boldsymbol{r})\,\mathrm{d}\boldsymbol{r}=0. \tag{9.77}$$

The electrostatic potential is given by

$$\phi(\boldsymbol{r})=\frac{1}{4\pi\varepsilon_0}\int\frac{n(\boldsymbol{r}')}{|\boldsymbol{r}-\boldsymbol{r}'|}\,\mathrm{d}\boldsymbol{r}'. \tag{9.78}$$

The electric field satisfies

$$\boldsymbol{E}=-\mathrm{grad}\,\phi. \tag{9.79}$$

From (B.11)

$$\int\frac{\exp(\mathrm{i}\boldsymbol{\kappa}'\,.\,\boldsymbol{r}')}{|\boldsymbol{r}-\boldsymbol{r}'|}\,\mathrm{d}\boldsymbol{r}'=\frac{4\pi}{\kappa'^2}\exp(\mathrm{i}\boldsymbol{\kappa}'\,.\,\boldsymbol{r}). \tag{9.80}$$

Thus

$$\phi(\boldsymbol{r})=\frac{1}{4\pi\varepsilon_0}\frac{1}{2\pi^2}\int\frac{1}{\kappa'^2}n(\boldsymbol{\kappa}')\exp(\mathrm{i}\boldsymbol{\kappa}'\,.\,\boldsymbol{r})\,\mathrm{d}\boldsymbol{\kappa}'. \tag{9.81}$$

We also have the result

$$\boldsymbol{p}\exp(\mathrm{i}\boldsymbol{k}\,.\,\boldsymbol{r})=\frac{\hbar}{\mathrm{i}}\,\mathrm{grad}\{\exp(\mathrm{i}\boldsymbol{k}\,.\,\boldsymbol{r})\}=\hbar\boldsymbol{k}\exp(\mathrm{i}\boldsymbol{k}\,.\,\boldsymbol{r}). \tag{9.82}$$

Note that the operator $(\boldsymbol{E}\times\boldsymbol{p})$ in (9.71) should have been written as $\frac{1}{2}\{(\boldsymbol{E}\times\boldsymbol{p})-(\boldsymbol{p}\times\boldsymbol{E})\}$. However

$$\boldsymbol{p}\times\boldsymbol{E}=-(\boldsymbol{E}\times\boldsymbol{p})+\frac{\hbar}{\mathrm{i}}\,\mathrm{curl}\,\boldsymbol{E}, \tag{9.83}$$

and curl $\boldsymbol{E}=0$, since $\boldsymbol{E}$ is due to a charge distribution.

From (9.79), (9.81), and (9.82)

$$\begin{aligned}&\int\exp(-\mathrm{i}\boldsymbol{k}'\,.\,\boldsymbol{r})(\boldsymbol{E}\times\boldsymbol{p})\exp(\mathrm{i}\boldsymbol{k}\,.\,\boldsymbol{r})\,\mathrm{d}\boldsymbol{r}\\&\quad=-\frac{1}{4\pi\varepsilon_0}\frac{\hbar}{2\pi^2}\int\exp(\mathrm{i}\boldsymbol{\kappa}\,.\,\boldsymbol{r})\,\mathrm{d}\boldsymbol{r}\,\mathrm{grad}_r\left\{\int\frac{n(\boldsymbol{\kappa}')}{\kappa'^2}\exp(\mathrm{i}\boldsymbol{\kappa}'\,.\,\boldsymbol{r})\,\mathrm{d}\boldsymbol{\kappa}'\right\}\times\boldsymbol{k}.\end{aligned} \tag{9.84}$$

This is evaluated from the relations

$$\mathrm{grad}_r\{\exp(\mathrm{i}\boldsymbol{\kappa}'\,.\,\boldsymbol{r})\}=\mathrm{i}\boldsymbol{\kappa}'\exp(\mathrm{i}\boldsymbol{\kappa}'\,.\,\boldsymbol{r}), \tag{9.85}$$

$$\int\exp(\mathrm{i}\boldsymbol{\kappa}\,.\,\boldsymbol{r})\exp(\mathrm{i}\boldsymbol{\kappa}'\,.\,\boldsymbol{r})\,\mathrm{d}\boldsymbol{r}=(2\pi)^3\delta(\boldsymbol{\kappa}+\boldsymbol{\kappa}'). \tag{9.86}$$

Further, since the scattering is elastic – the neutron does not have sufficient energy to change the spatial wavefunctions of the charges – we have

$$\kappa = 2k \sin \tfrac{1}{2}\theta, \tag{9.87}$$

and

$$\frac{\boldsymbol{\kappa} \times \boldsymbol{k}}{\kappa^2} = -\frac{\boldsymbol{k}' \times \boldsymbol{k}}{\kappa^2} = -\tfrac{1}{2}\hat{\boldsymbol{n}} \cot \tfrac{1}{2}\theta, \tag{9.88}$$

where θ is the angle between $\boldsymbol{k}$ and $\boldsymbol{k}'$, and $\hat{\boldsymbol{n}}$ is a unit vector in the direction of $(\boldsymbol{k}' \times \boldsymbol{k})$. Using these relations with (9.71) and (9.84) we obtain the result

$$\frac{m}{2\pi\hbar^2} \int \exp(-\mathrm{i}\boldsymbol{k}' . \boldsymbol{r}) V_{\mathrm{SO}}(\boldsymbol{r}) \exp(\mathrm{i}\boldsymbol{k} . \boldsymbol{r}) \,\mathrm{d}\boldsymbol{r}$$

$$= -\frac{\mu_0}{4\pi} \frac{\gamma\mu_{\mathrm{N}}}{\hbar} n(-\boldsymbol{\kappa})\mathrm{i} \cot(\tfrac{1}{2}\theta)\boldsymbol{\sigma} . \hat{\boldsymbol{n}}. \tag{9.89}$$

We may put

$$\frac{\mu_0}{4\pi} \frac{\gamma\mu_{\mathrm{N}}}{\hbar} = \frac{\gamma r_{\mathrm{p}}}{2e}, \tag{9.90}$$

where

$$r_{\mathrm{p}} = \frac{\mu_0}{4\pi} \frac{e^2}{m_{\mathrm{p}}} \tag{9.91}$$

is the proton analogue of r_0, the classical radius of the electron (see p. 133).

The matrix element for the Foldy interaction is evaluated directly from the relation

$$\operatorname{div} \boldsymbol{E} = \frac{1}{\varepsilon_0} n(\boldsymbol{r}). \tag{9.92}$$

From (9.72) and (9.75)

$$\frac{m}{2\pi\hbar^2} \int \exp(-\mathrm{i}\boldsymbol{k}' . \boldsymbol{r}) V_{\mathrm{F}}(\boldsymbol{r}) \exp(\mathrm{i}\boldsymbol{k} . \boldsymbol{r}) \,\mathrm{d}\boldsymbol{r} = \frac{\gamma r_{\mathrm{p}}}{2e} n(-\boldsymbol{\kappa}). \tag{9.93}$$

The matrix element for the combined potential $V_{\mathrm{E}}(\boldsymbol{r})$ is therefore

$$\frac{m}{2\pi\hbar^2} \int \exp(-\mathrm{i}\boldsymbol{k}' . \boldsymbol{r}) V_{\mathrm{E}}(\boldsymbol{r}) \exp(\mathrm{i}\boldsymbol{k} . \boldsymbol{r}) \,\mathrm{d}\boldsymbol{r}$$

$$= \frac{\gamma r_{\mathrm{p}}}{2e} n(-\boldsymbol{\kappa})\{1 - \mathrm{i} \cot(\tfrac{1}{2}\theta)\boldsymbol{\sigma} . \hat{\boldsymbol{n}}\}. \tag{9.94}$$

The quantity $\tfrac{1}{2}\gamma r_{\mathrm{p}}$ is the equivalent of the scattering length for the spin-orbit and Foldy interactions. The values of the atomic constants give

$$\tfrac{1}{2}\gamma r_{\mathrm{p}} = 1.468 \times 10^{-18} \text{ m}. \tag{9.95}$$

r_p is related to r_0 by

$$\frac{r_p}{r_0}=\frac{m_e}{m_p}. \tag{9.96}$$

Thus the cross-sections for electric interactions are less than those for magnetic interactions by a factor of the order of $(1/1836)^2$.

If the nuclei in the crystal are taken as fixed at their equilibrium positions, the matrix element $\langle\lambda|n(-\boldsymbol{\kappa})|\lambda\rangle$ for a Bravais crystal is given by

$$\langle\lambda|n(-\boldsymbol{\kappa})|\lambda\rangle=e\{Z-f(\boldsymbol{\kappa})\}\sum_l \exp(\mathrm{i}\boldsymbol{\kappa}\,.\,\boldsymbol{l}), \tag{9.97}$$

where
$$f(\boldsymbol{\kappa})=\int \rho(\boldsymbol{r}')\exp(\mathrm{i}\boldsymbol{\kappa}\,.\,\boldsymbol{r}')\,\mathrm{d}\boldsymbol{r}'. \tag{9.98}$$

Z is the charge on the nucleus, and $\rho(\boldsymbol{r}')$ is the density of all the electrons – paired and unpaired – in the atom at the position $\boldsymbol{r}'$ relative to the nucleus at the origin. The form factor $f(\boldsymbol{\kappa})$ is the one encountered in X-ray scattering. Notice that since $n(-\boldsymbol{\kappa})$ is zero for $\boldsymbol{\kappa}=0$ – see (9.77) – there is no electric scattering in the forward direction.

We now consider Bragg scattering due to the nuclear and electric interactions. From (9.18), (9.94), and (9.97) the scattering length operator is

$$\bar{b}+F_{\mathrm{E}}(\boldsymbol{\kappa})+\boldsymbol{\sigma}\,.\,\{B\boldsymbol{I}-\mathrm{i}\cot(\tfrac{1}{2}\theta)F_{\mathrm{E}}(\boldsymbol{\kappa})\hat{\boldsymbol{n}}\}, \tag{9.99}$$

where
$$F_{\mathrm{E}}(\boldsymbol{\kappa})=\tfrac{1}{2}\gamma r_p\{Z-f(\boldsymbol{\kappa})\}. \tag{9.100}$$

For randomly oriented nuclei, the results of (9.20) give the following scattering lengths for the 4 spin-state transitions:

$$\begin{aligned}
b_{uu}&=\bar{b}+F_{\mathrm{E}}(\boldsymbol{\kappa})-\mathrm{i}\cot(\tfrac{1}{2}\theta)F_{\mathrm{E}}(\boldsymbol{\kappa})\hat{n}_z,\\
b_{vv}&=\bar{b}+F_{\mathrm{E}}(\boldsymbol{\kappa})+\mathrm{i}\cot(\tfrac{1}{2}\theta)F_{\mathrm{E}}(\boldsymbol{\kappa})\hat{n}_z,\\
b_{uv}&=-\mathrm{i}\cot(\tfrac{1}{2}\theta)F_{\mathrm{E}}(\boldsymbol{\kappa})(\hat{n}_x+\mathrm{i}\hat{n}_y),\\
b_{vu}&=-\mathrm{i}\cot(\tfrac{1}{2}\theta)F_{\mathrm{E}}(\boldsymbol{\kappa})(\hat{n}_x-\mathrm{i}\hat{n}_y).
\end{aligned} \tag{9.101}$$

From these results we may derive an expression for the coherent elastic cross-section for a non-Bravais crystal. The nuclear scattering length $\bar{b}$ is replaced by

$$F_{\mathrm{N}}(\boldsymbol{\kappa})=\sum_d \bar{b}_d\exp(\mathrm{i}\boldsymbol{\kappa}\,.\,\boldsymbol{d}), \tag{9.102}$$

and $F_{\mathrm{E}}(\boldsymbol{\kappa})$ becomes

$$F_{\mathrm{E}}(\boldsymbol{\kappa})=\tfrac{1}{2}\gamma r_p\sum_d\{Z_d-f_d(\boldsymbol{\kappa})\}\exp(\mathrm{i}\boldsymbol{\kappa}\,.\,\boldsymbol{d}). \tag{9.103}$$

Suppose the incident neutron beam is polarised with a fraction f in the state u. Then the coherent elastic cross-section is proportional to

$$f\{|b_{uu}|^2+|b_{uv}|^2\}+(1-f)\{|b_{vv}|^2+|b_{vu}|^2\}. \qquad (9.104)$$

These results give

$$\left(\frac{d\sigma}{d\Omega}\right)_{\text{coh el}}$$
$$=N\frac{(2\pi)^3}{v_0}\sum_{\tau}\delta(\boldsymbol{\kappa}-\boldsymbol{\tau})[|F_N(\boldsymbol{\kappa})|^2+|F_E(\boldsymbol{\kappa})|^2+\cot^2(\tfrac{1}{2}\theta)|F_E(\boldsymbol{\kappa})|^2$$
$$+2\,\mathrm{Re}\{F_N(\boldsymbol{\kappa})F_E^*(\boldsymbol{\kappa})\}-2\cot(\tfrac{1}{2}\theta)\boldsymbol{P}\,.\,\hat{\boldsymbol{n}}\,\mathrm{Im}\{F_N(\boldsymbol{\kappa})F_E^*(\boldsymbol{\kappa})\}], \qquad (9.105)$$

where the polarisation $\boldsymbol{P}$ is related to f by (9.2). The first three terms in the square brackets correspond to nuclear, Foldy, and spin-orbit scattering respectively. The fourth term represents interference between the nuclear and Foldy scattering. The fifth term represents interference between the nuclear and spin-orbit scattering. It is the only polarisation-dependent term in the cross-section.

The polarisation-dependent term has been detected by Shull (1963) who measured the flipping ratio for the (110) reflection in vanadium for scattering to the left and to the right of the incident beam. The directions of $\hat{\boldsymbol{n}}$ are opposed in the two cases. He found, in accordance with the form of (9.105), that the flipping ratio was $1+\delta$ for one direction of $\hat{\boldsymbol{n}}$ and $1-\delta$ for the other, the value of δ being 0.003 ± 0.001.

Examples

9.1 Show that, for neutrons whose wavelength is large compared to the distance between the protons of a hydrogen molecule, the total scattering cross-sections for ortho- and parahydrogen are

$$\sigma_{\text{ortho}}=\frac{4\pi}{9}\{(3b^++b^-)^2+2(b^+-b^-)^2\},$$

$$\sigma_{\text{para}}=\frac{4\pi}{9}(3b^++b^-)^2,$$

where b^+ and b^- are the triplet and singlet scattering lengths respectively for the proton. (The two proton spins are parallel in orthohydrogen, and antiparallel in parahydrogen. Hint: Use the operator form of the scattering length.)

9.2 MnF_2 has a body-centred tetragonal lattice with unit cell dimensions a, a, c. The atoms occupy the following positions (in units of a, a, c)

Mn $(0, 0, 0)$, $(\frac{1}{2}, \frac{1}{2}, \frac{1}{2})$.

F $(u, u, 0)$, $(1-u, 1-u, 0)$, $(\frac{1}{2}+u, \frac{1}{2}-u, \frac{1}{2})$, $(\frac{1}{2}-u, \frac{1}{2}+u, \frac{1}{2})$. $u = 0.31$.

In the magnetically ordered phase, the spins of the Mn ions at $(0, 0, 0)$ and $(\frac{1}{2}, \frac{1}{2}, \frac{1}{2})$ are in opposite directions. At which reciprocal lattice points does nuclear–magnetic interference occur in this phase?

APPENDIX A

The Dirac delta function

A.1 Definition and basic properties

The Dirac delta function $\delta(x)$ is a function defined to have the following properties:

$$\begin{aligned} \delta(x) &= 0 \qquad x \neq 0, \\ \delta(x) &= \infty \qquad x = 0, \\ \int_{-\infty}^{\infty} \delta(x)\,\mathrm{d}x &= 1. \end{aligned} \tag{A.1}$$

$\delta(x)$ is not a proper mathematical function but is nevertheless a very useful tool in mathematical physics. Note that the function gives a meaningful result only after the integration process.

From the definition we have

$$\int_{-\infty}^{\infty} f(x)\delta(a-x)\,\mathrm{d}x = f(a), \tag{A.2}$$

$$\delta(cx) = \frac{1}{c}\,\delta(x), \tag{A.3}$$

where c is a positive constant,

$$\delta(x) = \delta(-x). \tag{A.4}$$

A.2 Representation in terms of an infinite integral

There are several ways of representing the delta function as the limit of a proper mathematical function. The most useful one for our purposes is in terms of an infinite integral. Consider the function

$$\begin{aligned} f(x) = \int_{-k_0}^{k_0} \exp(\mathrm{i}kx)\,\mathrm{d}k &= \frac{1}{\mathrm{i}x}\{\exp(\mathrm{i}k_0x) - \exp(-\mathrm{i}k_0x)\} \\ &= \frac{2}{x}\sin k_0x. \end{aligned} \tag{A.5}$$

$f(x)$ is shown for a particular value of k_0 in Fig. A.1. The total area under the curve is

$$2\int_{-\infty}^{\infty}\frac{1}{x}\sin k_0 x\,\mathrm{d}x = 2\int_{-\infty}^{\infty}\frac{\sin y}{y}\,\mathrm{d}y = 2\pi. \tag{A.6}$$

Now consider the function $f(x)$ as k_0 increases. The height of the peak at $x=0$ increases, and the first zero occurs at smaller and smaller values of x. But the value of $\int_{-\infty}^{\infty} f(x)\,\mathrm{d}x$ remains constant. So, as $k_0 \to \infty$ the function $f(x)/2\pi$ becomes more and more like a δ function, i.e.

$$\delta(x) = \frac{1}{2\pi}\int_{-\infty}^{\infty}\exp(\mathrm{i}kx)\,\mathrm{d}k. \tag{A.7}$$

Similarly

$$\delta(k) = \frac{1}{2\pi}\int_{-\infty}^{\infty}\exp(\mathrm{i}kx)\,\mathrm{d}x, \tag{A.8}$$

$$\delta(\hbar\omega) = \frac{1}{\hbar}\delta(\omega) = \frac{1}{2\pi\hbar}\int_{-\infty}^{\infty}\exp(\mathrm{i}\omega t)\,\mathrm{d}t, \tag{A.9}$$

Fig. A.1 Plot of the function $(2/x)\sin k_0x$.

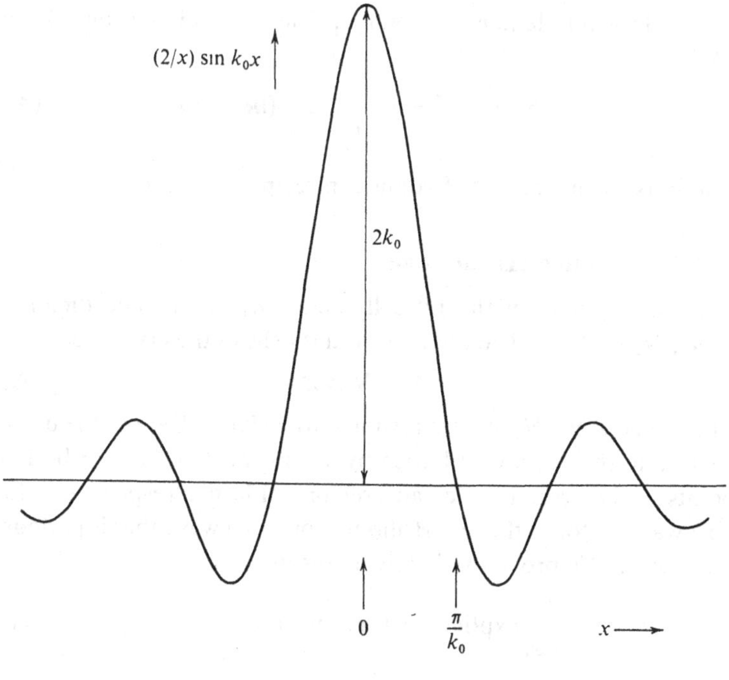

and

$$\delta(E_\lambda - E_{\lambda'} + \hbar\omega) = \frac{1}{2\pi\hbar}\int_{-\infty}^{\infty} \exp\{i(E_{\lambda'} - E_\lambda)t/\hbar\} \exp(-i\omega t)\, dt, \tag{A.10}$$

where E_λ and $E_{\lambda'}$ are constants.

A.3 Three-dimensional delta function

If $\boldsymbol{r}$ is a vector with components x, y, z, we define the three-dimensional δ-function by

$$\delta(\boldsymbol{r}) = \delta(x)\delta(y)\delta(z). \tag{A.11}$$

$\delta(\boldsymbol{r})$ has the following properties:

$$\begin{aligned} &\delta(\boldsymbol{r}) = 0 \qquad \boldsymbol{r} \neq 0, \\ &\delta(\boldsymbol{r}) = \infty \qquad \boldsymbol{r} = 0, \\ &\int_{\substack{\text{all}\\ \text{space}}} \delta(\boldsymbol{r})\, d\boldsymbol{r} = 1, \\ &\int_{\substack{\text{all}\\ \text{space}}} f(\boldsymbol{r})\delta(\boldsymbol{r} - \boldsymbol{r}_0)\, d\boldsymbol{r} = f(\boldsymbol{r}_0), \end{aligned} \tag{A.12}$$

where $d\boldsymbol{r}$ is an element of volume. The three-dimensional form of (A.7) is

$$\delta(\boldsymbol{r}) = \frac{1}{(2\pi)^3}\int_{\substack{\text{all recip}\\ \text{space}}} \exp(i\boldsymbol{\kappa}\,.\,\boldsymbol{r})\, d\boldsymbol{\kappa}, \tag{A.13}$$

where $d\boldsymbol{\kappa}$ is an element of volume in reciprocal space.

A.4 Lattice integrals and sums

Consider a crystal with unit-cell vectors $\boldsymbol{a}_1$, $\boldsymbol{a}_2$, $\boldsymbol{a}_3$ and dimensions N_1a_1, N_2a_2, N_3a_3. The number of unit cells in the crystal is

$$N = N_1N_2N_3. \tag{A.14}$$

The integers N_1, N_2, N_3 are assumed to be large. Denote the unit-cell vectors in the reciprocal lattice by $\boldsymbol{\tau}_1$, $\boldsymbol{\tau}_2$, $\boldsymbol{\tau}_3$. Let $\boldsymbol{l}$ and $\boldsymbol{\tau}$ be lattice points in the crystal lattice and reciprocal lattice respectively. Let $\boldsymbol{q}$ be a wavevector in the first Brillouin zone of a wave that is periodic in the crystal. We prove the following results:

$$\int_{\text{cell}} \exp\{i(\boldsymbol{\tau} - \boldsymbol{\tau}')\,.\,\boldsymbol{r}\}\, d\boldsymbol{r} = v_0\delta_{\boldsymbol{\tau}\boldsymbol{\tau}'}, \tag{A.15}$$

$$\int_{\text{cell}} \exp\{\mathrm{i}\boldsymbol{\kappa} \cdot (\boldsymbol{l}-\boldsymbol{l}')\}\,\mathrm{d}\boldsymbol{\kappa} = \frac{(2\pi)^3}{v_0}\delta_{ll'}, \tag{A.16}$$

$$\sum_{l} \exp(\mathrm{i}\boldsymbol{\kappa} \cdot \boldsymbol{l}) = \frac{(2\pi)^3}{v_0}\sum_{\boldsymbol{\tau}} \delta(\boldsymbol{\kappa}-\boldsymbol{\tau}), \tag{A.17}$$

$$\sum_{l} \exp\{\mathrm{i}(\boldsymbol{q}-\boldsymbol{q}') \cdot \boldsymbol{l}\} = N\delta_{\boldsymbol{qq}'}. \tag{A.18}$$

v_0 is the volume of the unit cell in the crystal lattice, and $\int_{\text{cell}}$ means integrate over the unit cell of the crystal or reciprocal lattice.

To prove (A.15) put $\boldsymbol{\tau}'=0$ and let

$$\boldsymbol{\tau} = t_1\boldsymbol{\tau}_1 + t_2\boldsymbol{\tau}_2 + t_3\boldsymbol{\tau}_3, \tag{A.19}$$

$$\boldsymbol{r} = r_1\boldsymbol{a}_1 + r_2\boldsymbol{a}_2 + r_3\boldsymbol{a}_3, \tag{A.20}$$

where t_1, t_2, t_3 are integers. Then

$$\int_{\text{cell}} \exp(\mathrm{i}\boldsymbol{\tau} \cdot \boldsymbol{r})\,\mathrm{d}\boldsymbol{r} = 0, \tag{A.21}$$

unless $\boldsymbol{\tau}=0$. This is readily seen by considering a one-dimensional component of the integral along one side of the unit cell. If t_i ($i = 1, 2, 3$) is an integer other than zero,

$$\int_0^1 \exp(2\pi\mathrm{i}t_i r_i)\,\mathrm{d}r_i = \frac{1}{2\pi\mathrm{i}t_i}\{\exp(2\pi\mathrm{i}t_i)-1\} = 0. \tag{A.22}$$

If $\boldsymbol{\tau}=0$,

$$\int_{\text{cell}} \exp(\mathrm{i}\boldsymbol{\tau} \cdot \boldsymbol{r})\,\mathrm{d}\boldsymbol{r} = v_0, \tag{A.23}$$

and (A.15) follows immediately.

The volume of the unit cell in the reciprocal lattice is $(2\pi)^3/v_0$. The result in (A.16) then follows from the previous reasoning – only the notation is different.

To prove (A.17) put

$$S = \sum_{l} \exp(\mathrm{i}\boldsymbol{\kappa} \cdot \boldsymbol{l}), \tag{A.24}$$

$$\boldsymbol{l} = l_1\boldsymbol{a}_1 + l_2\boldsymbol{a}_2 + l_3\boldsymbol{a}_3, \tag{A.25}$$

$$\boldsymbol{\kappa} = \kappa_1\boldsymbol{\tau}_1 + \kappa_2\boldsymbol{\tau}_2 + \kappa_3\boldsymbol{\tau}_3. \tag{A.26}$$

The sum over l in (A.24) is over all values of the integers l_1, l_2, l_3. Then

$$S = \sum_{l_1 l_2 l_3} \exp\{2\pi\mathrm{i}(\kappa_1 l_1 + \kappa_2 l_2 + \kappa_3 l_3)\} = S_1 S_2 S_3, \tag{A.27}$$

where

$$S_i = \sum_{l_i=-(N_i-1)/2}^{(N_i-1)/2} \exp(2\pi\mathrm{i}\kappa_i l_i) = \frac{\sin(N_i\pi\kappa_i)}{\sin(\pi\kappa_i)}. \tag{A.28}$$

For large values of N_i the function $\sin(N_i\pi\kappa_i)/\sin(\pi\kappa_i)$ is highly peaked as κ_i varies, and is effectively zero unless

$$\kappa_i = n_i, \tag{A.29}$$

where n_i is an integer. Thus S is effectively zero unless κ_1, κ_2, κ_3 are simultaneously integers, i.e. unless $\boldsymbol{\kappa}$ is a vector in the reciprocal lattice. Thus

$$S = \sum_l \exp(\mathrm{i}\boldsymbol{\kappa} \,.\, \boldsymbol{l}) = c \sum_{\boldsymbol{\tau}} \delta(\boldsymbol{\kappa} - \boldsymbol{\tau}). \tag{A.30}$$

To determine the constant c we integrate both sides of (A.30) over a unit cell in the reciprocal lattice. The integral on the right-hand side is equal to c. From (A.16) the integral on the left-hand side is equal to $(2\pi)^3/v_0$, which gives the required result.

Similar reasoning gives (A.18). Since $\boldsymbol{q}$ represents a wave periodic in the crystal,

$$\boldsymbol{q} = q_1\boldsymbol{\tau}_1 + q_2\boldsymbol{\tau}_2 + q_3\boldsymbol{\tau}_3, \tag{A.31}$$

where

$$q_i = \frac{n_i}{N}, \tag{A.32}$$

with n_i an integer ($|n_i| < N_i/2$). Put

$$S = \sum_l \exp(\mathrm{i}\boldsymbol{q} \,.\, \boldsymbol{l}) = S_1 S_2 S_3, \tag{A.33}$$

where

$$S_i = \sum_{l_i = -(N_i-1)/2}^{(N_i-1)/2} \exp(2\pi \mathrm{i} q_i l_i) = \frac{\sin(n_i\pi)}{\sin(n_i\pi/N_i)}. \tag{A.34}$$

Thus

$$\begin{aligned} S_i &= 0 \qquad \text{for } n_i \neq 0, \\ &= N_i \qquad \text{for } n_i = 0, \end{aligned} \tag{A.35}$$

and

$$\begin{aligned} \sum_l \exp(\mathrm{i}\boldsymbol{q} \,.\, \boldsymbol{l}) &= 0 \qquad \text{for } \boldsymbol{q} \neq 0, \\ &= N \qquad \text{for } \boldsymbol{q} = 0, \end{aligned} \tag{A.36}$$

which is equivalent to (A.18).

APPENDIX B

Fourier transforms

B.1 Definitions

Let $f(x)$ be a function defined by

$$f(x)=\int_{-\infty}^{\infty} g(k)\exp(\mathrm{i}kx)\,\mathrm{d}k. \tag{B.1}$$

Then $g(k)$ is said to be the *Fourier transform* of $f(x)$. $g(k)$ is given explicitly by

$$g(k)=\frac{1}{2\pi}\int_{-\infty}^{\infty} f(x)\exp(-\mathrm{i}kx)\,\mathrm{d}x. \tag{B.2}$$

This relation is proved as follows.

$$\begin{aligned}
&\frac{1}{2\pi}\int_{-\infty}^{\infty} f(x)\exp(-\mathrm{i}k'x)\,\mathrm{d}x \\
&\quad=\frac{1}{2\pi}\int_{-\infty}^{\infty}\int_{-\infty}^{\infty} g(k)\exp\{\mathrm{i}(k-k')x\}\,\mathrm{d}k\,\mathrm{d}x \\
&\quad=\int_{-\infty}^{\infty} g(k)\delta(k-k')\,\mathrm{d}k=g(k').
\end{aligned} \tag{B.3}$$

A three-dimensional function $f(\boldsymbol{r})$ has a Fourier transform $g(\boldsymbol{k})$. The two functions are related by

$$f(\boldsymbol{r})=\int_{\substack{\text{all } \boldsymbol{k}\\ \text{space}}} g(\boldsymbol{k})\exp(\mathrm{i}\boldsymbol{k}\,.\,\boldsymbol{r})\,\mathrm{d}\boldsymbol{k}, \tag{B.4}$$

$$g(\boldsymbol{k})=\frac{1}{(2\pi)^3}\int_{\substack{\text{all}\\ \text{space}}} f(\boldsymbol{r})\exp(-\mathrm{i}\boldsymbol{k}\,.\,\boldsymbol{r})\,\mathrm{d}\boldsymbol{r}. \tag{B.5}$$

B.2 Fourier transforms of some functions

The following relations are used in the text:

$$\int_{-\infty}^{\infty} \exp(-x^2/2\sigma^2)\exp(ikx)\,dx = (2\pi\sigma^2)^{1/2}\exp(-k^2\sigma^2/2), \quad \text{(B.6)}$$

$$\int \exp(-r^2/2\sigma^2)\exp(i\boldsymbol{\kappa}\,.\,\boldsymbol{r})\,d\boldsymbol{r} = (2\pi\sigma^2)^{3/2}\exp(-\kappa^2\sigma^2/2), \quad \text{(B.7)}$$

$$\int \exp(-\kappa^2\sigma^2/2)\exp(-i\boldsymbol{\kappa}\,.\,\boldsymbol{r})\,d\boldsymbol{\kappa} = (2\pi/\sigma^2)^{3/2}\exp(-r^2/2\sigma^2), \quad \text{(B.8)}$$

$$\int \frac{\hat{\boldsymbol{R}}}{R^2}\exp(i\boldsymbol{\kappa}\,.\,\boldsymbol{R})\,d\boldsymbol{R} = 4\pi i\frac{\hat{\boldsymbol{\kappa}}}{\kappa}, \quad \text{(B.9)}$$

$$\int \frac{1}{q^2}\exp(i\boldsymbol{q}\,.\,\boldsymbol{R})\,d\boldsymbol{q} = \frac{2\pi^2}{R}, \quad \text{(B.10)}$$

$$\int \frac{1}{R}\exp(i\boldsymbol{\kappa}\,.\,\boldsymbol{R})\,d\boldsymbol{R} = \frac{4\pi}{\kappa^2}, \quad \text{(B.11)}$$

$$\operatorname{curl}\left(\frac{\boldsymbol{s}\times\hat{\boldsymbol{R}}}{R^2}\right) = \frac{1}{2\pi^2}\int \hat{\boldsymbol{q}}\times(\boldsymbol{s}\times\hat{\boldsymbol{q}})\exp(i\boldsymbol{q}\,.\,\boldsymbol{R})\,d\boldsymbol{q}. \quad \text{(B.12)}$$

The three-dimensional integrals are over all space. We outline the proof of these relations.

$$\text{(B.6)} \quad \int_{-\infty}^{\infty} \exp(-x^2/2\sigma^2)\exp(ikx)\,dx$$
$$= \exp(-k^2\sigma^2/2)\int_{-\infty}^{\infty} \exp\left\{-\frac{1}{2}\left(\frac{x}{\sigma} - ik\sigma\right)^2\right\}dx$$
$$= (2\pi\sigma^2)^{1/2}\exp(-k^2\sigma^2/2).$$

(B.7) The integral may be written as the product of $\int_{-\infty}^{\infty}\exp(-x^2/2\sigma^2)\exp(i\kappa_x x)\,dx$ and two similar integrals for y and z. Each of these integrals is given by (B.6), and the result follows.

(B.8) The relation follows from (B.4) and (B.7).

(B.9) Take the polar axis in the direction of $\boldsymbol{\kappa}$. In the integration with respect to the azimuthal angle ϕ, only the component of $\boldsymbol{R}$ in the direction of $\boldsymbol{\kappa}$ contributes. Thus

$$\int \frac{\hat{\boldsymbol{R}}}{R^2}\exp(i\boldsymbol{\kappa}\,.\,\boldsymbol{R})\,d\boldsymbol{R} = 2\pi\hat{\boldsymbol{\kappa}}\int_0^{\infty} dR\int_{-1}^{1}\cos\theta\exp(i\kappa R\cos\theta)\,d(\cos\theta)$$
$$= 4\pi i\frac{\hat{\boldsymbol{\kappa}}}{\kappa}\int_0^{\infty}\left(\frac{\sin y}{y^2} - \frac{\cos y}{y}\right)dy = 4\pi i\frac{\hat{\boldsymbol{\kappa}}}{\kappa}.$$

(B.10) Take the polar axis in the direction of $\boldsymbol{R}$.

$$\int \frac{1}{q^2} \exp(\mathrm{i}\boldsymbol{q} \,.\, \boldsymbol{R})\, \mathrm{d}\boldsymbol{q} = 2\pi \int_0^\infty \mathrm{d}q \int_{-1}^{1} \exp(\mathrm{i}qR \cos\theta)\, \mathrm{d}(\cos\theta)$$

$$= 4\pi \int_0^\infty \frac{\sin(qR)}{qR}\, \mathrm{d}q = \frac{2\pi^2}{R}.$$

(B.11) The relation follows from (B.5) and (B.10).

(B.12) From the definitions of grad and curl we have the following equations.

$$\frac{\hat{\boldsymbol{R}}}{R^2} = -\mathrm{grad}\,\frac{1}{R}, \qquad \mathrm{grad}\{\exp(\mathrm{i}\boldsymbol{q} \,.\, \boldsymbol{R})\} = \mathrm{i}\boldsymbol{q} \exp(\mathrm{i}\boldsymbol{q} \,.\, \boldsymbol{R}), \quad \text{(B.13)}$$

$$\mathrm{curl}\{(\boldsymbol{s} \times \boldsymbol{q}) \exp(\mathrm{i}\boldsymbol{q} \,.\, \boldsymbol{R})\} = \mathrm{i}\boldsymbol{q} \times (\boldsymbol{s} \times \boldsymbol{q}) \exp(\mathrm{i}\boldsymbol{q} \,.\, \boldsymbol{R}). \quad \text{(B.14)}$$

The grad and curl operators operate only on $\boldsymbol{R}$. Now

$$\mathrm{curl}\left(\frac{\boldsymbol{s} \times \hat{\boldsymbol{R}}}{R^2}\right) = -\mathrm{curl}\left(\boldsymbol{s} \times \mathrm{grad}\,\frac{1}{R}\right)$$

$$= -\frac{1}{2\pi^2} \int \frac{1}{q^2}\, \mathrm{curl}[\boldsymbol{s} \times \mathrm{grad}\{\exp(\mathrm{i}\boldsymbol{q} \,.\, \boldsymbol{R})\}]\, \mathrm{d}\boldsymbol{q} \quad \text{(from (B.10))}$$

$$= \frac{1}{2\pi^2} \int \hat{\boldsymbol{q}} \times (\boldsymbol{s} \times \hat{\boldsymbol{q}}) \exp(\mathrm{i}\boldsymbol{q} \,.\, \boldsymbol{R})\, \mathrm{d}\boldsymbol{q}.$$

B.3 Fourier representation of a periodic function

If $f(\boldsymbol{r})$ is a function periodic in the unit cell of a crystal it may be expressed as

$$f(\boldsymbol{r}) = \frac{1}{v_0} \sum_{\boldsymbol{\tau}} g(\boldsymbol{\tau}) \exp(-\mathrm{i}\boldsymbol{\tau} \,.\, \boldsymbol{r}), \quad \text{(B.15)}$$

where v_0 is the volume of the unit cell in the crystal, and $\boldsymbol{\tau}$ is a vector in the reciprocal lattice. To obtain $g(\boldsymbol{\tau})$, multiply by $\exp(\mathrm{i}\boldsymbol{\tau}' \,.\, \boldsymbol{r})$, integrate over the unit cell of the crystal, and use (A.15).

$$\int_{\text{cell}} f(\boldsymbol{r}) \exp(\mathrm{i}\boldsymbol{\tau}' \,.\, \boldsymbol{r})\, \mathrm{d}\boldsymbol{r}$$

$$= \frac{1}{v_0} \sum_{\boldsymbol{\tau}} g(\boldsymbol{\tau}) \int_{\text{cell}} \exp\{\mathrm{i}(\boldsymbol{\tau}' - \boldsymbol{\tau}) \,.\, \boldsymbol{r}\}\, \mathrm{d}\boldsymbol{r} = g(\boldsymbol{\tau}'). \quad \text{(B.16)}$$

Note that $g(\boldsymbol{\tau})$ is the Fourier transform of the function that is equal to $f(\boldsymbol{r})$ within the $l = 0$ unit cell of the crystal and is zero outside this cell.

APPENDIX C

Some results for linear operators and matrix elements

We derive here some results, relevant to the present work, for linear operators and matrix elements in quantum mechanics. Let u_j and u_k be members of a finite family of functions of generalised† position $\boldsymbol{r}$. We assume firstly that they are complete, i.e. that any function of physical interest can be expressed as a linear combination of the us, and secondly that they are orthonormal, i.e. that they satisfy the relation

$$\begin{aligned} \int u_j^* u_k \, \mathrm{d}\boldsymbol{r} &= 1, \qquad j=k, \\ &= 0, \qquad j \neq k. \end{aligned} \tag{C.1}$$

The asterisk denotes the complex conjugate. Let A, B, C... be linear operators, which in general do not commute with each other.

(i) The *Hermitian conjugate* of A is denoted by A^+ and defined by

$$\begin{aligned} \int u_j^* A u_k \, \mathrm{d}\boldsymbol{r} &= \int (A^+ u_j)^* u_k \, \mathrm{d}\boldsymbol{r} \\ &= \left\{ \int u_k^* A^+ u_j \, \mathrm{d}\boldsymbol{r} \right\}^*. \end{aligned} \tag{C.2}$$

In Dirac notation (C.2) is written

$$\langle j|A|k\rangle = \langle k|A^+|j\rangle^*. \tag{C.3}$$

It follows from the definition that

$$(A^+)^+ = A. \tag{C.4}$$

† The variable $\boldsymbol{r}$ stands for all the position and spin variables of the physical system under discussion. The notation $\int \mathrm{d}\boldsymbol{r}$ in the present section stands for the integration over all space with respect to the position variables, and the sum over all the spin variables. A fuller discussion of the topics in Appendices C to F will be found in a number of textbooks on quantum mechanics. See for example Merzbacher (1970).

If $f(x)$ is a real function, the Hermitian conjugate of the operator $f(A)$ is

$$[f(A)]^+ = f(A^+). \tag{C.5}$$

If $f(x)$ is a complex function, it can be written as

$$f(x) = u(x) + \mathrm{i}v(x), \tag{C.6}$$

where $u(x)$ and $v(x)$ are real functions. Then

$$[f(A)]^+ = [u(A) + \mathrm{i}v(A)]^+ = u(A^+) - \mathrm{i}v(A^+). \tag{C.7}$$

If $A = A^+$, the operator is said to be *Hermitian*. All operators corresponding to physical quantities, e.g. position, momentum, energy, etc., are Hermitian. Examples of non-Hermitian operators in the present book are the annihilation and creation operators a and a^+, and the raising and lowering operators S^+ and S^-. Each one is the Hermitian conjugate of its partner.

(ii) From (C.2)

$$\int u_i^* A B u_k \,\mathrm{d}\boldsymbol{r} = \int (A^+ u_i)^* B u_k \,\mathrm{d}\boldsymbol{r} = \int (B^+ A^+ u_i)^* u_k \,\mathrm{d}\boldsymbol{r}, \tag{C.8}$$

i.e.

$$(AB)^+ = B^+ A^+. \tag{C.9}$$

Similarly

$$(ABC)^+ = C^+ B^+ A^+, \tag{C.10}$$

and so on.

(iii) For any pair of operators

$$\sum_j \left(\int u_i^* A u_j \,\mathrm{d}\boldsymbol{r}\right)\left(\int u_j^* B u_k \,\mathrm{d}\boldsymbol{r}\right) = \int u_i^* A B u_k \,\mathrm{d}\boldsymbol{r}, \tag{C.11}$$

or, in Dirac notation,

$$\sum_j \langle i|A|j\rangle\langle j|B|k\rangle = \langle i|AB|k\rangle. \tag{C.12}$$

The sum is over all the functions in the family u_j. This result is known as the *closure relation*.

To prove (C.11) we first express Bu_k in terms of the us themselves, i.e. we put

$$Bu_k = \sum_j c_{jk} u_j. \tag{C.13}$$

To find the coefficients c_{jk}, multiply by a particular $u_{j'}^*$ and integrate over all space. Since the u_j form an orthonormal set

$$\begin{aligned} \int u_{j'}^* B u_k \,\mathrm{d}\boldsymbol{r} &= \sum_j c_{jk} \int u_{j'}^* u_j \,\mathrm{d}\boldsymbol{r} \\ &= c_{j'k}. \end{aligned} \tag{C.14}$$

Thus
$$Bu_k = \sum_j \left(\int u_j^* B u_k \, d\mathbf{r} \right) u_j. \tag{C.15}$$

Now
$$\sum_j \left(\int u_i^* A u_j \, d\mathbf{r} \right)\left(\int u_j^* B u_k \, d\mathbf{r} \right)$$
$$= \int u_i^* A \sum_j \left(\int u_j^* B u_k \, d\mathbf{r} \right) u_j \, d\mathbf{r}$$
$$= \int u_i^* A B u_k \, d\mathbf{r}. \tag{C.16}$$

The last line follows from (C.15).

(iv)
$$\sum_k \langle k|AB|k\rangle = \sum_k \langle k|BA|k\rangle. \tag{C.17}$$

This result follows from (C.12).
$$\sum_k \langle k|AB|k\rangle = \sum_k \sum_j \langle k|A|j\rangle\langle j|B|k\rangle. \tag{C.18}$$
$$\sum_k \langle k|BA|k\rangle = \sum_k \sum_j \langle k|B|j\rangle\langle j|A|k\rangle. \tag{C.19}$$

The two right-hand sides are equal, since they differ only in the j, k labelling.

(v) The thermal average of the operator AB is defined in (2.58) to be
$$\langle AB\rangle = \sum_\lambda p_\lambda \langle \lambda|AB|\lambda\rangle. \tag{C.20}$$

From (C.3) and (C.9)
$$\langle AB\rangle^* = \sum_\lambda p_\lambda \langle \lambda|B^+A^+|\lambda\rangle = \langle B^+A^+\rangle. \tag{C.21}$$

APPENDIX D

Heisenberg operators

D.1 The Schrödinger and Heisenberg pictures of quantum mechanics

Quantum mechanics may be formulated in two equivalent ways, known as the *Schrödinger* and *Heisenberg* pictures. In most introductions to the subject the Schrödinger picture is used. Here the operator A corresponding to a physical quantity $\mathscr{A}$ is constant in time, and so therefore are its eigenfunctions. The function $\phi(t)$ representing the state of the physical system under consideration varies with time according to the equation

$$\frac{\partial\phi(t)}{\partial t}=\frac{1}{\mathrm{i}\hbar}H\phi(t), \tag{D.1}$$

where H is the Hamiltonian operator of the system.

In the Heisenberg picture the state function ϕ remains fixed in time. The Heisenberg operator $A(t)$ corresponding to the Schrödinger operator A is

$$A(t)=T^{+}AT, \tag{D.2}$$

where

$$T=\exp(-\mathrm{i}Ht/\hbar). \tag{D.3}$$

An operator of the form $\exp(\alpha H)$, where α is a number, is defined by expanding the exponential, i.e.

$$\exp(\alpha H)=1+\alpha H+\frac{\alpha^2}{2!}H^2+\dots \tag{D.4}$$

The expectation value of $\mathscr{A}$ is given by

$\langle\phi(t)|A|\phi(t)\rangle$ on the Schrödinger picture,

$\langle\phi|A(t)|\phi\rangle$ on the Heisenberg picture.

It can be shown from (D.1) and (D.2) that the two expressions are equal, as they must be, since the expectation value of an observable is a physically measured quantity and cannot depend on the picture used.

Eq. (D.1) does not apply in the Heisenberg picture. Instead we have an equation giving the time variation of the operator $A(t)$. From (D.2) and (D.3)

$$\frac{d}{dt}A(t)=\frac{i}{\hbar}HA(t)-\frac{i}{\hbar}A(t)H+T^{+}\frac{\partial A}{\partial t}T. \qquad \text{(D.5)}$$

If A does not vary explicitly with time

$$\frac{d}{dt}A(t)=\frac{i}{\hbar}[H, A(t)]. \qquad \text{(D.6)}$$

D.2 Properties of Heisenberg operators

(i) If A commutes with H, the Heisenberg and Schrödinger operators are the same at all times. For

$$A(t)=T^{+}AT=AT^{+}T=A. \qquad \text{(D.7)}$$

(ii) The Hermitian conjugate of the Heisenberg operator corresponding to A is the Heisenberg operator corresponding to A^{+}. This follows from (C.10) and (D.2).

$$[A(t)]^{+}=(T^{+}AT)^{+}=T^{+}A^{+}T=A^{+}(t). \qquad \text{(D.8)}$$

It follows that if A is Hermitian, so is $A(t)$.

(iii) $$[A(t)]^{2}=T^{+}ATT^{+}AT=T^{+}A^{2}T, \qquad \text{(D.9)}$$

and in general $$[A(t)]^{n}=T^{+}A^{n}T. \qquad \text{(D.10)}$$

Thus

$$\begin{aligned}\exp\{\alpha A(t)\} &= 1+\alpha A(t)+\frac{\alpha^{2}}{2!}[A(t)]^{2}+\dots \\ &= T^{+}\left(1+\alpha A+\frac{\alpha^{2}}{2!}A^{2}+\dots\right)T \\ &= T^{+}\exp(\alpha A)T. \end{aligned} \qquad \text{(D.11)}$$

(iv) If $A(t_1)$ and $B(t_2)$ are two Heisenberg operators, the value of $\langle\lambda|A(t_1)B(t_2)|\lambda\rangle$ depends not on t_1 and t_2 separately, but only on the difference t_2-t_1. To see this we note that for any operator C

$$\langle\lambda|\exp(iHt/\hbar)C|\lambda\rangle=\exp(iE_{\lambda}t/\hbar)\langle\lambda|C|\lambda\rangle, \qquad \text{(D.12)}$$

where E_{λ} is the eigenvalue of H for the state λ. Then

$$\begin{aligned}&\langle\lambda|A(t_1)B(t_2)|\lambda\rangle \\ &\quad=\langle\lambda|\exp(iHt_1/\hbar)A\exp(-iHt_1/\hbar)\exp(iHt_2/\hbar)B\exp(-iHt_2/\hbar|\lambda\rangle \\ &\quad=\exp(iE_{\lambda}t_1\hbar)\langle\lambda|A\exp\{iH(t_2-t_1)/\hbar\}B\exp(-iHt_2/\hbar)|\lambda\rangle \\ &\quad=\langle\lambda|A\exp\{iH(t_2-t_1)/\hbar\}B\exp\{-iH(t_2-t_1)/\hbar\}|\lambda\rangle \\ &\quad=\langle\lambda|A(0)B(t_2-t_1)|\lambda\rangle. \end{aligned} \qquad \text{(D.13)}$$

(v) $$\langle A(0)B(t)\rangle = \langle B(t)A(\mathrm{i}\hbar\beta)\rangle. \tag{D.14}$$

In this equation $\langle\ \rangle$ denotes the thermal average at temperature T (see p. 20) and $\beta = 1/k_B T$. To prove the result we have

$$\begin{aligned}
&\langle A(0)B(t)\rangle \\
&= \frac{1}{Z}\sum_\lambda \exp(-E_\lambda\beta)\langle\lambda|A(0)B(t)|\lambda\rangle \\
&= \frac{1}{Z}\sum_\lambda \langle\lambda|AT^+BT\exp(-H\beta)|\lambda\rangle \\
&= \frac{1}{Z}\sum_\lambda \langle\lambda|T^+BT\exp(-H\beta)A|\lambda\rangle \qquad \text{(from C.17)} \\
&= \frac{1}{Z}\sum_\lambda \exp(-E_\lambda\beta)\langle\lambda|\exp(H\beta)T^+BT\exp(-H\beta)A|\lambda\rangle \\
&= \langle B(t-\mathrm{i}\hbar\beta)A(0)\rangle = \langle B(t)A(\mathrm{i}\hbar\beta)\rangle. \qquad \text{(D.15)}
\end{aligned}$$

The last step follows from (D.13).

APPENDIX E

The harmonic oscillator in quantum mechanics

E.1 Annihilation and creation operators

Consider a particle of mass M undergoing simple harmonic motion in one dimension. Let the force on it when it is displaced a distance Q from its equilibrium position be $-kQ$, where k is a constant. The angular frequency of oscillation ω is given by

$$\omega^2 = \frac{k}{M}. \tag{E.1}$$

The energy is

$$E = \frac{1}{2M}P^2 + \tfrac{1}{2}kQ^2, \tag{E.2}$$

where P is the momentum of the particle. Thus the Hamiltonian is

$$H = \frac{1}{2M}(P^2 + M^2\omega^2 Q^2). \tag{E.3}$$

If P and Q are taken to be the quantum mechanical operators corresponding to momentum and position respectively, (E.3) gives the operator form of the Hamiltonian.

The operator for momentum is

$$P = -\mathrm{i}\hbar \frac{\mathrm{d}}{\mathrm{d}Q}. \tag{E.4}$$

Thus

$$[P, Q] = -\mathrm{i}\hbar. \tag{E.5}$$

We define operators a and a^+ by

$$a = (2M\hbar\omega)^{-1/2}(M\omega Q + \mathrm{i}P), \qquad a^+ = (2M\hbar\omega)^{-1/2}(M\omega Q - \mathrm{i}P). \tag{E.6}$$

From (E.3), (E.5), and (E.6) the following results may be deduced

(i)
$$aa^+ = \frac{H}{\hbar\omega} + \frac{1}{2}, \qquad a^+a = \frac{H}{\hbar\omega} - \frac{1}{2}, \tag{E.7}$$

$$[a, a^+] = 1. \tag{E.8}$$

(ii)
$$[H, a] = -\hbar\omega a, \qquad [H, a^+] = \hbar\omega a^+. \tag{E.9}$$

(iii) The eigenvalues of H have the form

$$E_n = (n + \tfrac{1}{2})\hbar\omega, \qquad n = 0, 1, 2, \ldots \tag{E.10}$$

(iv) If ψ_n is a normalised eigenfunction of H corresponding to the eigenvalue E_n, then

$$a^+\psi_n = (n+1)^{1/2}\psi_{n+1}, \qquad a\psi_n = n^{1/2}\psi_{n-1}. \tag{E.11}$$

We can arrange the eigenfunctions of the Hamiltonian on a ladder (Fig. E.1), the position on the ladder being proportional to the eigenvalue, i.e. the energy. The operators a and a^+ are known as *ladder operators.* Eq. (E.11) shows that a^+ converts the function ψ_n into the function one rung up the ladder. It is known as a *creation* operator, as it creates a quantum of energy. Similarly, the operator a, which knocks the eigenfunction one rung down the ladder, is known as an *annihilation* operator. Note that $a\psi_0 = 0$, expressing the fact that the state ψ_0 is on the bottom rung of the ladder.

Fig. E.1 Ladder arrangement of the eigenfunctions for the harmonic oscillator.

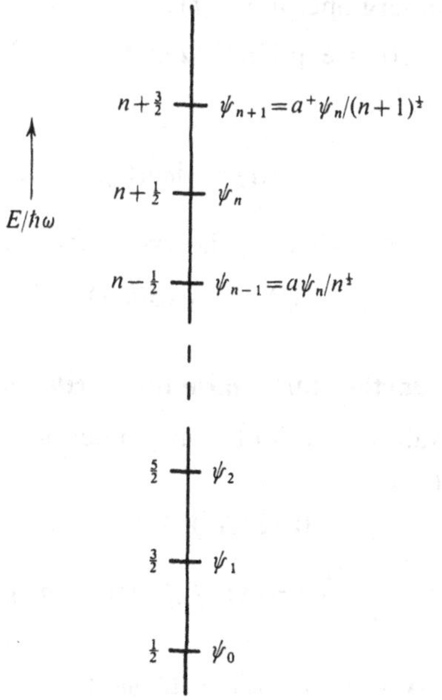

(v) From (E.11)

$$aa^+\psi_n=(n+1)\psi_n, \qquad a^+a\psi_n=n\psi_n. \tag{E.12}$$

Thus
$$\langle n|aa^+|n\rangle=n+1, \qquad \langle n|a^+a|n\rangle=n. \tag{E.13}$$

If the operator A is a product of the as and a^+s in some arbitrary order, then

$$\langle n|A|n\rangle=0, \tag{E.14}$$

unless the numbers of as and a^+s in the product are equal. For if this condition is not satisfied, $A\psi_n$ is an eigenfunction other than ψ_n, and the orthogonal property of the eigenfunctions makes the matrix element zero. Furthermore, if the physical system is equivalent to a number of independent harmonic oscillators, e.g. a crystal with harmonic interatomic forces, then an eigenfunction $|\lambda\rangle$ of the Hamiltonian of the overall system is a product of eigenfunctions of the Hamiltonians of the individual oscillators. If A is a product of the as and a^+s for the individual oscillators, then

$$\langle\lambda|A|\lambda\rangle=0, \tag{E.15}$$

unless the numbers of as and a^+s are equal for each of the oscillators separately. This is a very useful property of the annihilation and creation operators, and is used to evaluate expressions such as $\langle\lambda|U^2|\lambda\rangle$ in (3.55) and $\langle\lambda|UV|\lambda\rangle$ in (3.105).

(vi) The Heisenberg operator $a(t)$ is

$$a(t)=\exp(\mathrm{i}Ht/\hbar)a\exp(-\mathrm{i}Ht/\hbar). \tag{E.16}$$

From (D.6) and (E.9)

$$\frac{\mathrm{d}}{\mathrm{d}t}a(t)=-\mathrm{i}\omega a(t). \tag{E.17}$$

Thus
$$a(t)=a(0)\exp(-\mathrm{i}\omega t)=a\exp(-\mathrm{i}\omega t). \tag{E.18}$$

Similarly
$$a^+(t)=a^+\exp(\mathrm{i}\omega t). \tag{E.19}$$

E.2 Probability function for a harmonic oscillator

We prove the result for the probability function $f(Q)$ given in (3.19) and (3.20).† From (E.6)

$$Q=(\hbar/2M\omega)^{1/2}(a+a^+), \tag{E.20}$$

$$\frac{\mathrm{d}}{\mathrm{d}Q}=\frac{\mathrm{i}}{\hbar}P=(M\omega/2\hbar)^{1/2}(a-a^+). \tag{E.21}$$

† The proof here follows that of Landau and Lifshitz (1958).

From (3.15) and (3.16)

$$f=\frac{1}{Z}\sum_n \exp(-E_n\beta)\psi_n^2, \tag{E.22}$$

where $E_n=(n+\frac{1}{2})\hbar\omega$. Therefore

$$\frac{df}{dQ}=\frac{2}{Z}\sum_n \exp(-E_n\beta)\psi_n\frac{d\psi_n}{dQ}. \tag{E.23}$$

From (E.11) and (E.21)

$$\psi_n\frac{d}{dQ}\psi_n=(M\omega/2\hbar)^{1/2}\{n^{1/2}\psi_n\psi_{n-1}-(n+1)^{1/2}\psi_n\psi_{n+1}\}. \tag{E.24}$$

In the summation over n in (E.23) a specific $\psi_n\psi_{n+1}$ term occurs twice, once multiplied by $\exp(-E_n\beta)$, and once multiplied by

$$\exp(-E_{n+1}\beta)=\exp(-\hbar\omega\beta)\exp(-E_n\beta). \tag{E.25}$$

Therefore $$\frac{df}{dQ}=-(2M\omega/\hbar)^{1/2}\{1-\exp(-\hbar\omega\beta)\}S, \tag{E.26}$$

where $$S=\frac{1}{Z}\sum_n \exp(-E_n\beta)(n+1)^{1/2}\psi_n\psi_{n+1}. \tag{E.27}$$

Similarly $$Qf=(\hbar/2M\omega)^{1/2}\{1+\exp(-\hbar\omega\beta)\}S. \tag{E.28}$$

Therefore $$\frac{df/dQ}{Qf}=-\frac{2M\omega}{\hbar}\frac{1-\exp(-\hbar\omega\beta)}{1+\exp(-\hbar\omega\beta)}=-\frac{1}{\sigma^2}, \tag{E.29}$$

where $$\sigma^2=\frac{\hbar}{2M\omega}\coth(\tfrac{1}{2}\hbar\omega\beta). \tag{E.30}$$

The solution of (E.29) is

$$f=C\exp(-Q^2/2\sigma^2), \tag{E.31}$$

where C is a constant. Its value, obtained from (3.17) and (I.11), is

$$C=(2\pi\sigma^2)^{-1/2}. \tag{E.32}$$

From (E.31), (E.32), and (I.12), the thermal average of Q^2 is

$$\langle Q^2\rangle=\int_{-\infty}^{\infty}Q^2 f(Q)\,dQ=\sigma^2, \tag{E.33}$$

and the thermal average of exp Q is

$$\begin{aligned}\langle\exp Q\rangle&=C\int_{-\infty}^{\infty}\exp Q\exp(-Q^2/2\sigma^2)\,dQ\\&=C\exp(\tfrac{1}{2}\sigma^2)\int_{-\infty}^{\infty}\exp\left\{-\frac{1}{2}\left(\frac{Q}{\sigma}-\sigma\right)^2\right\}dQ\\&=\exp(\tfrac{1}{2}\sigma^2).\end{aligned} \tag{E.34}$$

Thus
$$\langle \exp Q \rangle = \exp\{\tfrac{1}{2}\langle Q^2 \rangle\}. \tag{E.35}$$

E.3 Value of $\langle n \rangle$

From (3.15) and (E.10)

$$\langle n \rangle = \sum_n p_n n = \sum_{n=0}^{\infty} n\, \mathrm{e}^{-nx} \Big/ \sum_{n=0}^{\infty} \mathrm{e}^{-nx}, \tag{E.36}$$

where
$$x = \hbar\omega\beta. \tag{E.37}$$

Now
$$\sum_{n=0}^{\infty} \mathrm{e}^{-nx} = 1/(1-\mathrm{e}^{-x}). \tag{E.38}$$

Differentiating this expression with respect to x gives

$$\sum_{n=0}^{\infty} n\, \mathrm{e}^{-nx} = \mathrm{e}^{-x}/(1-\mathrm{e}^{-x})^2. \tag{E.39}$$

Thus
$$\langle n \rangle = 1/(\mathrm{e}^x - 1), \tag{E.40}$$

$$\langle n+1 \rangle = \mathrm{e}^x/(\mathrm{e}^x - 1), \tag{E.41}$$

and
$$\langle 2n+1 \rangle = (\mathrm{e}^x + 1)/(\mathrm{e}^x - 1) = \coth \tfrac{1}{2}x. \tag{E.42}$$

For $x \ll 1$,
$$\coth x = \frac{1}{x} + \frac{x}{3} - \frac{x^3}{45} + \dots \tag{E.43}$$

APPENDIX F

Angular momentum in quantum mechanics

F.1 Raising and lowering operators

Let S_x, S_y, S_z be the operators corresponding to the x, y, z components of angular momentum in units of $\hbar$. The angular momentum may be of any kind – spin, orbital, or the resultant angular momentum of a system of particles. The operator

$$\boldsymbol{S}^2 = S_x^2 + S_y^2 + S_z^2 \tag{F.1}$$

corresponds to the square of the magnitude of the angular momentum.

It is a fundamental result in quantum mechanics that S_x, S_y, and S_z commute with $\boldsymbol{S}^2$, but not with each other. The commutations relations are

$$[\boldsymbol{S}^2, S_x] = [\boldsymbol{S}^2, S_y] = [\boldsymbol{S}^2, S_z] = 0, \tag{F.2}$$

$$[S_x, S_y] = \mathrm{i}S_z, \text{ etc.} \tag{F.3}$$

We define operators

$$S^+ = S_x + \mathrm{i}S_y, \qquad S^- = S_x - \mathrm{i}S_y. \tag{F.4}$$

The following results may be deduced from the commutation relations

$$S^+S^- = \boldsymbol{S}^2 - S_z^2 + S_z, \tag{F.5}$$

$$S^-S^+ = \boldsymbol{S}^2 - S_z^2 - S_z, \tag{F.6}$$

$$[S_z, S^+] = S^+, \qquad [S_z, S^-] = -S^-. \tag{F.7}$$

Since $\boldsymbol{S}^2$ and S_z commute we can find functions which are eigenfunctions of both operators. The following results may be deduced from (F.5) to (F.7):

(i) The eigenvalues of $\boldsymbol{S}^2$ have the form $S(S+1)$, where $S = 0, \frac{1}{2}, 1, \frac{3}{2}, 2, \ldots$

(ii) There are $(2S+1)$ eigenfunctions of $\boldsymbol{S}^2$ corresponding to the eigenvalue $S(S+1)$. They are eigenfunctions of S_z with eigenvalues

M, where

$$M = S, S-1, \ldots, -S. \tag{F.8}$$

(iii) If $|S, M\rangle$ is a normalised eigenfunction of $\boldsymbol{S}^2$ and S_z with respective eigenvalues $S(S+1)$ and M, then

$$\begin{aligned} S^+|S, M\rangle &= \{(S-M)(S+M+1)\}^{1/2}|S, M+1\rangle \\ S^-|S, M\rangle &= \{(S+M)(S-M+1)\}^{1/2}|S, M-1\rangle. \end{aligned} \tag{F.9}$$

For a fixed value of S, the set of eigenfunctions $|S, M\rangle$ may be regarded as arranged on a ladder with $2S+1$ rungs. The operator S^+ converts the function $|S, M\rangle$ into the function one rung *up* the ladder, while the operator S^- converts it into the function one rung *down*. From (F.9)

$$S^+|S, S\rangle = S^-|S, -S\rangle = 0, \tag{F.10}$$

which is consistent with the fact that the ladder is bounded at $M = \pm S$. Like a and a^+ for the harmonic oscillator, the operators S^+ and S^- are known as *ladder* operators. S^+ is referred to as the *raising* operator, and S^- as the *lowering* operator.

F.2 Pauli spin operators

For $S = \frac{1}{2}$ there are only 2 eigenfunctions for S_z. Denote them by u and v, i.e.

$$|\tfrac{1}{2}, \tfrac{1}{2}\rangle \equiv u, \qquad |\tfrac{1}{2}, -\tfrac{1}{2}\rangle \equiv v. \tag{F.11}$$

Colloquially u and v are referred to as the 'spin up' and 'spin down' states respectively. The states are normalised and orthogonal, i.e.

$$\begin{aligned} \langle u|u\rangle &= \langle v|v\rangle = 1 \\ \langle u|v\rangle &= \langle v|u\rangle = 0. \end{aligned} \tag{F.12}$$

From (F.9)

$$\begin{aligned} S^+u &= 0, \qquad & S^-u &= v, \\ S^+v &= u, \qquad & S^-v &= 0. \end{aligned} \tag{F.13}$$

The Pauli spin operators are defined by

$$\sigma_x = 2S_x, \text{ etc.} \tag{F.14}$$

They are used only for the case $S = \frac{1}{2}$. From (F.4) and (F.14)

$$\sigma_x = (S^+ + S^-), \qquad \sigma_y = -\mathrm{i}(S^+ - S^-). \tag{F.15}$$

From (F.13) and (F.15)

$$\begin{aligned} \sigma_x u &= v, & \sigma_y u &= \mathrm{i}v, & \sigma_z u &= u, \\ \sigma_x v &= u, & \sigma_y v &= -\mathrm{i}u, & \sigma_z v &= -v. \end{aligned} \tag{F.16}$$

APPENDIX G

Normal modes of crystals

A theoretical treatment of normal modes of vibration for atoms in a crystal may be found in a number of textbooks (see for example Ghatak and Kothari, 1972). We give here an outline of the theory, showing first how the modes arise in classical mechanics and then how they are quantised.

G.1 One-dimensional crystal

Several of the basic properties of normal modes can be obtained from a simple one-dimensional crystal. Consider a line of identical particles of mass M, distance a apart. Suppose the particles can move only at right angles to the line. Let the displacement of the nth particle be u_n (Fig. G.1). Assume that the forces are harmonic, i.e. the force on the nth particle has the form

$$F_n = \alpha_0 u_n + \alpha_1(u_{n-1} + u_{n+1}) + \alpha_2(u_{n-2} + u_{n+2}) + \ldots, \qquad \text{(G.1)}$$

where α_0, α_1, $\alpha_2, \ldots$ are constants known as *force constants*. Although in principle the forces may extend to a large number of neighbours, it is usually assumed that the force constants become small for distant neighbours.

Fig. G.1 Transverse displacements of the atoms in a one-dimensional crystal.

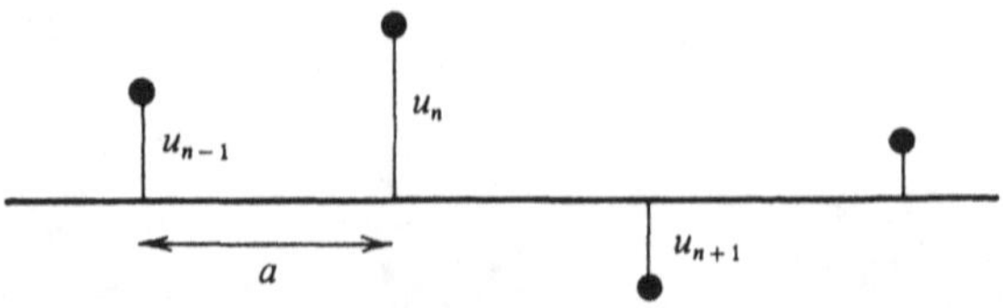

The equation of motion for the nth atom is

$$F_n = M\ddot{u}_n, \tag{G.2}$$

and it can be shown that a solution of this equation has the form

$$u_n(t) = A_q \exp\{i(qna - \omega t)\}. \tag{G.3}$$

Eq. (G.3) represents a sinusoidal wave of wavenumber $q = 2\pi/\text{wavelength}$, angular frequency ω, and amplitude A_q, running through the lattice. It is known as a *normal mode.*

For a given M and a set of α values, ω may be calculated as a function of q; the relation is known as the *dispersion relation.* It is shown in Example 3.1 that the dispersion relation in the present case is

$$\omega^2 = \frac{4}{M}\sum_\nu \alpha_\nu \sin^2(\tfrac{1}{2}\nu q a), \tag{G.4}$$

where α_ν is the force constant for the νth neighbour. An example of the relation is shown in Fig. G.2. It can be seen from (G.4) that ω is a periodic function of q, i.e.

$$\omega(q) = \omega\left(q + \frac{2\pi}{a}\right). \tag{G.5}$$

Fig. G.2 Dispersion relation for a one-dimensional crystal with one atom per unit cell.

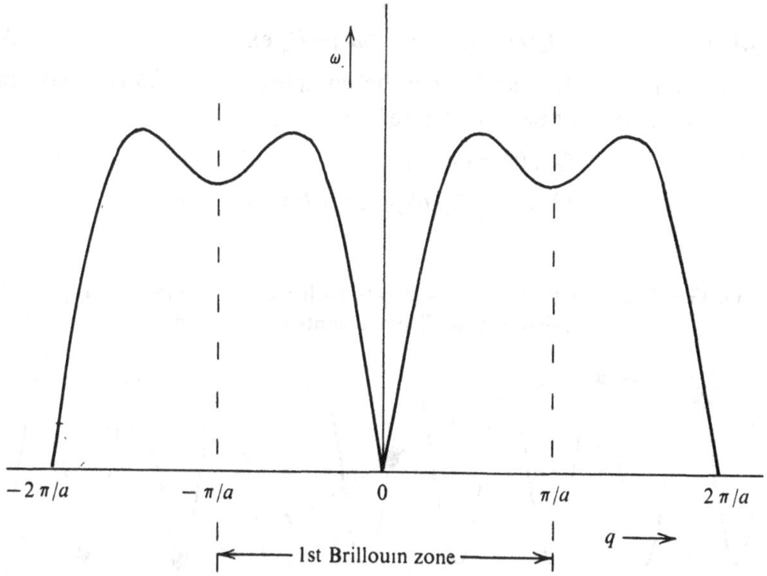

This must be the case because the normal modes are physically defined only by their displacements at the lattice points, and the waves with wavenumbers q and $q+(2\pi/a)$ have the same displacements at the lattice points. This is shown in Fig. G.3. We may therefore restrict our attention to those modes whose q values lie in a range $2\pi/a$, and it is physically reasonable to select the range with the smallest values of q, namely $|q|\leqslant \pi/a$. This range is known as the *1st Brillouin zone* for a one-dimensional lattice.

If the crystal is finite with N atoms, we further restrict the qs to waves that have an integral number of wavelengths in the length $L=Na$, i.e.

$$q=0, \pm\frac{2\pi}{L}, \pm\frac{4\pi}{L}, \ldots, \pm\frac{N}{2}\frac{2\pi}{L}. \tag{G.6}$$

Thus the q values form a one-dimensional lattice with N points in the 1st Brillouin zone.

Any linear combination of equations like (G.3) is a solution of the equations of motion. We may also have normal modes of the form

$$u_n(t)=B_q \exp\{\mathrm{i}(qna+\omega t)\}, \tag{G.7}$$

which is a wave travelling in the direction opposite to the wave in (G.3). So the most general solution can be written as

$$u_n(t)=N^{-1/2}\sum_q \exp(\mathrm{i}qna)Q_q(t), \tag{G.8}$$

where
$$Q_q(t)=A_q \exp(-\mathrm{i}\omega t)+B_q \exp(\mathrm{i}\omega t). \tag{G.9}$$

The quantities A_q and B_q may be complex. But $u_n(t)$ is a physical quantity and must be real. Therefore

$$Q_{-q}(t)=Q_q^*(t), \tag{G.10}$$

i.e.
$$Q_{-q}(t)=A_q^* \exp(\mathrm{i}\omega t)+B_q^* \exp(-\mathrm{i}\omega t). \tag{G.11}$$

Fig. G.3 Diagram showing how waves with wavenumbers q and $q+2\pi/a$ give the same displacements of the atoms.

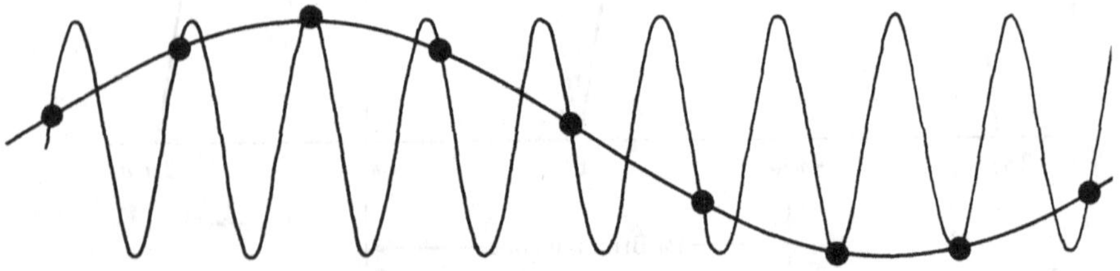

The right-hand side of (G.8) contains $2N$ arbitrary parameters. To see this we note that there are $N/2$ positive values of q. Each of these has two complex quantities, A_q and B_q, representing 4 parameters. The negative values of q give no more parameters since A_{-q} and B_{-q} are fixed by (G.11). Thus there is a total of $2N$ parameters, which is the correct number to specify the position and velocity of each of the N particles. That is to say, given the position and velocity of each particle in the crystal at time zero we can calculate the values of all the A_q and B_q. Provided the forces are harmonic, the expression in (G.8) will then correctly give the position and velocity of each of the particles at any later time.

It may be shown (see Example 3.1) that the kinetic energy T and potential energy V of the system are given by

$$T = \tfrac{1}{2}M\sum_n \dot{u}_n^2 = \tfrac{1}{2}M\sum_q \dot{Q}_q\dot{Q}_{-q} = \tfrac{1}{2}M\sum_q |\dot{Q}_q|^2, \tag{G.12}$$

$$V = \tfrac{1}{2}\sum_{nn'} \alpha_{n'}(u_n - u_{n+n'})^2 = \tfrac{1}{2}M\sum_q \omega_q^2 Q_q Q_{-q} = \tfrac{1}{2}M\sum_q \omega_q^2 |Q_q|^2. \tag{G.13}$$

G.2 Three-dimensional crystal

For a three-dimensional crystal the normal modes are plane waves. Each mode has a wavevector $\boldsymbol{q}$ whose magnitude is 2π divided by the wavelength. The atoms in a plane perpendicular to $\boldsymbol{q}$ have the same displacement, which varies sinusoidally in the direction of $\boldsymbol{q}$. For a Bravais crystal there are three normal modes corresponding to each $\boldsymbol{q}$. They are characterised by the *polarisation index* $j = 1, 2, 3$. Each mode has its own frequency $\omega_{\boldsymbol{q}j}$ and polarisation vector $\boldsymbol{e}_{\boldsymbol{q}j}$. The latter is a unit vector specifying the direction of the displacement of the atoms. The three $\boldsymbol{e}_{\boldsymbol{q}j}$ corresponding to the same $\boldsymbol{q}$ are orthogonal. They are not simply related to the direction of $\boldsymbol{q}$ except in conditions of high symmetry. For example, in a cubic crystal, if $\boldsymbol{q}$ is along a [100] or [111] axis, $\boldsymbol{e}_{\boldsymbol{q}j}$ is either in the direction of $\boldsymbol{q}$ (longitudinal mode) or perpendicular to $\boldsymbol{q}$ (transverse mode).

As for the one-dimensional crystal we specify that the normal modes should be periodic in the crystal. This causes the permitted $\boldsymbol{q}$s to form a lattice in reciprocal space with unit cell of volume $(2\pi)^3/V$, where V is the volume of the crystal. Two waves whose wavevectors differ by a vector in the reciprocal lattice give identical displacements at the lattice points in the crystal. It is therefore sufficient to consider only those $\boldsymbol{q}$ points that lie in the 1st Brillouin zone. This is the volume enclosed by planes that perpendicularly bisect the lines from the origin in reciprocal space to the neighbouring reciprocal lattice points. The 1st Brillouin zone contains N $\boldsymbol{q}$ points, where N is the number of unit cells or atoms in the crystal. Thus there are $3N$ normal modes, which is the correct number to represent in three dimensions the positions and velocities of the N atoms.

The displacement of the atoms at the lattice site l may be written as the sum of displacements due to the normal modes, thus

$$\boldsymbol{u}_l(t) = N^{-1/2} \sum_s \exp(\mathrm{i}\boldsymbol{q} \cdot \boldsymbol{l}) \boldsymbol{e}_s Q_s(t), \tag{G.14}$$

where s stands for the double index $\boldsymbol{q}$, j. The quantity $Q_s(t)$ has the same form as $Q_q(t)$ in (G.9), and

$$Q_s(t) = Q^*_{-s}(t), \tag{G.15}$$

where $-s$ stands for $-\boldsymbol{q}, j$. The expressions for the kinetic energy T and potential energy V of the crystal are

$$T = \tfrac{1}{2} M \sum_s \dot{Q}_s \dot{Q}_{-s}, \tag{G.16}$$

$$V = \tfrac{1}{2} M \sum_s Q_s Q_{-s} \omega_s^2. \tag{G.17}$$

Note that

$$\omega_s = \omega_{-s}, \quad \text{and} \quad \boldsymbol{e}_s = \boldsymbol{e}_{-s}. \tag{G.18}$$

G.3 Non-Bravais crystal

For a three-dimensional crystal with r atoms per unit cell, there are $3r$ normal modes corresponding to each $\boldsymbol{q}$. Each mode has its own frequency and polarisation vector. The latter has $3r$ components, r trios which give the directions of the displacements of the r atoms. The components may be complex, indicating that the atoms do not vibrate in phase. As $q \to 0$, the frequencies of three of the modes tend to zero, while the frequencies of the other $3r-3$ tend to finite values. The former are known as *acoustic* modes and correspond to motions in which (for small q) the atoms in the unit cell vibrate in phase. The latter, known as *optic* modes, correspond to vibrations with phase differences between the atoms in the cell. An example of the dispersion relations of a crystal with two atoms per unit cell is given in Fig. 3.14.

The reciprocal lattice, and hence the 1st Brillouin zone, are defined by the unit cell in the crystal – its contents are irrelevant. The number of unit cells in the crystal is N. For waves periodic in the crystals there are N $\boldsymbol{q}$ points in the 1st Brillouin zone, and hence a total of $3rN$ normal modes.

The displacement $\boldsymbol{u}\begin{pmatrix} l \\ d \end{pmatrix}t$ of atom d $(=1, 2, \ldots, r)$ in unit cell l may be written as

$$\boldsymbol{u}\begin{pmatrix} l \\ d \end{pmatrix}t = (M_d N)^{-1/2} \sum_s \exp(\mathrm{i}\boldsymbol{q} \cdot \boldsymbol{l}) \boldsymbol{e}_{ds} \tilde{Q}_s(t), \qquad \text{(G.19)}$$

where M_d is the mass of atom d. The tilde in $\tilde{Q}_s(t)$ is to distinguish this quantity (of dimension $\mathrm{LM}^{1/2}$) from $Q_s(t)$ (of dimension L) in (G.14). $\boldsymbol{e}_{ds}$ is that part of the polarisation vector that gives the displacement and relative phase of atom d for the mode s. Since the displacements $\boldsymbol{u}\begin{pmatrix} l \\ d \end{pmatrix}t$ are real, we have the relations

$$\tilde{Q}_s(t) = \tilde{Q}^*_{-s}(t), \qquad e^{\alpha}_{ds} = (e^{\alpha}_{d-s})^*, \qquad \text{(G.20)}$$

where e^{α}_{ds} is the component of $\boldsymbol{e}_{ds}$ along the α axis ($\alpha = x, y, z$). The components of the polarisation vectors corresponding to the same $\boldsymbol{q}$ are orthonormal, i.e.

$$\sum_d \sum_\alpha (e^{\alpha}_{d\boldsymbol{q}j})^* e^{\alpha}_{d\boldsymbol{q}j'} = \delta_{jj'}. \qquad \text{(G.21)}$$

The kinetic and potential energies of the crystal are given by

$$T = \tfrac{1}{2} \sum_s \dot{\tilde{Q}}_s \dot{\tilde{Q}}_{-s}, \qquad \text{(G.22)}$$

$$V = \tfrac{1}{2} \sum_s \tilde{Q}_s \tilde{Q}_{-s} \omega_s^2. \qquad \text{(G.23)}$$

G.4 Quantisation of the normal modes

We consider first a Bravais crystal. From (G.16) and (G.17) the Lagrangian of the crystal is

$$L = T - V = \tfrac{1}{2} M \sum_s (\dot{Q}_s \dot{Q}_{-s} - Q_s Q_{-s} \omega_s^2). \qquad \text{(G.24)}$$

Then the momentum P_s conjugate to Q_s is

$$P_s = \frac{\partial L}{\partial \dot{Q}_s} = M \dot{Q}_{-s}. \qquad \text{(G.25)}$$

The Hamiltonian of the crystal may now be written as

$$H = T + V = \sum_s \frac{1}{2M} P_s P_{-s} + \tfrac{1}{2} M Q_s Q_{-s} \omega_s^2, \qquad \text{(G.26)}$$

which is a function of a set of displacements and their conjugate momenta. We may therefore replace these quantities by their usual quantum mechanical operators to obtain the operator form of the Hamiltonian. The operators corresponding to Q_s and P_s obey the commutation relations

$$[Q_s, Q_{s'}] = [P_s, P_{s'}] = 0,$$
$$[P_s, Q_{s'}] = -i\hbar\delta_{ss'}. \tag{G.27}$$

The annihilation and creation operators for the mode s are defined by

$$a_s = (2M\hbar\omega_s)^{-1/2}(M\omega_s Q_s + iP_{-s}),$$
$$a_s^+ = (2M\hbar\omega_s)^{-1/2}(M\omega_s Q_{-s} - iP_s). \tag{G.28}$$

Note that

$$Q_s^+ = Q_{-s}, \quad \text{and} \quad P_s^+ = P_{-s}. \tag{G.29}$$

Eqs. (G.27) and (G.28) give

$$[a_s, a_{s'}^+] = \delta_{ss'}. \tag{G.30}$$

From (G.26), (G.27), and (G.28) we have

$$H = \sum_s \hbar\omega_s(a_s^+ a_s + \tfrac{1}{2}). \tag{G.31}$$

Comparing (G.31) with (E.7) we see that the Hamiltonian of the crystal is the sum of the Hamiltonians for the harmonic oscillators that represent the normal modes. This is an important result and is a direct consequence of the independence of the normal modes, which in turn is a consequence of harmonic forces. Because the overall Hamiltonian has the form of (G.31), its eigenfunctions are products of the eigenfunctions of the individual Hamiltonians. The quantum of energy $\hbar\omega_s$ is known as a *phonon*.

We require an expression for $\boldsymbol{u}(l)$ in terms of the operators a_s and a_s^+. From (G.28)

$$Q_s = (\hbar/2M\omega_s)^{1/2}(a_s + a_{-s}^+). \tag{G.32}$$

Inserting this in (G.14) we have

$$\boldsymbol{u}_l = (\hbar/2MN)^{1/2} \sum_s \boldsymbol{e}_s \omega_s^{-1/2}(a_s + a_{-s}^+)\exp(i\boldsymbol{q}\,.\,\boldsymbol{l}). \tag{G.33}$$

Now

$$\sum_s \boldsymbol{e}_s \omega_s^{-1/2} a_{-s}^+ \exp(i\boldsymbol{q}\,.\,\boldsymbol{l}) = \sum_s \boldsymbol{e}_s \omega_s^{-1/2} a_s^+ \exp(-i\boldsymbol{q}\,.\,\boldsymbol{l}). \tag{G.34}$$

This follows from the fact that the s values range over the 1st Brillouin zone, so that for every $\boldsymbol{q}, j$ term there is a corresponding term $-\boldsymbol{q}, j$. From (G.33) and (G.34)

$$\boldsymbol{u}_l = (\hbar/2MN)^{1/2} \sum_s \boldsymbol{e}_s \omega_s^{-1/2} \{a_s \exp(\mathrm{i}\boldsymbol{q} \cdot \boldsymbol{l}) + a_s^+ \exp(-\mathrm{i}\boldsymbol{q} \cdot \boldsymbol{l})\}. \tag{G.35}$$

For a non-Bravais crystal we define the annihilation and creation operators by

$$\begin{aligned} a_s &= (2\hbar\omega_s)^{-1/2}(\omega_s \tilde{Q}_s + \mathrm{i}\tilde{P}_{-s}), \\ a_s^+ &= (2\hbar\omega_s)^{-1/2}(\omega_s \tilde{Q}_{-s} - \mathrm{i}\tilde{P}_s), \end{aligned} \tag{G.36}$$

where

$$\tilde{P}_s = \frac{\partial L}{\partial \dot{\tilde{Q}}_s}. \tag{G.37}$$

Then

$$\tilde{Q}_s = (\hbar/2\omega_s)^{1/2}(a_s + a_{-s}^+). \tag{G.38}$$

Eqs. (G.19) and (G.38) give

$$\boldsymbol{u}\binom{l}{d} = (\hbar/2M_d N)^{1/2} \sum_s \omega_s^{-1/2} \{\boldsymbol{e}_{ds} a_s \exp(\mathrm{i}\boldsymbol{q} \cdot \boldsymbol{l}) + \boldsymbol{e}_{ds}^* a_s^+ \exp(-\mathrm{i}\boldsymbol{q} \cdot \boldsymbol{l})\}. \tag{G.39}$$

APPENDIX H

The proofs of two results for magnetic scattering

H.1 Relation between $Q_{\perp L}$ and the Fourier transform of the orbital magnetisation

We prove† the result

$$Q_{\perp L} = \frac{i}{\hbar\kappa}\sum_i \exp(i\boldsymbol{\kappa}\,.\,\boldsymbol{r}_i)(\boldsymbol{p}_i \times \hat{\boldsymbol{\kappa}}) = -\frac{1}{2\mu_B}\hat{\boldsymbol{\kappa}} \times \{\boldsymbol{M}_L(\boldsymbol{\kappa}) \times \hat{\boldsymbol{\kappa}}\}. \quad \text{(H.1)}$$

We need the following results from vector calculus. If $\phi(\boldsymbol{r})$ is any scalar field that vanishes at infinity, and $\boldsymbol{M}(\boldsymbol{r})$ is any vector field that vanishes at infinity, then

$$\int \{\text{grad}\,\phi(\boldsymbol{r})\} \exp(i\boldsymbol{\kappa}\,.\,\boldsymbol{r})\,d\boldsymbol{r} = -i\boldsymbol{\kappa}\int \phi(\boldsymbol{r}) \exp(i\boldsymbol{\kappa}\,.\,\boldsymbol{r})\,d\boldsymbol{r}, \quad \text{(H.2)}$$

$$\int \{\text{curl}\,\boldsymbol{M}(\boldsymbol{r})\} \exp(i\boldsymbol{\kappa}\,.\,\boldsymbol{r})\,d\boldsymbol{r} = -i\boldsymbol{\kappa} \times \int \boldsymbol{M}(\boldsymbol{r}) \exp(i\boldsymbol{\kappa}\,.\,\boldsymbol{r})\,d\boldsymbol{r}. \quad \text{(H.3)}$$

The integrals are taken over all space. Eq. (H.2) follows from the relations

$$\begin{aligned}
&\text{grad}\{\phi(\boldsymbol{r}) \exp(i\boldsymbol{\kappa}\,.\,\boldsymbol{r})\} \\
&\quad = \phi(\boldsymbol{r})\,\text{grad}\{\exp(i\boldsymbol{\kappa}\,.\,\boldsymbol{r})\} + \exp(i\boldsymbol{\kappa}\,.\,\boldsymbol{r})\,\text{grad}\,\phi(\boldsymbol{r}) \\
&\quad = i\boldsymbol{\kappa}\phi(\boldsymbol{r}) \exp(i\boldsymbol{\kappa}\,.\,\boldsymbol{r}) + \exp(i\boldsymbol{\kappa}\,.\,\boldsymbol{r})\,\text{grad}\,\phi(\boldsymbol{r}), \quad \text{(H.4)}
\end{aligned}$$

$$\int \text{grad}\{\phi(\boldsymbol{r}) \exp(i\boldsymbol{\kappa}\,.\,\boldsymbol{r})\}\,d\boldsymbol{r} = 0. \quad \text{(H.5)}$$

Eq. (H.3) follows from the relations

$$\begin{aligned}
&\text{curl}\{\boldsymbol{M}(\boldsymbol{r}) \exp(i\boldsymbol{\kappa}\,.\,\boldsymbol{r})\} \\
&\quad = \{\text{curl}\,\boldsymbol{M}(\boldsymbol{r})\} \exp(i\boldsymbol{\kappa}\,.\,\boldsymbol{r}) + [\text{grad}\{\exp(i\boldsymbol{\kappa}\,.\,\boldsymbol{r})\} \times \boldsymbol{M}(\boldsymbol{r})], \quad \text{(H.6)}
\end{aligned}$$

$$\int \text{curl}\{\boldsymbol{M}(\boldsymbol{r}) \exp(i\boldsymbol{\kappa}\,.\,\boldsymbol{r})\}\,d\boldsymbol{r} = 0. \quad \text{(H.7)}$$

† The proof here follows that of Steinsvoll *et al.* (1967).

Since $\exp(\mathrm{i}\boldsymbol{\kappa}\,.\,\boldsymbol{r}_i)$ and $\boldsymbol{p}_i$ do not commute, we put

$$\frac{\mathrm{i}}{\hbar\kappa}\sum_i \exp(\mathrm{i}\boldsymbol{\kappa}\,.\,\boldsymbol{r}_i)(\boldsymbol{p}_i\times\hat{\boldsymbol{\kappa}})$$
$$=\frac{\mathrm{i}}{2\hbar\kappa}\left[\sum_i\{\exp(\mathrm{i}\boldsymbol{\kappa}\,.\,\boldsymbol{r}_i)\boldsymbol{p}_i+\boldsymbol{p}_i\exp(\mathrm{i}\boldsymbol{\kappa}\,.\,\boldsymbol{r}_i)\}\right]\times\hat{\boldsymbol{\kappa}}. \qquad \text{(H.8)}$$

Now

$$\frac{1}{2}\left[\sum_i \exp(\mathrm{i}\boldsymbol{\kappa}\,.\,\boldsymbol{r}_i)\boldsymbol{p}_i+\boldsymbol{p}_i\exp(\mathrm{i}\boldsymbol{\kappa}\,.\,\boldsymbol{r}_i)\}\right]$$
$$=\frac{1}{2}\int\sum_i\{\delta(\boldsymbol{r}-\boldsymbol{r}_i)\boldsymbol{p}_i+\boldsymbol{p}_i\delta(\boldsymbol{r}-\boldsymbol{r}_i)\}\exp(\mathrm{i}\boldsymbol{\kappa}\,.\,\boldsymbol{r})\,\mathrm{d}\boldsymbol{r}$$
$$=-\frac{m_{\mathrm{e}}}{e}\int \boldsymbol{j}(\boldsymbol{r})\exp(\mathrm{i}\boldsymbol{\kappa}\,.\,\boldsymbol{r})\,\mathrm{d}\boldsymbol{r}, \qquad \text{(H.9)}$$

where $$\boldsymbol{j}(\boldsymbol{r})=-\frac{e}{2m_{\mathrm{e}}}\sum_i\{\delta(\boldsymbol{r}-\boldsymbol{r}_i)\boldsymbol{p}_i+\boldsymbol{p}_i\delta(\boldsymbol{r}-\boldsymbol{r}_i)\}. \qquad \text{(H.10)}$$

$\boldsymbol{j}(\boldsymbol{r})$ is the operator corresponding to the orbital current density. It can be expressed in the form

$$\boldsymbol{j}(\boldsymbol{r})=\operatorname{curl}\boldsymbol{M}_{\mathrm{L}}(\boldsymbol{r})+\operatorname{grad}\phi(\boldsymbol{r}). \qquad \text{(H.11)}$$

Then

$$\int \boldsymbol{j}(\boldsymbol{r})\exp(\mathrm{i}\boldsymbol{\kappa}\,.\,\boldsymbol{r})\,\mathrm{d}\boldsymbol{r}$$
$$=\int\{\operatorname{curl}\boldsymbol{M}_{\mathrm{L}}(\boldsymbol{r})+\operatorname{grad}\phi(\boldsymbol{r})\}\exp(\mathrm{i}\boldsymbol{\kappa}\,.\,\boldsymbol{r})\,\mathrm{d}\boldsymbol{r}$$
$$=-\mathrm{i}\left\{\boldsymbol{\kappa}\times\int\boldsymbol{M}_{\mathrm{L}}(\boldsymbol{r})\exp(\mathrm{i}\boldsymbol{\kappa}\,.\,\boldsymbol{r})\,\mathrm{d}\boldsymbol{r}\right\}-\mathrm{i}\boldsymbol{\kappa}\int\phi(\boldsymbol{r})\exp(\mathrm{i}\boldsymbol{\kappa}\,.\,\boldsymbol{r})\,\mathrm{d}\boldsymbol{r}, \qquad \text{(H.12)}$$

from (H.2) and (H.3). When the last expression is substituted into (H.8), the second term gives zero, since $\boldsymbol{\kappa}\times\hat{\boldsymbol{\kappa}}=0$. Thus

$$\frac{\mathrm{i}}{\hbar\kappa}\sum_i\exp(\mathrm{i}\boldsymbol{\kappa}\,.\,\boldsymbol{r}_i)(\boldsymbol{p}_i\times\hat{\boldsymbol{\kappa}})=-\frac{1}{2\mu_{\mathrm{B}}}\hat{\boldsymbol{\kappa}}\times\{\boldsymbol{M}_{\mathrm{L}}(\boldsymbol{\kappa})\times\hat{\boldsymbol{\kappa}}\}, \qquad \text{(H.13)}$$

where $$\boldsymbol{M}_{\mathrm{L}}(\boldsymbol{\kappa})=\int\boldsymbol{M}_{\mathrm{L}}(\boldsymbol{r})\exp(\mathrm{i}\boldsymbol{\kappa}\,.\,\boldsymbol{r})\,\mathrm{d}\boldsymbol{r}. \qquad \text{(H.14)}$$

H.2 Spin-only scattering for a localised model

We prove the result

$$\langle\lambda'|\exp(\mathrm{i}\boldsymbol{\kappa}\cdot\boldsymbol{R}_{ld})\sum_{\nu}\exp(\mathrm{i}\boldsymbol{\kappa}\cdot\boldsymbol{r}_{\nu})\boldsymbol{s}_{\nu}|\lambda\rangle$$

$$=F_{d}(\boldsymbol{\kappa})\langle\lambda'|\exp(\mathrm{i}\boldsymbol{\kappa}\cdot\boldsymbol{R}_{ld})\boldsymbol{S}_{ld}|\lambda\rangle. \qquad \text{(H.15)}$$

Put

$$f_{\nu}=\exp(\mathrm{i}\boldsymbol{\kappa}\cdot\boldsymbol{r}_{\nu}), \qquad \text{(H.16)}$$

$$g_{ld}=\exp(\mathrm{i}\boldsymbol{\kappa}\cdot\boldsymbol{R}_{ld}). \qquad \text{(H.17)}$$

Then the left-hand side of (H.15) is

$$\langle\lambda'|g_{ld}\sum_{\nu}f_{\nu}\boldsymbol{s}_{\nu}|\lambda\rangle=\sum_{\lambda''}\sum_{\nu}\langle\lambda'|f_{\nu}|\lambda''\rangle\langle\lambda''|g_{ld}\boldsymbol{s}_{\nu}|\lambda\rangle \qquad \text{(H.18)}$$

$$=\sum_{\lambda''}\langle\lambda'|f_{\nu}|\lambda''\rangle\langle\lambda''|g_{ld}\sum_{\nu}\boldsymbol{s}_{\nu}|\lambda\rangle. \qquad \text{(H.19)}$$

Eq. (H.18) comes from the closure relation (C.12). The step from (H.18) to (H.19) is justified by the fact that, since the electron space states are symmetric or antisymmetric, the matrix element $\langle\lambda'|f_{\nu}|\lambda''\rangle$ is independent of ν (see Example 7.1).

Now the states $|\lambda'\rangle$ and $|\lambda''\rangle$ differ from $|\lambda\rangle$, and hence from each other, only in the functions giving the orientations of the electron spins and the positions of the nuclei. The operator f_{ν} depends only on the space variables of the electrons. Therefore, since the electron spin states and the nuclear position states are orthogonal, the matrix element $\langle\lambda'|f_{\nu}|\lambda''\rangle$ is zero unless $\lambda''=\lambda'$. We then have

$$\langle\lambda'|f_{\nu}|\lambda'\rangle=\langle\lambda|f_{\nu}|\lambda\rangle$$

$$=\int s_{d}(\boldsymbol{r})\exp(\mathrm{i}\boldsymbol{\kappa}\cdot\boldsymbol{r})\,\mathrm{d}\boldsymbol{r}=F_{d}(\boldsymbol{\kappa}). \qquad \text{(H.20)}$$

where $s_{d}(\boldsymbol{r})$ is the normalised spin density of the unpaired electrons.

Finally

$$\sum_{\nu}\boldsymbol{s}_{\nu}=\boldsymbol{S}_{ld}. \qquad \text{(H.21)}$$

Inserting these results in (H.19) gives (H.15).

APPENDIX I

Some mathematical results

I.1 Theorem on operators

If A and B are two operators such that

$$AB - BA = c, \tag{I.1}$$

where c is a number (i.e. not an operator), then

$$\exp A \exp B = \exp(A+B)\exp(\tfrac{1}{2}c). \tag{I.2}$$

To prove this result we first prove by induction that

$$AB^n - B^nA = cnB^{n-1}. \tag{I.3}$$

Suppose (I.3) is true for a particular value of n. Multiply by B on the right

$$AB^{n+1} - B^nAB = cnB^n. \tag{I.4}$$

From (I.1)

$$B^nAB = cB^n + B^{n+1}A. \tag{I.5}$$

Therefore

$$AB^{n+1} - B^{n+1}A = c(n+1)B^n. \tag{I.6}$$

So if (I.3) is true for n, it is also true for $n+1$. It is clearly true for $n=1$. Hence it is true for all n.

If λ is a number

$$\begin{aligned} A\exp(\lambda B) - \exp(\lambda B)A &= \sum_{n=0}^{\infty}\frac{\lambda^n}{n!}(AB^n - B^nA) \\ &= \lambda c \sum_{n=1}^{\infty}\frac{(\lambda B)^{n-1}}{(n-1)!} \\ &= \lambda c \exp(\lambda B). \end{aligned} \tag{I.7}$$

In the middle line we have used (I.3). Define a function

$$f(\lambda) = \exp(\lambda A)\exp(\lambda B)\exp\{-\lambda(A+B)\}. \tag{I.8}$$

Then

$$\begin{aligned}\frac{df}{d\lambda} &= \exp(\lambda A)A\exp(\lambda B)\exp\{-\lambda(A+B)\}\\ &\quad+\exp(\lambda A)\exp(\lambda B)B\exp\{-\lambda(A+B)\}\\ &\quad-\exp(\lambda A)\exp(\lambda B)(A+B)\exp\{-\lambda(A+B)\}\\ &= \lambda cf. \end{aligned} \tag{I.9}$$

The last line follows from (I.7). The solution of (I.9) is

$$f(\lambda)=\exp(\tfrac{1}{2}c\lambda^2). \tag{I.10}$$

Equating the right-hand sides of (I.8) and (I.10), and putting $\lambda = 1$, gives the result (I.2).

I.2 Values of integrals

$$\int_{-\infty}^{\infty} \exp(-x^2/2\sigma^2)\,dx = (2\pi)^{1/2}\sigma \tag{I.11}$$

$$\int_{-\infty}^{\infty} x^2\exp(-x^2/2\sigma^2)\,dx = (2\pi)^{1/2}\sigma^3 \tag{I.12}$$

$$\int_{-\infty}^{\infty} x^4\exp(-x^2/2\sigma^2)\,dx = 3(2\pi)^{1/2}\sigma^5 \tag{I.13}$$

$$\int_{-\infty}^{\infty} \frac{\sin x}{x}\,dx = \pi. \tag{I.14}$$

SOLUTIONS TO EXAMPLES

3.1 (*a*) The equation of motion for atom n is

$$M\ddot{u}_n = \alpha_0 u_n + \alpha_1(u_{n+1} + u_{n-1}) + \alpha_2(u_{n+2} + u_{n-2}) + \dots .$$

If all the atoms are given equal displacements the force is zero. Therefore

$$\alpha_0 + 2\alpha_1 + 2\alpha_2 + \dots = 0.$$

Take a single q term in (G.8), substitute in the equation of motion, and use the relation $\ddot{Q}_q = -\omega^2 Q_q$.

$$\omega^2 M = -\sum_\nu \alpha_\nu \{\exp(\mathrm{i}\nu qa) + \exp(-\mathrm{i}\nu qa) - 2\} = 4\sum_\nu \alpha_\nu \sin^2(\tfrac{1}{2}\nu qa).$$

(*b*) Use (G.8) and (A.18).

$$T = \tfrac{1}{2}M\sum_n \dot{u}_n^2 = \tfrac{1}{2}M\sum_{qq'}\frac{1}{N}\sum_n \exp\{\mathrm{i}(q+q')na\}\dot{Q}_q\dot{Q}_{q'}$$

$$= \tfrac{1}{2}M\sum_q \dot{Q}_q\dot{Q}_{-q},$$

$$V = \tfrac{1}{2}\sum_n\sum_\nu \alpha_\nu(u_{n+\nu} - u_n)^2$$

$$= \tfrac{1}{2}\sum_{qq'}\sum_\nu \alpha_\nu \frac{1}{N}\sum_n \exp\{\mathrm{i}(q+q')na\}\{\exp(\mathrm{i}\nu qa) - 1\}$$

$$\times\{\exp(\mathrm{i}\nu q'a) - 1\}Q_qQ_{q'}$$

$$= \tfrac{1}{2}\sum_q\sum_\nu 4\alpha_\nu \sin^2(\tfrac{1}{2}\nu qa)Q_qQ_{-q} = \tfrac{1}{2}M\sum_q \omega_q^2 Q_qQ_{-q}.$$

(*c*) Use (G.9).

$$|\dot{Q}_q|^2 = \omega_q^2\{|A_q|^2 + |B_q|^2 - C_q(t)\},$$

$$\omega^2|Q_q|^2 = \omega_q^2\{|A_q|^2 + |B_q|^2 + C_q(t)\},$$

where $$C_q(t) = 2\,\mathrm{Re}\{A_q^* B_q \exp(2\mathrm{i}\omega_q t)\}.$$

The average of $C_q(t)$ over one cycle is zero. Therefore

$$\bar{T} = \bar{V} = \tfrac{1}{2}M \sum_q (|A_q|^2 + |B_q|^2)\omega_q^2.$$

3.2 As $\hbar\omega\beta \to 0$, $\sigma^2 \to 1/M\omega^2\beta$ in (3.20). Classically we have a particle moving in a potential $V = \frac{1}{2}M\omega^2 Q^2$. The probability of the particle having a displacement Q to $Q + \mathrm{d}Q$ is given by the Boltzmann distribution

$$f(Q) \propto \int_V^\infty \exp(-E\beta)\,\mathrm{d}E,$$

where E is the energy of the particle. The integral is evaluated at constant Q. Put $E = V + x$. Then

$$f(Q) \propto \exp(-V\beta) \int_0^\infty \exp(-x\beta)\,\mathrm{d}x.$$

The integral is independent of Q. Therefore

$$f(Q) \propto \exp(-\tfrac{1}{2}M\omega^2\beta Q^2),$$

which is a Gaussian with the same σ as above.

3.3 Consider (3.66). For $\hbar\omega_{\mathrm{m}}\beta \gg 1$, $\coth(\frac{1}{2}\hbar\omega\beta) \approx 1$ for most of the range of integration. Therefore

$$2W \propto \int_0^{\omega_{\mathrm{m}}} Z(\omega)\,\mathrm{d}\omega/\omega.$$

For $\hbar\omega_{\mathrm{m}}\beta \ll 1$, $\coth(\frac{1}{2}\hbar\omega\beta) \approx 2/\hbar\omega\beta$. Therefore

$$2W \propto \int_0^{\omega_m} Z(\omega)\,\mathrm{d}\omega/\omega^2.$$

3.4 For a Debye frequency spectrum $Z(\omega) = 3\omega^2/\omega_{\mathrm{m}}^3$, $\hbar\omega_{\mathrm{m}} = k_{\mathrm{B}}\theta_{\mathrm{D}}$. Use (3.67) and the results of the last example.

For $T \ll \theta_{\mathrm{D}}$, $$\langle u^2 \rangle = \frac{9\hbar}{2M\omega_{\mathrm{m}}^3} \int_0^{\omega_{\mathrm{m}}} \omega\,\mathrm{d}\omega = \frac{9}{4}\frac{\hbar^2}{M}\frac{1}{k_{\mathrm{B}}\theta_{\mathrm{D}}}.$$

For $T \gg \theta_{\mathrm{D}}$, $$\langle u^2 \rangle = \frac{9}{M\beta\omega_{\mathrm{m}}^3} \int_0^{\omega_{\mathrm{m}}} \mathrm{d}\omega = \frac{9\hbar^2}{M}\frac{T}{k_{\mathrm{B}}\theta_{\mathrm{D}}^2}.$$

3.5 (*a*) Use the results of the last example.

T/K	$\langle u^2 \rangle^{1/2}$/Å	$2W$	$\exp(-2W)$
20	0.07	0.016	0.984
1000	0.26	0.202	0.817

(*b*) The intensity of the peak is proportional to $\exp(-2W)$. $0.817/0.984 = 0.831$.

3.6 For coherent elastic scattering the amplitude of the neutron wave is proportional to

$$\sum_l \exp(\mathrm{i}\boldsymbol{\kappa}\,.\,\boldsymbol{R}_l)$$

$$=\sum_l \exp(\mathrm{i}\boldsymbol{\kappa}\,.\,\boldsymbol{l})\exp[\tfrac{1}{2}\mathrm{i}\boldsymbol{\kappa}\,.\,\boldsymbol{A}\{\exp(\mathrm{i}\boldsymbol{q}\,.\,\boldsymbol{l})+\exp(-\mathrm{i}\boldsymbol{q}\,.\,\boldsymbol{l})\}]$$

$$\approx\sum_l \exp(\mathrm{i}\boldsymbol{\kappa}\,.\,\boldsymbol{l})+\tfrac{1}{2}\mathrm{i}\boldsymbol{\kappa}\,.\,\boldsymbol{A}\exp\{\mathrm{i}(\boldsymbol{\kappa}+\boldsymbol{q})\,.\,\boldsymbol{l}\}+\tfrac{1}{2}\mathrm{i}\boldsymbol{\kappa}\,.\,\boldsymbol{A}\exp\{\mathrm{i}(\boldsymbol{\kappa}-\boldsymbol{q})\,.\,\boldsymbol{l}\}.$$

The first term in the last expression gives scattering when $\boldsymbol{\kappa}=\boldsymbol{\tau}$, the second when $\boldsymbol{\kappa}=\boldsymbol{\tau}-\boldsymbol{q}$, and the third when $\boldsymbol{\kappa}=\boldsymbol{\tau}+\boldsymbol{q}$.

Let the velocities of the incident and scattered neutrons in the laboratory frame be $\boldsymbol{v}$ and $\boldsymbol{v}'$. If the velocity of the crystal in the laboratory frame is $\boldsymbol{c}$ then, for elastic scattering in the crystal frame, $|\boldsymbol{v}-\boldsymbol{c}|=|\boldsymbol{v}'-\boldsymbol{c}|$; whence $v^2-v'^2=2(\boldsymbol{v}-\boldsymbol{v}')\,.\,\boldsymbol{c}$. Thus

$$\frac{\hbar^2}{2m}(k^2-k'^2)=\frac{m}{2}(v^2-v'^2)=m(\boldsymbol{v}-\boldsymbol{v}')\,.\,\boldsymbol{c}=\hbar\boldsymbol{\kappa}\,.\,\boldsymbol{c}$$

$$=\pm\hbar\boldsymbol{q}\,.\,\boldsymbol{c}\ \text{(from the result of the first part with } \boldsymbol{\tau}=0)$$

$$=\pm\hbar\omega_s.$$

3.7 From Fig. S.1*a*

$$\omega=\frac{1}{\hbar}(E-E')=\frac{\hbar}{2m}(k^2-k'^2)=-\frac{\hbar}{2m}(q^2+2kq\cos\phi),$$

where q is taken positive. P is a point on the scattering surface if $\omega=\pm\omega_s=\pm cq$. (The upper sign corresponds to phonon emission, and the lower sign to phonon absorption.) Thus

$$\mp cq=\frac{\hbar}{2m}(q^2+2kq\cos\phi).$$

Fig. S.1 Diagram for the solution of Example 3.7. The scattering surface near the origin is shown in (*b*) for $c<v$ and in (*c*) for $c>v$.

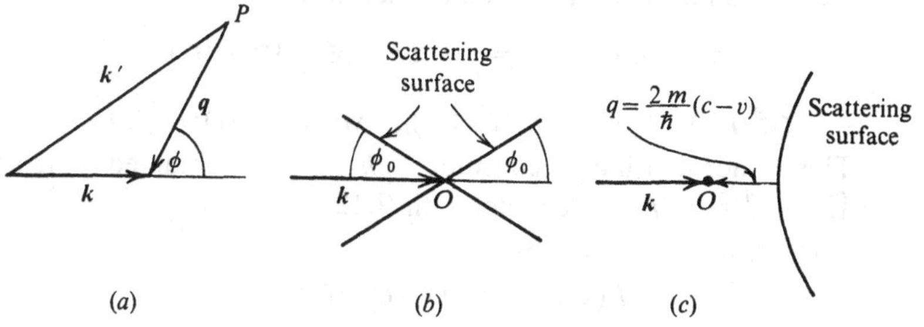

$q=0$ is always a solution of this equation. If $q \neq 0$,

$$\mp c = \frac{\hbar q}{2m} + v \cos \phi,$$

where $v = \hbar k/m$ is the velocity of the incident neutron.

If $c < v$, the scattering surface for small q is a pair of cones with apex at the origin, axis along $\boldsymbol{k}$, and semi-angle $\phi_0 = \cos^{-1}(c/v)$, see Fig. S.1*b*. If $c > v$, there is no solution for small q. The point on the scattering surface nearest the origin has $q = (2m/\hbar)(c-v)$, and corresponds to phonon absorption (Fig. S.1*c*). These results are demonstrated in the scattering surfaces shown in Fig. 3.12*b*. The velocity of the incident neutrons lies between the maximum sound velocities of the highest and middle frequency polarisation branches. Thus the scattering surfaces of the two lowest frequencies emerge from the origin; that of the highest frequency does not.

3.8 Consider the expression for the coherent one-phonon absorption cross-section given in (3.113). As $k \to 0$ (with $\boldsymbol{k}'$ constant), $\boldsymbol{\kappa} \to -\boldsymbol{k}'$, and everything in the expression tends to a constant except the term $1/k$.

The physical interpretation of this result is that the probability of the neutron being scattered into a fixed element of $\boldsymbol{k}'$ space is the product of two terms. One is the probability of the neutron being scattered per unit time, which is given by the constant part of the cross-section. The other is the time the neutron spends in the crystal, which is proportional to $1/k$.

4.1 For a perfect gas $G_s^{cl}(\boldsymbol{r}, t)$ is spherically symmetric. If an atom is at the origin at time $t=0$, the probability of its being between r and $r+dr$ from the origin at time t is $4\pi r^2\, dr G_s^{cl}(\boldsymbol{r}, t)$. This is equal to $P(v)\,dv$, the probability of the atom having velocity of magnitude v to $v+dv$, where $v = r/t$. $P(v)$ is given by the Maxwellian distribution

$$P(v) = h_1 v^2 \exp(-\tfrac{1}{2}Mv^2\beta),$$

where h_1 is a normalising constant. Therefore

$$4\pi r^2\, dr\, G_s^{cl}(\boldsymbol{r}, t) = \frac{h_1}{t^3} r^2\, dr \exp(-\tfrac{1}{2}Mr^2\beta/t^2),$$

$$G_s^{cl}(\boldsymbol{r}, t) = h_2 \exp\{-r^2/2\sigma_{cl}^2(t)\}, \quad \text{where } \sigma_{cl}^2(t) = t^2/M\beta.$$

The normalising constant h_2 is obtained from $4\pi \int_0^\infty r^2 G_s^{cl}(\boldsymbol{r}, t)\, dr = 1$, together with (I.12).

4.2 (*a*) From (3.126) and (4.9)

$$I_s(\boldsymbol{\kappa}, t) = \exp\langle U^2 \rangle \exp\langle UV_0 \rangle,$$

where $\quad U = -\mathrm{i}\boldsymbol{\kappa}\,.\,\boldsymbol{u}(0), \qquad V_0 = \mathrm{i}\boldsymbol{\kappa}\,.\,\boldsymbol{u}(t).$

For a single harmonic oscillator (3.129) and (3.134) give

$$\langle U^2\rangle + \langle UV_0\rangle = -\frac{\hbar\kappa^2}{2M\omega}\{\coth(\tfrac{1}{2}\hbar\omega\beta)(1-\cos\omega t) - \mathrm{i}\sin\omega t\}.$$

The expression for $G_s(\boldsymbol{r}, t)$ follows from (4.10) and (B.8).

(*b*) From (4.43) $\tilde{\sigma}^2(t) = \sigma^2(t + \frac{1}{2}\mathrm{i}\hbar\beta)$. Substitute in the expression for $\sigma^2(t)$, and use the formulae for $\cos(\omega t + \mathrm{i}x)$ and $\sin(\omega t + \mathrm{i}x)$, where $x = \frac{1}{2}\hbar\omega\beta$.

(*c*) The classical form of $\sigma^2(t)$ is obtained by allowing $\frac{1}{2}\hbar\omega\beta \to 0$. For small x, $\coth x \to x^{-1}$, which gives the required result.

(*d*) $\langle r_{\mathrm{cl}}^2(t)\rangle = 4\pi \int_0^\infty G_s^{\mathrm{cl}}(r, t) r^4\,\mathrm{d}r = 3\sigma_{\mathrm{cl}}^2(t)$ from (I.13).

4.3 $\langle \exp(\mathrm{i}\boldsymbol{\kappa}\,.\,\boldsymbol{R}_l)\rangle = \exp(\mathrm{i}\boldsymbol{\kappa}\,.\,\boldsymbol{l})\langle \exp(\mathrm{i}\boldsymbol{\kappa}\,.\,\boldsymbol{u}_l)\rangle$

$$= \exp(\mathrm{i}\boldsymbol{\kappa}\,.\,\boldsymbol{l})\langle \exp U\rangle$$
$$= \exp(\mathrm{i}\boldsymbol{\kappa}\,.\,\boldsymbol{l})\exp\{\tfrac{1}{2}\langle U^2\rangle\} \quad \text{from (3.32).}$$

Thus

$$\sum_{ll'} \langle \exp(-\mathrm{i}\boldsymbol{\kappa}\,.\,\boldsymbol{R}_{l'})\rangle\langle \exp(\mathrm{i}\boldsymbol{\kappa}\,.\,\boldsymbol{R}_l)\rangle$$
$$= \exp\langle U^2\rangle \sum_{ll'} \exp(-\mathrm{i}\boldsymbol{\kappa}\,.\,\boldsymbol{l}')\exp(\mathrm{i}\boldsymbol{\kappa}\,.\,\boldsymbol{l})$$
$$= N\exp\langle U^2\rangle \sum_{l} \exp(\mathrm{i}\boldsymbol{\kappa}\,.\,\boldsymbol{l})$$
$$= N\frac{(2\pi)^3}{v_0}\exp(-2W)\sum_{\boldsymbol{\tau}}\delta(\boldsymbol{\kappa}-\boldsymbol{\tau}) \quad \text{from (3.49) and (A.17).}$$

4.4 (*a*) From (4.30) and (A.13)

$$\rho(\boldsymbol{r}) = \sum_l \delta\{\boldsymbol{r}-\boldsymbol{l}-\boldsymbol{u}(l)\} = (2\pi)^{-3}\sum_l \int \exp[\mathrm{i}\boldsymbol{\kappa}\,.\,\{\boldsymbol{r}-\boldsymbol{l}-\boldsymbol{u}(l)\}]\,\mathrm{d}\boldsymbol{\kappa},$$
$$\langle\rho(\boldsymbol{r})\rangle = (2\pi)^{-3}\sum_l \int \exp\{\mathrm{i}\boldsymbol{\kappa}\,.\,(\boldsymbol{r}-\boldsymbol{l})\}\langle \exp(\mathrm{i}\boldsymbol{\kappa}\,.\,\boldsymbol{u})\rangle\,\mathrm{d}\boldsymbol{\kappa},$$
$$\langle \exp(\mathrm{i}\boldsymbol{\kappa}\,.\,\boldsymbol{u})\rangle = \exp\{-\tfrac{1}{2}\langle(\boldsymbol{\kappa}\,.\,\boldsymbol{u})^2\rangle\} \quad \text{from (3.32)}$$
$$= \exp\{-\tfrac{1}{6}\kappa^2\langle u^2\rangle\} \quad \text{for a cubic crystal}$$
$$= \exp(-\tfrac{1}{4}\kappa^2\sigma^2).$$

These relations together with (B.8) give the required result.

(*b*) $G_s(\boldsymbol{r}, \infty)$ may be obtained from (4.122) with the expressions for $\langle\delta(\boldsymbol{r}'-\boldsymbol{R}_j)\rangle$ and $\langle\delta(\boldsymbol{r}'+\boldsymbol{r}-\boldsymbol{R}_j)\rangle$ given in (*a*). However a quicker method is to use the inverse relation of (4.123)

$$G_s(\boldsymbol{r}, \infty) = (2\pi)^{-3}\int \exp(-2W)\exp(-\mathrm{i}\boldsymbol{\kappa}\,.\,\boldsymbol{r})\,\mathrm{d}\boldsymbol{\kappa}.$$

Put $$2W=\tfrac{1}{3}\kappa^2\langle u^2\rangle=\tfrac{1}{2}\kappa^2\sigma^2,\quad \text{and use (B.8).}$$

(*c*) $$\langle r^2\rangle=\int r^2 G_s(\boldsymbol{r},\infty)\,\mathrm{d}\boldsymbol{r}$$
$$=4\pi(2\pi\sigma^2)^{-3/2}\int_0^\infty r^4\exp(-r^2/2\sigma^2)\,\mathrm{d}r$$
$$=3\sigma^2\ \text{(from (I.13))}=2\langle u^2\rangle.$$

Alternatively, if $\boldsymbol{u}_1$ and $\boldsymbol{u}_2$ are the displacements of the atom at times 0 and t. Then for large t
$$\langle r^2\rangle=\langle(\boldsymbol{u}_1-\boldsymbol{u}_2)^2\rangle=\langle u_1^2\rangle+\langle u_2^2\rangle-2\langle\boldsymbol{u}_1\,.\,\boldsymbol{u}_2\rangle$$
$$=2\langle u^2\rangle,\quad \text{since } \langle\boldsymbol{u}_1\,.\,\boldsymbol{u}_2\rangle=0.$$

4.5 For incoherent one-phonon scattering we have, from (3.136) and (4.13)
$$S_{\mathrm{i}}(\boldsymbol{\kappa},\omega)=\frac{\kappa^2}{4M}\exp(-2W)\frac{Z(\omega)}{\omega}\{\coth(\tfrac{1}{2}\hbar\omega\beta)\pm1\}.$$
For $\hbar\omega\beta\ll1$, $\coth(\tfrac{1}{2}\hbar\omega\beta)\pm1\approx2/\hbar\omega\beta$.

6.1 The neutron flux $\boldsymbol{j}_\nu$ is obtained from the wavefunction ψ_ν by the formula $\boldsymbol{j}_\nu=(\hbar/2\mathrm{i}m)(\psi_\nu^*\,\mathrm{grad}\,\psi_\nu-\psi_\nu\,\mathrm{grad}\,\psi_\nu^*)$. Insert ψ_ν from (6.41) and use the relation $\mathrm{grad}\{\exp(\mathrm{i}\boldsymbol{k}_\nu\,.\,\boldsymbol{r})\}=\mathrm{i}\boldsymbol{k}_\nu\exp(\mathrm{i}\boldsymbol{k}_\nu\,.\,\boldsymbol{r})$. This gives
$$\boldsymbol{j}_\nu=\frac{\hbar}{m}|A_\nu|^2\{\boldsymbol{k}_\nu+|\alpha_\nu|^2(\boldsymbol{k}_\nu-\boldsymbol{\tau})+\boldsymbol{B}\},$$
where $$\boldsymbol{B}=\mathrm{Re}\{\boldsymbol{k}_\nu\alpha_\nu^*\exp(\mathrm{i}\boldsymbol{\tau}\,.\,\boldsymbol{r})+(\boldsymbol{k}_\nu-\boldsymbol{\tau})\alpha_\nu\exp(-\mathrm{i}\boldsymbol{\tau}\,.\,\boldsymbol{r})\}.$$
The average of $\exp(\pm\mathrm{i}\boldsymbol{\tau}\,.\,\boldsymbol{r})$ over the unit cell is zero (see A.21).

6.2 Inserting the values given in (6.47) in (6.41) leads to $|\psi_\nu|^2=\tfrac{1}{4}|\exp(\mathrm{i}\boldsymbol{k}_\nu\,.\,\boldsymbol{r})\mp\exp\{\mathrm{i}(\boldsymbol{k}_\nu-\boldsymbol{\tau})\,.\,\boldsymbol{r}\}|^2=\tfrac{1}{2}(1\mp\cos\boldsymbol{\tau}\,.\,\boldsymbol{r})$. Thus $|\psi_\nu|^2$ has a minimum or a maximum at the atomic sites depending on whether we take the negative or positive sign. The negative sign here corresponds to the positive sign in (6.46), which corresponds to the larger value of k_ν, since G_τ has the same sign as $\bar{b}$ and is therefore positive. Note that ψ_ν is made up of two waves travelling in the directions of $\boldsymbol{k}_\nu$ and $\boldsymbol{k}_\nu-\boldsymbol{\tau}$, while $|\psi_\nu|^2$ is a standing wave with planes of equal amplitude perpendicular to $\boldsymbol{\tau}$.

7.1 $\langle\phi|f(\boldsymbol{r}_1)|\phi\rangle$
$$=\tfrac{1}{2}\int\{\psi_{\mathrm{a}}(\boldsymbol{r}_1)\psi_{\mathrm{b}}(\boldsymbol{r}_2)\pm\psi_{\mathrm{b}}(\boldsymbol{r}_1)\psi_{\mathrm{a}}(\boldsymbol{r}_2)\}^*$$
$$\times\{\psi_{\mathrm{a}}(\boldsymbol{r}_1)\psi_{\mathrm{b}}(\boldsymbol{r}_2)\pm\psi_{\mathrm{b}}(\boldsymbol{r}_1)\psi_{\mathrm{a}}(\boldsymbol{r}_2)\}f(\boldsymbol{r}_1)\,\mathrm{d}\boldsymbol{r}_1\,\mathrm{d}\boldsymbol{r}_2$$
$$=\tfrac{1}{2}\int\{|\psi_{\mathrm{a}}(\boldsymbol{r}_1)|^2+|\psi_{\mathrm{b}}(\boldsymbol{r}_1)|^2\}f(\boldsymbol{r}_1)\,\mathrm{d}\boldsymbol{r}_1.$$

This follows from the orthonormal properties of ψ_a and ψ_b, i.e.

$$\int \psi_a^*(\boldsymbol{r}_2)\psi_a(\boldsymbol{r}_2)\,d\boldsymbol{r}_2 = 1, \qquad \int \psi_a^*(\boldsymbol{r}_2)\psi_b(\boldsymbol{r}_2)\,d\boldsymbol{r}_2 = 0.$$

$\langle\phi|f(\boldsymbol{r}_2)|\phi\rangle$ has the same final form except that $\boldsymbol{r}_2$ is the variable instead of $\boldsymbol{r}_1$. The two integrals are therefore equal.

7.2 (*a*) When a magnetic field acts we have the following results:

For all l, $\langle S_0^x S_l^y\rangle = \langle S_0^x S_l^z\rangle = \langle S_0^y S_l^z\rangle = 0.$

For $l \neq 0$, $\langle S_0^x S_l^x\rangle = \langle S_0^y S_l^y\rangle = 0$, $\langle S_0^z S_l^z\rangle = \langle S^z\rangle^2.$

For $l = 0$, $\langle (S_0^x)^2\rangle = \langle (S_0^y)^2\rangle = \frac{1}{2}\{S(S+1) - \langle (S^z)^2\rangle\}$, $\langle (S_0^z)^2\rangle = \langle (S^z)^2\rangle.$

Inserting these values in (7.88) gives (7.95).

The magnetic quantum number M for the z direction takes values $S, S-1, \ldots, -S$. The energy for the state M is $-Mg\mu_B B$; thus the probability of the atom being in this state, given by the Boltzmann factor, is proportional to $\exp(Mu)$ where $u = g\mu_B B\beta$.

Put
$$Z = \sum_{M=-S}^{S} \exp(Mu) = \frac{\sinh\{(S+\frac{1}{2})u\}}{\sinh(\frac{1}{2}u)}.$$

Then
$$\langle S^z\rangle = \frac{\sum_{-S}^{S} M\exp(Mu)}{Z} = \frac{\partial Z/\partial u}{Z},$$

and
$$\langle (S^z)^2\rangle = \frac{\sum_{-S}^{S} M^2\exp(Mu)}{Z} = \frac{\partial^2 Z/\partial u^2}{Z}.$$

Evaluate $\partial Z/\partial u$ and $\partial^2 Z/\partial u^2$ from the expression for Z and substitute.

(*b*) As $u \to \infty$, $\coth\{(S+\frac{1}{2})u\} \to 1$ and $\coth(\frac{1}{2}u) \to 1$. For $u = \infty$, all the atoms are in the state $M = S$. So $\langle S^z\rangle = S$, and $\langle (S^z)^2\rangle = S^2$. As $x \to 0$, $\coth x \to x^{-1} + \frac{1}{3}x$.

8.1 In units of the side of the cube in the simple cubic lattice, the atoms at the corners of the primitive unit cell of the face-centred cubic lattice have coordinates (000), (111), $(\frac{1}{2}\frac{1}{2}0)$, $(\frac{1}{2}\frac{1}{2}1)$, $(0\frac{1}{2}\frac{1}{2})$, $(1\frac{1}{2}\frac{1}{2})$, $(\frac{1}{2}0\frac{1}{2})$, $(\frac{1}{2}1\frac{1}{2})$. The spins of the first four atoms point in one direction, and those of the second four point in the opposite direction. The magnetic structure factor is therefore proportional to

$$1 + \exp\{\pi i(t_1+t_2)\} - \exp\{\pi i(t_2+t_3)\} - \exp\{\pi i(t_3+t_1)\},$$

which is zero for integral values of t_1, t_2, t_3 unless, either t_1 and t_2 are even with t_3 odd, or t_1 and t_2 are odd with t_3 even.

8.2 The Hamiltonian is $H = -JA$, where $A = \sum_{ij}(S_i^+S_j^- + S_i^zS_j^z)$. The sum is taken over all nearest-neighbour pairs, with i, $i+1$ and $i+1$, i

counted separately. We have

$$\phi_n = |u_1 u_2 \ldots u_{n-1} v_n u_{n+1} \ldots u_N\rangle,$$

$$S_i^+ u_i = 0, \qquad S_i^- u_i = v_i, \qquad S_i^z u_i = \tfrac{1}{2} u_i,$$

$$S_i^+ v_i = u_i, \qquad S_i^- v_i = 0, \qquad S_i^z v_i = -\tfrac{1}{2} v_i.$$

Thus $$A\phi_n = \phi_{n-1} + \phi_{n+1} - 2\phi_n + \tfrac{1}{2}(N-1)\phi_n.$$

Let $\psi = \sum_n c_n \phi_n$ be an eigenfunction of H with energy E. Then

$$H\psi = E \sum_n c_n \phi_n = -J \sum_n c_n A\phi_n.$$

Equating the coefficients of ϕ_n gives

$$Ec_n = -J\{c_{n+1} + c_{n-1} - 2c_n + \tfrac{1}{2}(N-1)c_n\}.$$

Put $c_n = \exp(iqna)$. Then

$$E = 2J(1 - \cos qa) - \tfrac{1}{2}J(N-1).$$

The ground state $|g\rangle$ consists of all us. Thus

$$H|g\rangle = -\tfrac{1}{2}J(N-1)|g\rangle,$$

i.e. the energy of the ground state is $-\tfrac{1}{2}J(N-1)$.

8.3 From (8.61) to (8.64)

$$\hbar\omega = 2rJS\left\{1 - \frac{1}{r}\sum_{\boldsymbol{\rho}} \exp(i\boldsymbol{q}\cdot\boldsymbol{\rho})\right\}.$$

For $\boldsymbol{q}\cdot\boldsymbol{\rho} \ll 1$

$$1 - \frac{1}{r}\sum_{\boldsymbol{\rho}} \exp(i\boldsymbol{q}\cdot\boldsymbol{\rho}) \approx 1 - \frac{1}{r}\sum_{\boldsymbol{\rho}} \{1 + i\boldsymbol{q}\cdot\boldsymbol{\rho} - \tfrac{1}{2}(\boldsymbol{q}\cdot\boldsymbol{\rho})^2\}$$

$$= \frac{1}{2r}\sum_{\boldsymbol{\rho}} (\boldsymbol{q}\cdot\boldsymbol{\rho})^2.$$

Thus $$\hbar\omega = JS\sum_{\boldsymbol{\rho}} (\boldsymbol{q}\cdot\boldsymbol{\rho})^2 = \tfrac{1}{3}rJSd^2q^2$$

for a cubic lattice (simple, face-centred, or body-centred), where d is the distance to the nearest neighbour. For a simple cubic lattice $r = 6$, $d = a$; for a face-centred cubic lattice $r = 12$, $d = a/\sqrt{2}$; for a body-centred cubic lattice $r = 8$, $d = \sqrt{3}(a/2)$. For all three lattices, $D = \tfrac{1}{3}rJSd^2 = 2JSa^2$.

8.4 For scattering in the forward direction, $\boldsymbol{\tau} = 0$. From (8.65), (8.81), and (8.82)

$$k^2 + k'^2 - 2kk'\cos\theta = q^2, \qquad \frac{\hbar^2}{2m}(k^2 - k'^2) = \pm Dq^2.$$

These equations lead to a quadratic equation for k', whose solution is

$$\frac{k'}{k}=\{\cos\theta\pm(\alpha^2-\sin^2\theta)^{1/2}\}/(1\pm\alpha), \quad \text{where } \alpha=\hbar^2/2mD.$$

Scattering occurs only for real values of k', i.e. for $\theta<\theta_0=\sin^{-1}\alpha$.

It may be noted that this result is independent of the wavelength of the incident neutrons. The experimental values of D for iron and nickel (Fig. 8.9) give $\theta_0\leqslant 1°$.

8.5 From (8.37), (8.58), (8.71), and (8.80)

$$S-\langle S^z\rangle=\langle a_l^+a_l\rangle=\frac{1}{N}\sum_{\boldsymbol{q}}\langle b_{\boldsymbol{q}}^+b_{\boldsymbol{q}}\rangle=\frac{1}{N}\sum_{\boldsymbol{q}}\langle n_{\boldsymbol{q}}\rangle$$

$$=\frac{1}{N}\sum_{\boldsymbol{q}}\{\exp(\hbar\omega_{\boldsymbol{q}}\beta)-1\}^{-1}=\frac{v_0}{(2\pi)^3}\int\{\exp(\hbar\omega_{\boldsymbol{q}}\beta)-1\}^{-1}\,\mathrm{d}\boldsymbol{q}.$$

The integral is taken over the Brillouin zone. For small T (large β) the integrand is small except when $\omega_{\boldsymbol{q}}$ is small. Therefore we may use the quadratic dispersion relation $\hbar\omega_{\boldsymbol{q}}=Dq^2$ with $\mathrm{d}\boldsymbol{q}=4\pi q^2\,\mathrm{d}q$. Further, since the integrand is small at the zone boundary the upper limit of integration may be extended to infinity. Put $x=D\beta q^2$. Then

$$S-\langle S^z\rangle=\frac{v_0}{4\pi^2}\frac{1}{(D\beta)^{3/2}}\int_0^\infty\frac{x^{1/2}\,\mathrm{d}x}{\mathrm{e}^x-1}.$$

The integral is a pure number. Thus

$$S-\langle S^z\rangle\propto\beta^{-3/2}\propto T^{3/2}.$$

9.1 From (9.10), (9.13), and (9.14) the scattering length operator for the proton is

$$\mathscr{b}=A+B\boldsymbol{\sigma}\,.\,\boldsymbol{I}, \qquad A=\tfrac{3}{4}b^++\tfrac{1}{4}b^-, \qquad B=\tfrac{1}{2}(b^+-b^-),$$

where $\boldsymbol{I}$ is the spin of the proton. If the wavelength of the neutron is large compared to the distance between the two protons in the molecule, the two scattered neutron waves are in phase, and the scattering length for the molecule is $\mathscr{b}_{\text{mol}}=2A+B\boldsymbol{\sigma}\,.\,\boldsymbol{\mathscr{I}}$, where $\boldsymbol{\mathscr{I}}$ is the spin of the molecule. $\mathscr{I}=1$ for the ortho-molecule, and $\mathscr{I}=0$ for the para-molecule.

We require the value of $\langle\mathscr{b}_{\text{mol}}^2\rangle$ where $\langle\ \rangle$ stands for the average over the spin directions of the incident neutrons. Since the latter are unpolarised $\langle\boldsymbol{\sigma}\,.\,\boldsymbol{\mathscr{I}}\rangle=0$, and we have

$$\langle\mathscr{b}_{\text{mol}}^2\rangle=4A^2+B^2\langle(\boldsymbol{\sigma}\,.\,\boldsymbol{\mathscr{I}})^2\rangle,$$

$$\langle(\boldsymbol{\sigma}\,.\,\boldsymbol{\mathscr{I}})^2\rangle=\langle(\sigma_x\mathscr{I}_x+\sigma_y\mathscr{I}_y+\sigma_z\mathscr{I}_z)^2\rangle$$

$$=\langle\mathscr{I}_x^2+\mathscr{I}_y^2+\mathscr{I}_z^2\rangle=\mathscr{I}(\mathscr{I}+1).$$

(We have used the results $\langle\sigma_x\sigma_y\rangle = 0$ and $\sigma_x^2 = \sigma_y^2 = \sigma_z^2 = 1$). Thus

$$\langle \bar{b}^2_{\text{mol}}\rangle = \tfrac{1}{4}(3b^+ + b^-)^2 + \tfrac{1}{2}(b^+ - b^-)^2 \quad \text{ortho},$$

$$\langle \bar{b}^2_{\text{mol}}\rangle = \tfrac{1}{4}(3b^+ + b^-)^2 \quad \text{para}.$$

The above scattering lengths are for bound protons. The scattering length for a proton in a hydrogen molecule is (from 2.36) $\frac{2}{3}$ times the bound scattering length. Thus the total cross-section is

$$\sigma_{\text{tot}} = \frac{16\pi}{9}\langle \bar{b}^2_{\text{mol}}\rangle.$$

9.2 Nuclear–magnetic interference occurs at reciprocal lattice points at which the nuclear and magnetic structure factors are both non-zero. At the reciprocal lattice point $\boldsymbol{\tau} = t_1\boldsymbol{\tau}_1 + t_2\boldsymbol{\tau}_2 + t_3\boldsymbol{\tau}_3$ the nuclear structure factor is

$$F_{\text{N}}(\boldsymbol{\tau}) = \bar{b}_{\text{Mn}} g_{\text{Mn}} + \bar{b}_{\text{F}} g_{\text{F}},$$

where $g_{\text{Mn}} = 1 + \exp\{\pi\text{i}(t_1 + t_2 + t_3)\}$,

$$\begin{aligned} g_{\text{F}} = {} & \exp\{2\pi\text{i}u(t_1+t_2)\} + \exp\{2\pi\text{i}(1-u)(t_1+t_2)\} \\ & + \exp[2\pi\text{i}\{(\tfrac{1}{2}+u)t_1 + (\tfrac{1}{2}-u)t_2 + \tfrac{1}{2}t_3\}] \\ & + \exp[2\pi\text{i}\{(\tfrac{1}{2}-u)t_1 + (\tfrac{1}{2}+u)t_2 + \tfrac{1}{2}t_3\}]. \end{aligned}$$

Thus

$$\begin{aligned} F_{\text{N}}(\boldsymbol{\tau}) &= 2\bar{b}_{\text{Mn}} + 4\bar{b}_{\text{F}}\cos(2\pi u t_1)\cos(2\pi u t_2) \quad &\text{if } t_1+t_2+t_3 \text{ is even,} \\ &= -4\bar{b}_{\text{F}}\sin(2\pi u t_1)\sin(2\pi u t_2) \quad &\text{if } t_1+t_2+t_3 \text{ is odd.} \end{aligned}$$

The magnetic structure factor $F_{\text{M}}(\boldsymbol{\tau})$ is proportional to

$$\begin{aligned} 1 - \exp\{\pi\text{i}(t_1+t_2+t_3)\} &= 0 \quad &\text{if } t_1+t_2+t_3 \text{ is even,} \\ &= 2 \quad &\text{if } t_1+t_2+t_3 \text{ is odd.} \end{aligned}$$

The required condition is therefore $t_1 + t_2 + t_3$ odd, with t_1 and t_2 both non-zero.

BIBLIOGRAPHY

Books on the theory and applications of thermal neutron scattering

Bacon, G. E. 1975. *Neutron Diffraction*, 3rd ed., Oxford: Clarendon Press.

Boutin, H. and Yip, S. 1968. *Molecular Spectroscopy with Neutrons*, Cambridge, Massachusetts: M.I.T. Press.

Dachs, H. 1978. Editor, *Neutron Diffraction* (Topics in Current Physics, Vol. 6), Berlin: Springer-Verlag.

Egelstaff, P. A. 1965. Editor, *Thermal Neutron Scattering*, London: Academic Press.

Gurevich, I. I. and Tarasov, L. V. 1968. *Low Energy Neutron Physics*, Amsterdam: North-Holland.

Hughes, D. J. 1953. *Pile Neutron Research*, Cambridge, Massachusetts: Addison-Wesley.

Izyumov, Yu A. and Ozerov, R. P. 1966. *Magnetic Neutron Diffraction*, translated 1970, New York: Plenum Press.

Liquids by Neutron Scattering (Topics in Current Physics, Vol. 3), Berlin: *Liquids by Neutron Scatterin* (Topics in Current Physics, Vol. 3), Berlin: Springer-Verlag.

Marshall, W. and Lovesey, S. W. 1971. *Theory of Thermal Neutron Scattering*, Oxford: Clarendon Press.

Turchin, V. F. 1963. *Slow Neutrons*, translated 1965, Jerusalem: Israel Program for Scientific Translations.

Willis, B. T. M. 1970. Editor, *Thermal Neutron Diffraction*, Oxford University Press.

Willis, B. T. M. 1973. Editor, *Chemical Applications of Thermal Neutron Scattering*, Oxford University Press.

Proceedings of some conferences and symposia on thermal neutron scattering

The following symposia have been organised and the proceedings published by the International Atomic Energy Agency, Vienna:

Inelastic Scattering of Neutrons in Solids and Liquids: Vienna, 1960; Chalk River, 1962.

Inelastic Scattering of Neutrons: Bombay, 1964.

Neutron Inelastic Scattering: Copenhagen, 1968; Grenoble, 1972; Vienna, 1977.
New Methods and Techniques in Neutron Diffraction, Petten, 1975, Reactor Centrum Nederland.
Neutron Scattering, Gatlinburg, 1976, Oak Ridge National Laboratory and the U.S. Energy Research and Development Administration.

Guide to the literature

Larose, A. and Vanderwal, J. 1974. Compilers, *Scattering of Thermal Neutrons – A Bibliography (1932–1974)*, New York: IFI/Plenum Press.

REFERENCES

Abragam, A., Bacchella, G. L., Glättli, H., Mériel, P., Piesvaux, J., and Pinot, M. 1975. *J. Physique*, **36**, L-263.

Allen, G. and Higgins, J. S. 1973. *Rep. Prog. Phys.*, **36**, 1073.

Bacon, G. E. 1975. *Neutron Diffraction*, 3rd ed., Oxford: Clarendon Press.

Bacon, G. E. and Lowde, R. D. 1948. *Acta Cryst.*, **1**, 303.

Batterman, B. W. and Cole, H. 1964. *Rev. Mod. Phys.*, **36**, 681.

Bauspiess, W., Bonse, U., Rauch, H., and Treimer, W. 1974. *Z. Phys.*, **271**, 177.

Bloch, F. 1932. *Z. Phys.*, **74**, 295.

Blume, M. 1963. *Phys. Rev. Lett.*, **10**, 489.

Bonse, U. and Graeff, W. 1977. *Topics in Applied Physics*, Vol. **22**, Ch. 4, Berlin: Springer-Verlag.

Boutin, H. and Yip, S. 1968. *Molecular Spectroscopy with Neutrons*, Cambridge, Massachusetts: M.I.T. Press.

Breit, G. 1947. *Phys. Rev.*, **71**, 215.

Brockhouse, B. N. 1960. *Bull. Amer. Phys. Soc.*, **5**, 462.

Brockhouse, B. N., Becka, L. N., Rao, K. R., and Woods, A. D. B. 1963. *Inelastic Scattering of Neutrons in Solids and Liquids*, IAEA, Vienna, Vol. II, p. 23.

Brugger, R. M. 1965 in *Thermal Neutron Scattering*, ed. Egelstaff, P.A., London: Academic Press, p. 53.

Burgy, M. T., Ringo, G. R., and Hughes, D. J. 1951. *Phys. Rev.*, **84**, 1160.

Carneiro, K. 1976. *Phys. Rev.*, **A14**, 517.

Cochran, W. 1973. *The Dynamics of Atoms in Crystals*, London: Arnold.

Cooke, J. F. 1976. Conference on Neutron Scattering, Gatlinburg, Oak Ridge National Laboratory and U.S. ERDA, p. 723.

Cooper, B. R. 1968. *Solid State Physics*, **21**, 393.

Copley, J. R. D. and Lovesey, S. W. 1975. *Rep. Prog. Phys.*, **38**, 461.

Cowley, R. A. 1968. *Rep. Prog. Phys.*, **31**, 123.

Croxton, C. A. 1974. *Liquid State Physics – A Statistical Mechanical Introduction*, Cambridge University Press.

Dyson, F. J. 1956. *Phys. Rev.*, **102**, 1217.

Egelstaff, P. A. 1967. *An Introduction to the Liquid State*, London: Academic Press.

Enderby, J. E. 1968 in *Physics of Simple Liquids*, ed. Temperley, H. N. V., Rowlinson, J. S., and Rushbrooke, G. S., Amsterdam: North-Holland, p. 611.

Ernst, M. H., Hauge, E. H., and van Leeuwen, J. M. J. 1970. *Phys. Rev. Lett.*, **25**, 1254.

Faber, T. E. 1972. *Theory of Liquid Metals*, Cambridge University Press.

Fermi, E. 1936. *Ricerca Scientifica*, **7**, 13.

Foldy, L. L. 1958. *Rev. Mod. Phys.*, **30**, 471.

Ghatak, A. K. and Kothari, L. S. 1972. *An Introduction to Lattice Dynamics*, London: Addison-Wesley.

Gläser, W., Carvalho, F., and Ehret, G. 1965. *Inelastic Scattering of Neutrons*, IAEA, Vienna, Vol. I, p. 99.

Goldberger, M. L. and Seitz, F. 1947. *Phys. Rev.*, **71**, 294.

Hayter, J. B. 1976. Conference on Neutron Scattering, Gatlinburg, Oak Ridge National Laboratory and U.S. ERDA, p. 1074.

Herpin, A., Mériel, P., and Villain, J. 1959. *Comptes Rendus*, **249**, 1334.

Herring, C. 1966. *Magnetism*, Vol. 4, ed. Rado, G. T. and Suhl, H., New York: Academic Press.

Holstein, T. and Primakoff, H. 1940. *Phys. Rev.*, **58**, 1098.

Houmann, J. G., and Møller, H. B. 1976. Conference on Neutron Scattering, Gatlinburg, Oak Ridge National Laboratory and U.S. ERDA, p. 743.

Iyengar, P. K. 1965 in *Thermal Neutron Scattering*, ed. Egelstaff, P. A., London: Academic Press, p. 97.

Izuyama, T., Kim, D. J., and Kubo, R. 1963. *J. Phys. Soc. Japan*, **18**, 1025.

James, R. W. 1963. *Solid State Physics*, **15**, 53.

Johnston, D. F. 1966. *Proc. Phys. Soc.*, **88**, 37.

Kato, N. 1958. *Acta Cryst.*, **11**, 885.

Keffer, F. 1966. *Handbuch der Physik*, Vol. **18/2**, Berlin: Springer-Verlag.

Koehler, W. C., Child, H. R., Nicklow, R. M., Smith, H. G., Moon, R. M., and Cable, J. W. 1970. *Phys. Rev. Lett.*, **24**, 16.

Koester, L. 1977. *Springer Tracts in Modern Physics*, Vol. **80**, Berlin: Springer-Verlag.

Koester, L., Knopf, K., and Waschkowski, W. 1974. *Z. Phys.*, **271**, 201.

Landau, L. D., and Lifshitz, E. M. 1958. *Statistical Physics*, London: Pergamon Press.

Levesque, D. and Verlet, L. 1970. *Phys. Rev.*, **A2**, 2514.

Lowde, R. D. 1965. *Jour. App. Phys.*, **36**, 884.

Maier-Leibnitz, H. 1962. *Z. angew. Phys.*, **14**, 738.

Maier-Leibnitz, H. 1972. *Neutron Inelastic Scattering*, IAEA, Vienna, p. 681.

Marshall, W., and Lovesey, S. W. 1971. *Theory of Thermal Neutron Scattering*, Oxford: Clarendon Press.

Merzbacher, E. 1970. *Quantum Mechanics*, 2nd ed., New York: Wiley.

Mook, H. A. 1966. *Phys. Rev.*, **148**, 495.
Moon, R. M. and Koehler, W. C. 1969. *Phys. Rev.*, **181**, 883.
Moon, R. M., Riste, T., and Koehler, W. C. 1969. *Phys. Rev.*, **181**, 920.
Nagamiya, T. 1967. *Solid State Physics*, **20**, 305.
Nijboer, B. R. A., and Rahman, A. 1966. *Physica*, **32**, 415.
Page, D. I. 1973 in *Chemical Applications of Thermal Neutron Scattering*, ed. Willis, B. T. M., Oxford University Press, p. 173.
Placzek, G. 1952. *Phys. Rev.*, **86**, 377.
Placzek, G. 1954. *Phys. Rev.*, **93**, 895.
Placzek, G. 1957. *Phys. Rev.*, **105**, 1240.
Placzek, G., and Van Hove, L. 1955. *Nuov. Cim.*, **1**, 233.
Powles, J. G. 1973. *Adv. Phys.*, **22**, 1.
Pryde, J. A. 1966. *The Liquid State*, London: Hutchinson University Library.
Pynn, R., and Squires, G. L. 1972. *Proc. Roy. Soc.*, **A326**, 347.
Rahman, A., Singwi, K. S., and Sjölander, A. 1962. *Phys. Rev.*, **126**, 986.
Rauch, H. and Petrascheck, D. 1978. *Topics in Current Physics*, Vol. **6**, Ch. 9, Berlin: Springer-Verlag.
Rauch, H., Zeilinger, A., Badurek, G., Wilfing, A., Bauspiess, W., and Bonse, U. 1975. *Phys. Lett.*, **54A**, 425.
Reif, F. 1965. *Fundamentals of Statistical and Thermal Physics*, New York: McGraw-Hill.
Scatturin, V., Corliss, L., Elliott, N., and Hastings, J. 1961. *Acta Cryst.*, **14**, 19.
Schofield, P. 1960. *Phys. Rev. Lett.*, **4**, 239.
Shirane, G., Minkiewicz, V. J., and Nathans, R. 1968. *Jour. App. Phys.*, **39**, 383.
Shull, C. G. 1963. *Phys. Rev. Lett.*, **10**, 297.
Shull, C. G. 1968. *Phys. Rev. Lett.*, **21**, 1585.
Shull, C. G., and Oberteuffer, J. A. 1972. *Phys. Rev. Lett.*, **29**, 871.
Shull, C. G. 1973. *J. Appl. Cryst.*, **6**, 257.
Sköld, K., Rowe, J. M., Ostrowski, G., and Randolph, P. D. 1972. *Phys. Rev.*, **A6**, 1107.
Squires, G. L. 1956. *Phys. Rev.*, **103**, 304.
Squires, G. L. 1966. *Proc. Phys. Soc.*, **88**, 919.
Squires, G. L. 1976. *Contemp. Phys.*, **17**, 411.
Stassis, C., and Oberteuffer, J. A. 1974. *Phys. Rev.*, **B10**, 5192.
Steinsvoll, O., Shirane, G., Nathans, R., and Blume, M. 1967. *Phys. Rev.*, **161**, 499.
Stringfellow, M. W. 1968. *J. Phys. C*, **1**, 950.
Tucciarone, A., Lau, H. Y., Corliss, L. M., Delapalme, A., and Hastings, J. M. 1971. *Phys. Rev.* **B4**, 3206.
Van Hove, L. 1954. *Phys. Rev.*, **95**, 249.
Venkataraman, G., Feldkamp, L. A., and Sahni, V. C. 1975. *Dynamics of Perfect Crystals*, Cambridge, Massachusetts: M.I.T. Press.
Werner, S. A., Colella, R., Overhauser, A. W. and Eagen, C. F. 1975. *Phys. Rev. Lett.*, **35**, 1053.

Williams, W. G. 1975. Conference on New Methods and Techniques in Neutron Diffraction, Petten, Reactor Centrum Nederland, p. 103.

Wood, W. W. 1968 in *Physics of Simple Liquids*, ed. Temperley, H. N. V., Rowlinson, J. S., and Rushbrooke, G. S., Amsterdam: North-Holland, p. 115.

Woods, A. D. B. and Cowley, R. A. 1973. *Rep. Prog. Phys.*, **36**, 1135.

Woolfson, M. M. 1970. *An Introduction to X-ray Crystallography*, Cambridge University Press.

Yarnell, J. L., Katz, M. J., Wenzel, R. G., and Koenig, S. H. 1973. *Phys. Rev.*, **A7**, 2130.

GLOSSARY OF SYMBOLS

The numbers in square brackets are the page numbers where the symbol is first defined or reintroduced.

$A = \{(I+1)b^+ + Ib^-\}/(2I+1)$ [174]
$\boldsymbol{a}_1, \boldsymbol{a}_2, \boldsymbol{a}_3$ unit-cell vectors of crystal lattice [25]
a, a^+ annihilation and creation operators [27, 157, Appendix E]
a_τ coefficient in Bloch function [117]

$B = (b^+ - b^-)/(2I+1)$ [174]
b scattering length [8]
b^+, b^- scattering lengths for nucleus–neutron system with spins $I+\frac{1}{2}, I-\frac{1}{2}$ [23, 173]
b_q, b_q^+ operators in the Fourier representation of a_l, a_l^+ [158]
$\hat{b}$ scattering length operator [173]
$\bar{b}$ coherent scattering length [21]

$C_l = \frac{1}{2}\gamma r_0 g F(\boldsymbol{\kappa})\hat{\boldsymbol{\kappa}} \times \{\langle \boldsymbol{S}_l \rangle \times \hat{\boldsymbol{\kappa}}\}$ [182]
c velocity of sound [60, 233]
c velocity of light [83, 190]

D diffusion constant [103, Chapter 5]
D stiffness constant for spin waves [161, Chapter 8]
$\mathrm{d}\Omega$ element of solid angle [6]
$\mathrm{d}^2\sigma/\mathrm{d}\Omega\, \mathrm{d}E'$ partial differential cross-section [6]
$\mathrm{d}\sigma/\mathrm{d}\Omega$ differential cross-section [6]
$\boldsymbol{d}$ equilibrium position of atom d within unit cell [36]

E, E' initial and final energies of neutron [5, 6]

E_λ, $E_{\lambda'}$ initial and final energies of scattering system [14]
E_n energy of particle in quantum state n [27]
$E_r = \hbar^2\kappa^2/2M$ recoil energy [71, 74]
e elementary charge
$\boldsymbol{e}_s$ polarisation vector of normal mode s [27]
$\boldsymbol{e}_{ds}$ polarisation vector of normal mode s for atom d [37, 223]

$F_N(\boldsymbol{\kappa})$ nuclear unit-cell structure factor [37]
$F_{\boldsymbol{\tau}} = F_N(\boldsymbol{\tau})$ [117]
$F_d(\boldsymbol{\kappa})$ magnetic form factor [138]
$F_M(\boldsymbol{\kappa})$ magnetic unit-cell structure factor [150]
$F_E(\boldsymbol{\kappa})$ electrostatic unit-cell structure factor [193]
$\mathscr{F}(\boldsymbol{\kappa})$ magnetic unit-cell vector structure factor [149]
$f(Q)$ probability function [28, 213]

$G(\boldsymbol{r}, t)$ time-dependent pair-correlation function [61]
$G_s(\boldsymbol{r}, t)$ self time-dependent pair-correlation function [62]
$G^{cl}(\boldsymbol{r}, t)$ classical form of $G(\boldsymbol{r}, t)$ [64]
$\tilde{G}(\boldsymbol{r}, t)$ Fourier transform in space of $\tilde{I}(\boldsymbol{\kappa}, t)$ [67]
$G_{\boldsymbol{\tau}} = 4\pi F_{\boldsymbol{\tau}}/v_0$ [117]
$g_s = (\hbar/2MN\omega_s)^{1/2}\boldsymbol{\kappa}\,.\,\boldsymbol{e}_s$ [29]
$g(\boldsymbol{r})$ static pair-distribution function [65]
g Landé splitting factor [139]

H Hamiltonian of scattering system
$\hbar$ = Planck constant/2π
$h_s = (\hbar/2MN\omega_s)^{1/2}\boldsymbol{\kappa}\,.\,\boldsymbol{e}_s \exp\{i(\boldsymbol{q}\,.\,\boldsymbol{l} - \omega_s t)\}$ [29]

I spin quantum number of nucleus [8]
$I(\boldsymbol{\kappa}, t)$ intermediate function [61]
$I_s(\boldsymbol{\kappa}, t)$ self intermediate function [62]
$\tilde{I}(\boldsymbol{\kappa}, t) = I(\boldsymbol{\kappa}, t + \frac{1}{2}i\hbar\beta)$ [67]
$\boldsymbol{I}$ spin angular momentum of nucleus [174]

$J(\boldsymbol{l} - \boldsymbol{l}')$ exchange integral in Heisenberg Hamiltonian [156]
$\mathscr{J}(\boldsymbol{q}) = \sum_{\boldsymbol{\rho}} J(\boldsymbol{\rho}) \exp(i\boldsymbol{q}\,.\,\boldsymbol{\rho})$ [159]
j polarisation index [27, 37, Appendix G]

k_B Boltzmann constant
$\boldsymbol{k}$, $\boldsymbol{k}'$ initial and final wavevectors of neutron [10]

$\boldsymbol{k}_0$ wavevector of incident neutrons outside crystal [117, Chapter 6]
$\boldsymbol{k}_1, \boldsymbol{k}_2$ wavevectors of neutrons near incident direction inside crystal [120, Chapter 6]

$\boldsymbol{l}$ vector in crystal lattice [25]
l_1, l_2, l_3 trio of integers, $\boldsymbol{l} = l_1\boldsymbol{a}_1 + l_2\boldsymbol{a}_2 + l_3\boldsymbol{a}_3$ [25]

M mass of atom [27]
M_d mass of atom in position d of unit cell [37]
$\boldsymbol{M}_{\mathrm{S}}(\boldsymbol{\kappa})$ Fourier transform of spin magnetisation $\boldsymbol{M}_{\mathrm{S}}(\boldsymbol{r})$ [134]
$\boldsymbol{M}_{\mathrm{L}}(\boldsymbol{\kappa})$ Fourier transform of orbital magnetisation $\boldsymbol{M}_{\mathrm{L}}(\boldsymbol{r})$ [134]
$\boldsymbol{M}(\boldsymbol{\kappa}) = \boldsymbol{M}_{\mathrm{S}}(\boldsymbol{\kappa}) + \boldsymbol{M}_{\mathrm{L}}(\boldsymbol{\kappa})$ [134]
M quantum number for z component of angular momentum [156]
m mass of neutron
m_{e} mass of electron
m_{p} mass of proton

N number of unit cells in crystal [Chapter 3, Appendix G]
N number of nuclei in scattering system [Chapters 4 and 5]
N_{m} number of magnetic unit cells in crystal [150]
N_1, N_2, N_3 numbers of atoms along sides of crystal [198]
n_s quantum number of normal mode s [30]
n number of particles in subsystem of liquid [88]
n neutron refractive index [110, Chapter 6]
n spin deviation [157]
$n_{\boldsymbol{q}}$ quantum number of spin wave $\boldsymbol{q}$ [162]
$n(\boldsymbol{r})$ charge density in atom [190]
$\hat{\boldsymbol{n}}$ unit vector in direction of $\boldsymbol{k}' \times \boldsymbol{k}$ [192]

P number of neutrons scattered per second in Bragg peak [40]
$\boldsymbol{P}$ polarisation vector of beam of neutrons [172]
P momentum variable and operator [210, Appendix E]
p_λ probability of scattering system being in state λ [19]
$p(\omega)$ velocity frequency function [99]
$\boldsymbol{p}$ momentum of electron [130, Appendix H.1]
p_σ probability that neutron is in spin state σ [136]
$\boldsymbol{p}$ momentum of neutron [190, Section 9.5]

Q position variable and operator [27, 210, Appendix E]
$\boldsymbol{Q}_\perp = \sum_i \exp(i\boldsymbol{\kappa}\,.\,\boldsymbol{r}_i)\{\hat{\boldsymbol{\kappa}}\times(\boldsymbol{s}_i\times\hat{\boldsymbol{\kappa}}) + i(\boldsymbol{p}_i\times\hat{\boldsymbol{\kappa}})/\hbar\kappa\}$ [132]
$\boldsymbol{Q}_{\perp S}$ spin part of $\boldsymbol{Q}_\perp$ [133]
$\boldsymbol{Q}_{\perp L}$ orbital part of $\boldsymbol{Q}_\perp$ [134]
$\boldsymbol{Q}_S$ defined by $\boldsymbol{Q}_{\perp S} = \hat{\boldsymbol{\kappa}}\times(\boldsymbol{Q}_S\times\hat{\boldsymbol{\kappa}})$ [133]
$\boldsymbol{Q}_L$ defined by $\boldsymbol{Q}_{\perp L} = \hat{\boldsymbol{\kappa}}\times(\boldsymbol{Q}_L\times\hat{\boldsymbol{\kappa}})$ [134]
$\boldsymbol{Q} = \boldsymbol{Q}_S + \boldsymbol{Q}_L$ [135]
$\boldsymbol{Q}$ helix vector [154]
$Q_s(t)$ magnitude of displacement of normal mode s [222]
$\tilde{Q}_s(t) = M_d^{1/2} Q_s(t)$ [223]
$\boldsymbol{q}$ wavevector of normal mode [27, 221]
$\boldsymbol{q}$ variable in reciprocal space [130]

$\boldsymbol{R}_j$ position of jth nucleus [10]
$\boldsymbol{R}_{ld}$ position of atom d in unit cell l [36]
$\boldsymbol{R}$ position relative to electron [130, Chapter 7]
R flipping ratio [184]
$\boldsymbol{r}$ position of neutron [10, 131]
r number of atoms in unit cell [37, 222]
$\boldsymbol{r}$ space variable in scattering system [61, Chapter 4]
$\langle r^2(t)\rangle$ mean-square distance between positions of an atom at interval t [98]
$\boldsymbol{r}_i$ position of ith electron [131, Chapter 7]
$r_0 = \mu_0 e^2/4\pi m_e$ classical radius of electron [132]
$r_p = \mu_0 e^2/4\pi m_p$ [192]

$S(\boldsymbol{\kappa}, \omega)$ scattering function [61]
$S_i(\boldsymbol{\kappa}, \omega)$ incoherent scattering function [62]
$S_n(\boldsymbol{\kappa})$ nth moment of scattering function [73]
$S(\boldsymbol{\kappa}) = S_0(\kappa)$ structure factor [88]
$\tilde{S}(\boldsymbol{\kappa}, \omega)$ Fourier transform in time of $\tilde{I}(\boldsymbol{\kappa}, t)$ [67]
$\tilde{S}_i(\boldsymbol{\kappa}, \omega) = \exp(-\frac{1}{2}\hbar\omega\beta)S_i(\kappa, \omega)$ [96]
S quantum number for spin angular momentum [137]
$\boldsymbol{S}_{ld}$ spin angular momentum of atom l, d [138]
S_{ld}^{β} β component of $\boldsymbol{S}_{ld}$ [138]
$\langle S^{\eta}\rangle$ mean value of component of spin in direction $\hat{\boldsymbol{\eta}}$ [147]
S_l^+, S_l^- raising and lowering spin operators for atom l [157, 215]
s label for normal mode $\boldsymbol{q}, j$ [27, 222]
$\boldsymbol{s}$ spin angular momentum of electron [129]

$s_d(\boldsymbol{r})$ normalised density of unpaired electrons in atom d [138, 229]

T absolute temperature
$T = \exp(-\mathrm{i}Ht/\hbar)$ [Appendix D]
$T_l = A_l + \boldsymbol{\sigma} . \{B_l \boldsymbol{I}_l - \frac{1}{2}\gamma r_0 g F(\boldsymbol{\kappa}) \hat{\boldsymbol{\kappa}} = (\boldsymbol{S}_l \times \hat{\boldsymbol{\kappa}})\}$ [181]
t time variable
t_1, t_2, t_3 trio of integers $\boldsymbol{\tau} = t_1\boldsymbol{\tau}_1 + t_2\boldsymbol{\tau}_2 + t_3\boldsymbol{\tau}_3$ [150]
t quantum number for spin of nucleus–neutron system [174]
$\boldsymbol{t}$ spin angular momentum of nucleus–neutron system [174]

$U = -\mathrm{i}\boldsymbol{\kappa} . \boldsymbol{u}_0(0)$ [29]
$\boldsymbol{u}_l$ displacement of nucleus l [26]
$\boldsymbol{u}\binom{l}{d}$ displacement of atom d in unit cell l [36]
u, v spin up, spin down states of neutron [136, 171]

V nuclear potential of neutron and scattering system [10]
$V_j(\boldsymbol{r} - \boldsymbol{R}_j)$ nuclear potential of neutron and jth nucleus [14]
$V = \mathrm{i}\boldsymbol{\kappa} . \boldsymbol{u}_l(t)$ [29]
V volume of crystal [40]
$V_0 = \mathrm{i}\boldsymbol{\kappa} . \boldsymbol{u}_0(t)$ [54]
V_{m} magnetic potential of neutron and scattering system [131]
$V_{\mathrm{E}}(\boldsymbol{r})$ electrostatic potential of neutron and atom [190]
$V_{\mathrm{SO}}(\boldsymbol{r})$ spin-orbit potential [190]
$V_{\mathrm{F}}(\mathbf{r})$ Foldy potential [190]
$\boldsymbol{v}$ velocity of neutron [3]
v_0 volume of unit-cell in crystal lattice [25]
$\boldsymbol{v}(t)$ velocity of atom at time t [98]
$\langle \boldsymbol{v}(0) . \boldsymbol{v}(t) \rangle$ velocity autocorrelation function [98]
$v_{0\mathrm{m}}$ volume of magnetic unit-cell in crystal [150]

$W = \frac{1}{2}\langle(\boldsymbol{\kappa} . \boldsymbol{u})^2\rangle$. The quantity $\exp(-2W)$ is the Debye–Waller factor [32, 34]
$W_d = \frac{1}{2}\langle\{\boldsymbol{\kappa} . \boldsymbol{u}\binom{l}{d}\}^2\rangle$ [37]

$\boldsymbol{x}_j = \boldsymbol{r} - \boldsymbol{R}_j$ [14]
$x = \hbar\omega/E$ [91]

Y volume of normalisation box [12]

$Z = \sum_\lambda \exp(-E_\lambda \beta)$, partition function [19]
$Z(\omega)$ phonon density of states [35]

α_1, α_2 values of a_τ/a_0 for wavevectors $\boldsymbol{k}_1$ and $\boldsymbol{k}_2$ [120]

$\beta = 1/k_B T$ [19]

γ frictional force constant [108]
γ_c critical glancing angle [114]
$\gamma = 1.913$; magnetic moment of neutron $= -\gamma \mu_N \boldsymbol{\sigma}$ [129]

Δ_0 Pendellösung length [125]
$\delta(x)$ Dirac delta function [Appendix A]
δ_{ij} Kronecker delta

ε_0 permittivity of a vacuum [191]

$\hat{\boldsymbol{\eta}}$ unit vector in mean spin direction [147]

θ scattering angle [6]
θ_D Debye temperature [59]

$\boldsymbol{\kappa} = \boldsymbol{k} - \boldsymbol{k}'$ scattering vector [15]
$\boldsymbol{\kappa}_0$ value of $\boldsymbol{\kappa}$ for elastic scattering [79]
κ_T isothermal compressibility [90]

λ wavelength of neutron [3]
λ, λ' initial and final states of scattering system [10]

$\boldsymbol{\mu}_n$ magnetic dipole moment of neutron [129]
μ_N nuclear magneton [129]
μ_B Bohr magneton [129]
$\boldsymbol{\mu}_e$ magnetic dipole moment of electron [129]
$\mu_0 = 4\pi \times 10^{-7}$ Hm^{-1}; permeability of a vacuum [130, 192]

$\boldsymbol{\xi}$ wavevector of nucleus [70]
ξ extinction distance [126]

$\boldsymbol{\rho} = \boldsymbol{k} - \boldsymbol{\tau}$ [38]
$\rho(\boldsymbol{r}, t)$ particle density operator [65]
$\rho_{\boldsymbol{\kappa}}(t)$ Fourier transform of $\rho(\boldsymbol{r}, t)$ [65]
ρ mean number density [87, 110]

$\boldsymbol{\rho}_{\mathrm{S}}(\boldsymbol{r})$ electron vector spin density [133]
$\boldsymbol{\rho} = \boldsymbol{l} - \boldsymbol{l}'$ [159]

σ_{tot} total scattering cross-section [6]
$\sigma_{\text{coh}} = 4\pi(\bar{b})^2$ [22]
$\sigma_{\text{inc}} = 4\pi\{\overline{b^2} - (\bar{b})^2\}$ [22]
σ standard deviation of Gaussian function [28]
$\boldsymbol{\sigma}$ Pauli spin operator for neutron [129]
σ spin state of neutron [131]

$\boldsymbol{\tau}_1, \boldsymbol{\tau}_2, \boldsymbol{\tau}_3$ unit-cell vectors of reciprocal lattice [25]
$\boldsymbol{\tau}$ vector in reciprocal lattice [32]
$\boldsymbol{\tau}_{\text{m}}$ vector in magnetic reciprocal lattice [150]

Φ incident neutron flux [6]
$\phi(v)$ velocity distribution of incident neutron flux [2]

$\psi_{\boldsymbol{k}}, \psi_{\boldsymbol{k}'}$ initial and final wavefunctions of neutron [10]
ψ_n eigenfunction of Hamiltonian [27, Appendix E]
ψ angle between $\boldsymbol{k}$ and $\boldsymbol{\tau}$ [38]
ψ Bloch function [117]

ω defined by $\hbar\omega = E - E'$ [18]
ω_s angular frequency of normal mode s [27]
$\omega_{\boldsymbol{q}}$ angular frequency of spin wave $\boldsymbol{q}$ [160]

General symbols

$\langle A \rangle$ thermal average of operator A [20]
$\hat{\boldsymbol{R}}$ unit vector in the direction of vector $\boldsymbol{R}$ [130]
$(\ )_{\text{sp}}$ average over spin states of nucleus [175]
$\langle\ \rangle_{\text{iso}}$ average over isotopes in scattering system [175]

INDEX

Printed in the United States
By Bookmasters